MESOSCOPIC QUANTUM OPTICS

MESOSCOPIC QUANTUM OPTICS

by

Yoshihisa Yamamoto
Ataç İmamoḡlu

A Wiley-Interscience Publication
JOHN WILEY & SONS, INC.

New York • Chichester • Weinheim • Brisbane • Toronto • Singapore

This book is printed on acid-free paper. ∞

Copyright © 1999 by John Wiley & Sons, Inc. All rights reserved.

Published simultaneously in Canada.

No part of this publication may be reproduced, stored in a retrieval system, or transmitted in any form or by any means, electronic, mechanical, photocopying, recording, scanning, or otherwise, except as permitted under Sections 107 or 108 of the 1976 United States Copyright Act, without either the prior written permission of the Publisher, or authorization through payment of the appropriate per-copy fee to the Copyright Clearance Center, 222 Rosewood Drive, Danvers, MA 01923, (978) 750-8400, fax (978) 750-4744. Requests to the Publisher for permission should be addressed to the Permissions Department, John Wiley & Sons, Inc., 605 Third Avenue, New York, NY 10158-0012, (212) 850-6011, fax (212) 850-6008. E-Mail: PERMREQ@WILEY.COM.

For ordering and customer service, call 1-800-CALL-WILEY.

Library of Congress Cataloging-in-Publication Data:

Yamamoto, Yoshihisa.
    Mesoscopic quantum optics / by Yoshihisa Yamamoto, Ataç İmamoğlu.
       p.   cm.
    Includes bibliographical references.
    ISBN 0-471-14874-1 (alk. paper)
    1. Quantum optics. 2. Mesoscopic phenomena (Physics)
I. Imamoğlu, Ataç. II. Title.
QC446.2.Y36 1999
535—dc21   99-14000
                                                                                CIP

Printed in the United States of America

10 9 8 7 6 5 4 3 2 1

CONTENTS

Preface   xi

1 Basic Concepts   1
   1.1 Probability Interpretation of Quantum Mechanics   1
   1.2 Heisenberg Uncertainty Principle   4
      1.2.1 Commutator Bracket and the Minimum Uncertainty State   5
      1.2.2 Quantum Interference and Wave-Particle Duality   8
      1.2.3 Time Evolution of a Minimum Uncertainty State   9
      1.2.4 Coherent State and Squeezed State   11
   1.3 von Neumann's Projection Postulate   14
      1.3.1 Measurement Error and Back Action Noise   15
      1.3.2 Pauli's First-Kind and Second-Kind Measurements   18
      1.3.3 Density Operator and Liouville–von Neumann Equation   19
      1.3.4 General Description of Indirect Quantum Measurements   21
      1.3.5 Quantum Nondemolition Measurement   24
   1.4 Simultaneous Measurement of Two Conjugate Observables   25
   References   28

2 Canonical Quantization   29
   2.1 Minimum Action Principle and Lagrangian Formalism   29
   2.2 Quantization of the Electromagnetic Field   34
   2.3 Second Quantization of the Schrödinger Field   38
   References   42

3 Quantum States of the Electromagnetic Fields   44
   3.1 Photon Number Eigenstates   45
      3.1.1 Properties of Photon Number Eigenstates   46
      3.1.2 Generation of Photon Number Eigenstates   51
   3.2 Coherent States   53
      3.2.1 Properties of Coherent States   53
      3.2.2 Generation of Coherent States   58

3.3 Squeezed States 60
    3.3.1 Properties of Quadrature Amplitude Squeezed States 60
    3.3.2 Generation of Quadrature Amplitude Squeezed States 61
    3.3.3 Properties of Number-Phase Squeezed States 63
    3.3.4 Generation of Number-Phase Squeezed States 65
3.4 Correlated Twin Photons and Quantum Entanglement 67
    3.4.1 Generation of Arbitrary Quantum States by Projective Measurements 68
    3.4.2 Nonlocal Quantum Entanglement 70
3.5 Probability Distribution Functions for a Density Operator 71
References 74

4 Coherence of the Electromagnetic Fields 76

4.1 Photodetection 76
4.2 Young's Interference Experiment and First-Order Coherence 77
4.3 Hanbury–Brown–Twiss Experiment and Second-Order Coherence 81
4.4 Photon Counting 85
    4.4.1 Classical Theory of Photon Count Distribution 86
    4.4.2 Quantum Theory of Photon Count Distribution 88
4.5 Phase Operator of the Quantized Electromagnetic Field 89
4.6 Quantum Limits of Optical Interferometry 92
    4.6.1 Standard Quantum Limits of Optical Interferometry 92
    4.6.2 Squeezed State Interferometer 93
    4.6.3 Photon Number Eigenstate Interferometer 96
    4.6.4 Photonic de Broglie Wave Interferometer 97
References 100

5 Quantum States of Atoms 101

5.1 Angular Momentum Algebra 102
    5.1.1 Quantization of Angular Momentum 102
    5.1.2 Angular Momentum Operators and Eigenstates 103
5.2 Assembly of Two-Level Atoms 105
    5.2.1 Pauli Spin Operators 105
    5.2.2 Collective Angular Momentum Operators 106
    5.2.3 Angular Momentum Eigenstates (Dicke States) 108
    5.2.4 Coherent Atomic States (Bloch States) 109
References 113

CONTENTS vii

6 Interaction between Atoms and Fields   **114**
- 6.1 Atom-Field Interaction in the Length Gauge   114
- 6.2 Jaynes–Cummings Hamiltonian   118
- 6.3 Two-Level Atom and Single-Mode Photon Number State   119
  - 6.3.1 Vacuum Rabi Oscillation   119
  - 6.3.2 Normal Mode Splitting   121
- 6.4 N Two-Level Atoms and a Single-Mode Photon Number State   122
  - 6.4.1 Collective Rabi Oscillation and Normal Mode Splitting   122
  - 6.4.2 Dressed Fermions and Dressed Bosons   124
  - 6.4.3 Atomic Cavity Quantum Electrodynamics (QED)   125
  - 6.4.4 Mollow's Triplet   127
  - 6.4.5 Radiation Trapped State   128
- 6.5 Cummings Collapse and Revival   128
- 6.6 Two-Level Atoms with a Continuum of Radiation Fields   131
- 6.7 Superradiance   134
- References   136

7 Mathematical Methods for System-Reservoir Interaction   **137**
- 7.1 Noise Operator Method   139
  - 7.1.1 Field Damping by Field Reservoirs   139
  - 7.1.2 Einstein Relation between Drift and Diffusion Coefficients   144
- 7.2 Density Operator Method   145
  - 7.2.1 Derivation of the Master Equation   145
  - 7.2.2 Field Damping by Atomic Reservoirs   146
  - 7.2.3 Atom Damping by Field Reservoirs   150
  - 7.2.4 Field Damping by Field Reservoirs   151
- 7.3 The Fokker-Planck Equation   152
  - 7.3.1 Glauber-Sudarshan P Representation   152
  - 7.3.2 Stochastic Differential Equations   153
- 7.4 Quantum Regression Theorem   154
- References   161

8 Stochastic Wavefunction Methods   **162**
- 8.1 Monte Carlo Wavefunction Approach   163
  - 8.1.1 Description and Equivalence to Master Equation   164
  - 8.1.2 Two-Time Correlation Functions   166
  - 8.1.3 Two-Level Atom Driven by a Laser Field   167
  - 8.1.4 Single-Mode Cavity Driven by a Thermal Field   168

8.2 Quantum State Diffusion Model 171
References 174

9 Quantum Nondemolition Measurements — 175
9.1 QND Measurement of Photon Number 175
    9.1.1 Heisenberg Picture 175
    9.1.2 Schrödinger Picture 178
9.2 Experimental QND Measurements 180
9.3 QND Measurement of Quadrature Amplitudes 182
References 184

10 Semiconductor Bloch Equations — 186
10.1 Field Theory of Semiconductors 186
10.2 Semiconductor Bloch Equations 194
    10.2.1 Semiconductor Bloch Equations in the Non-interacting Limit 195
    10.2.2 Semiconductor Bloch Equations with Interactions 196
10.3 Excitons 197
10.4 Nonlinear Response of Excitons 198
References 200

11 Excitons and Polaritons — 202
11.1 Non-interacting Excitons 202
11.2 Bulk Exciton Polaritons 203
11.3 Quantum-Well Excitons and Cavity Polaritons 206
11.4 Excitons as Bosons 211
    11.4.1 Bosonization 211
    11.4.2 Bose Condensation of Excitons 212
    11.4.3 Stimulated Scattering of Excitons 213
11.5 Stimulated Scattering Experiment With Excitons 216
References 219

12 Coulomb Blockade and Squeezing — 220
12.1 Coulomb Blockade of Tunneling 220
12.2 Macroscopic Coulomb Blockade of Electron Injection 224
12.3 AC-Voltage-Driven Mesoscopic p-i-n Junctions 228
References 234

13 Current Noise in Mesoscopic and Macroscopic Conductors — 235

- 13.1 Suppression of Electrical Current Noise in Dissipative Conductors — 236
  - 13.1.1 Equilibrium and Nonequilibrium Transport Noise in Mesoscopic Conductors — 236
  - 13.1.2 Suppression of Nonequilibrium Partition Noise by Inelastic Scattering — 239
  - 13.1.3 Microscopic Model for Suppression of Nonequilibrium Partition Noise and Loss of Electron Wave Coherence — 242
- 13.2 Quantum Interference in Electron Collision — 246
- 13.3 Negative Correlation in Electron Partition — 250
- References — 252

14 Nonequilibrium Green's Function Formalism — 253

- 14.1 Green's Function and Self-Energy — 253
- 14.2 Correlation and Scattering Functions — 256
- 14.3 Current Flow — 257
  - 14.3.1 Noninteracting Transport — 257
  - 14.3.2 Strongly Interacting Transport — 258
  - 14.3.3 Conductance through a Single Site — 262
- References — 264

15 Quantum Statistical Properties of a Laser — 265

- 15.1 Density Operator Master Equation — 266
  - 15.1.1 Derivation of the Master Equation — 266
  - 15.1.2 Fast Dephasing Case — 269
  - 15.1.3 Lifetime Broadened Laser System — 269
  - 15.1.4 Threshold Characteristics — 271
  - 15.1.5 Photon Statistics — 273
  - 15.1.6 Spectral Linewidth — 276
- 15.2 Fokker-Planck Equation — 278
- 15.3 Langevin Equation — 280
  - 15.3.1 Derivation of the Langevin Equation — 280
  - 15.3.2 Linearization of the Langevin Equations — 281
  - 15.3.3 Quantum Noise at Well Above Threshold — 283
- 15.4 Sub-Poissonian Lasers — 284
  - 15.4.1 Standard Quantum Limit of the Output Field — 285
  - 15.4.2 Photon Number Noise and Phase Noise of Pump-Noise-Suppressed Lasers — 287
  - 15.4.3 Commutator Bracket Conservation — 288

15.5 Squeezing in Semiconductor Lasers 289
15.6 Observation of Mode Partition Noise 294
References 297

Index **299**

PREFACE

This textbook is an introduction to the quantum theory of light–matter interaction in mesoscopic systems, a rapidly developing field of great interest to scientists and engineers for its potential applications to future precision measurement, quantum communication, and quantum information processing technologies. Much of the contents of this book originated in graduate courses of lectures in quantum optics and mesoscopic physics that we have given over several years at Stanford University and University of California (UC) at Santa Barbara.

We have felt that graduate students and other beginning researchers in this field need an introductory book that provides fundamental mathematics and physical pictures through which to understand current research articles. Thus the goal of the book was to be selective in terms of subjects rather than comprehensive. We limit our efforts so that the material presented in this textbook could be covered in one semester course (or two quarters) rather than a full year. Thus the aim was to explain essential concepts in a mathematically rigorous but physical manner, rather than to present an exhaustive treatment of any one topic.

Throughout this textbook, we have emphasized that the theoretical predictions must be tested by experimental observations due to the problematic nature of quantum mechanics—in particular, the quantum theory of measurement. We have described the relevant recent experiments in Chapters 3, 9, 11, 12, 13 and 15, which treat advanced concepts and topics in mesoscopic quantum optics. The remaining chapters (2, 4, 5, 6, 7, 8, 10, and 14) are devoted to the mathematical methods.

We would like to thank Hermann Haus, Olle Nilsson, and Steve Harris for introducing us to this exciting field. We would also like to thank our colleagues and students at Nippon Telegraph and Telephone Corporation (NTT), Stanford University, and UC Santa Barbara who have contributed so much to the material of the book: in particular, Susumu Machida, Gunnar Björk, Nobuyuki Imoto, Stephen Friberg, Masahiro Kitagawa, Kimitaka Watanabe, Masahito Ueda, Joseph Jacobson, Isaac Chuang, Shuichiro Inoue, Fumiko Yamaguchi, Dmitri Nikonov, Rajeev Ram, Holger Schmidt, Orly Alter, Robert Liu, Jungsang Kim, Francesco Tassone, Hui Cao, Stanley Pau, Robin Huang, Oliver Benson, Shadong Jiang, Hirofumi Kan, Steven Kasapi, and Seema Lathi.

The completion of this book would not have been possible without the excellent and patient work of Minako Shioda, who performed an entire word processing.

Y. YAMAMOTO

A. İMAMOĞLU

1 Basic Concepts

The goal of this introductory chapter is to present an overview of the basic concepts of quantum optics and measurements. After a brief description of the standard probability interpretation of the Schrödinger wavefunction, the two basic assumptions of the quantum theory—namely, the Heisenberg uncertainty principle and von Neumann's projection postulate—are discussed. The texts by P. A. M. Dirac [1] and by W. H. Louisell [2] provide a general and excellent background of the material presented in this chapter. Several current topics in quantum optics, such as squeezed states, quantum entanglement, and quantum nondemolition measurements, are only briefly explained here. These subjects will be discussed in detail with recent experimental results in later chapters.

1.1. PROBABILITY INTERPRETATION OF QUANTUM MECHANICS

In classical physics the physical state of a system is uniquely determined by the dynamic variables of this system, such as the position r and the momentum p of a particle. If we have complete information about the physical state, we can tell the position and the momentum of the particle precisely. There is no fundamental limit as to how accurately we can measure these quantities. In quantum mechanics, however, such direct correspondence does not exist between the physical state of a system and the dynamic variables. The complete description of a quantum system is contained either in the Schrödinger wavefunction or the Dirac state vector. In Dirac formulation of quantum mechanics [1], a quantum system is described by an abstract state vector in Hilbert space. In order to connect the state vector $|\psi\rangle$ in Dirac formulation to the Schrödinger wavefunction $\psi(q)$ in the coordinate ($q-$) representation, one can project the system ket vector $|\psi\rangle$ onto the coordinate eigenbra vector $\langle q|$:

$$\psi(\mathbf{q}) \equiv \langle \mathbf{q} | \psi \rangle. \tag{1.1}$$

Therefore, the Schrödinger wavefunction $\psi(\mathbf{q})$ is the projection of the state vector $|\psi\rangle$ onto each vector $(q_1, q_2 \ldots q_n, \ldots)$ of the Hilbert space. An important difference between such a Hilbert space vector in quantum mechanics and an ordinary vector in classical physics is that those projections (Schrödinger wavefunction) $\psi(q_1), \psi(q_2) \ldots$ are complex numbers (c-numbers). This point is an important departure of quantum theory from its classical counterpart, as will be explained later in this chapter.

If the Schrödinger wavefunction $\psi(\mathbf{q})$ is known for all q_i values, it is said that we have the complete information about the state of this system. In this case, the system is described by a single ket vector and is called a *pure state*. The system state is now expanded linearly as follows:

$$|\psi\rangle = \sum_i |q_i\rangle\langle q_i|\psi\rangle$$
$$= \sum_i \psi(q_i)|q_i\rangle. \quad (1.2)$$

Here $|q_i\rangle\langle q_i|$ is a *projection operator* and satisfies the completeness relation

$$\sum_i |q_i\rangle\langle q_i| = \hat{I},$$

where $\hat{I}$ is an identity operator. The physical interpretation of this completeness relation of the projection operator is easily obtained if we recall that the sum of the probabilities of finding a particle in all positions $(q_1, q_2, \ldots q_n, \ldots)$ should be equal to one. Equation (1.2) implies that the particle exists simultaneously in different positions q_i with the probability amplitude $\psi(q_i)$. This simultaneous appearance of the particle at different positions is referred to as a *linear superposition state*, which is one of the most remarkable features of quantum mechanics. In quantum mechanics this probability amplitude $\psi(q_i)$ is a c-number and thus carries an amplitude and a phase simultaneously.

In order to connect the Schrödinger wavefunction $\psi(\mathbf{q})$ in (1.2) to the Schrödinger wavefunction $\varphi(p)$ in the momentum ($p-$) representation, one can project the same ket vector $|\psi\rangle$ onto the momentum eigenbra vector $\langle \mathbf{p}|$:

$$\varphi(\mathbf{p}) \equiv \langle \mathbf{p}|\psi\rangle. \quad (1.3)$$

Hence $\varphi(p)$ is the projection of the state vector $|\psi\rangle$ onto each momentum $(p_1, p_2, \ldots p_n, \ldots)$ in Hilbert space. The state vector $|\psi\rangle$ is thus expanded by the linear superposition of the momentum eigenstates:

$$|\psi\rangle \equiv \sum_i |p_i\rangle\langle p_i|\psi\rangle$$
$$= \sum_i \varphi(p_i)|p_i\rangle. \quad (1.4)$$

Here the completeness relation of the projection operator, $\sum_i |p_i\rangle\langle p_i| = \hat{I}$, is used. In this expression $\varphi(p_i)$ is interpreted as the probability amplitude that the particle has a particular momentum p_i.

The measured quantities of a quantum system, are represented by Hermitian operators acting on state vectors. For position and momentum, the corresponding operators are

$$\hat{q} \equiv \sum_i q_i |q_i\rangle\langle q_i|, \qquad (1.5)$$

$$\hat{p} \equiv \sum_i p_i |p_i\rangle\langle p_i|, \qquad (1.6)$$

where q_i and p_i are real numbers and $\hat{q}^\dagger = \hat{q}$ and $\hat{p}^\dagger = \hat{p}$. The Hermitian operators $\hat{q}$ and $\hat{p}$ are not real numbers even though actual measurement results always yield one of their eigenvalues, which are real numbers. They are often referred to as quantum ($q-$) numbers. Those Hermitian operators representing dynamical variables are called *observables*. In order to connect the quantum theory to actual experimental results, we have to operate the Hermitian operator, which represents a dynamical variable, upon the state vector that describes the physical state of a measured system. If the particle is in a state $|\psi\rangle$ and the position of this particle is measured, the measurement result can be any one of the eigenvalues q_i of the Hermitian operator $\hat{q}$ that satisfies the following eigenvalue relation:

$$\hat{q}|q_i\rangle = q_i|q_i\rangle. \qquad (1.7)$$

The probability of obtaining this specific result q_i is given by the squared Schrödinger wavefunction:

$$P(q_i) = |\langle q_i|\psi\rangle|^2 = |\psi(q_i)|^2. \qquad (1.8)$$

The average value of many measurements performed for an ensemble of identical systems, which are described by $|\psi\rangle$, is calculated by

$$\langle\hat{q}\rangle \equiv \langle\psi|\hat{q}|\psi\rangle = \sum_i q_i |\psi(q_i)|^2, \qquad (1.9a)$$

$$\langle\hat{q}^2\rangle \equiv \langle\psi|\hat{q}^2|\psi\rangle = \sum_i q_i^2 |\psi(q_i)|^2. \qquad (1.9b)$$

The direct one-to-one correspondence between the physical state and the measured quantities, which exists in classical physics, is absent in quantum mechanics. We have to deal with *vectors and operators* in Hilbert space to predict a measurement result. Even though we have complete information about the system—namely, the state is described by a single ket vector like (1.2) or (1.4)—we still cannot predict a single measurement result. This is often referred to as *lack of causality* in quantum measurements. Except in certain special cases (discussed subsequently), the result can be any eigenvalue of the measured observable. However, the statistics of many measurements performed for an ensemble of identical systems reproduce the Schrödinger wavefunction as suggested by (1.8). This is often called the *prob-*

ability interpretation of the Schrödinger wavefunction and is the only connection between the quantum theory and measurement results.

The only exception to this lack of direct correspondence between the physical state and the measurement result is the case when the system is prepared in one of the eigenstates of an observable and the measurement is performed on this observable. In such a case, the measurement result is completely deterministic.

The *reality* in quantum mechanics is significantly different from that in classical physics. Unless a measurement is performed on a certain observable for a given system and the measurement result is read out, all we have is an abstract *information* about the system. This observation highlights the importance of the measurement process for a quantum system, which brings the abstract information into the fact.

Perhaps the most important feature of this branch of science, quantum optics, is that it provides us an experimentally accessible and theoretically tractable paradigm with which we can study these fundamental properties of quantum mechanics. The finite measurement error and the unavoidable quantum fluctuations associated with a measurement process are at the heart of quantum optics.

1.2. HEISENBERG UNCERTAINTY PRINCIPLE

The Heisenberg uncertainty principle is usually interpreted as a statement about the properties of a state vector $|\psi\rangle$ or wavefunction $\psi(\mathbf{q})$ of a measured system in the standard quantum theory. Connection of the uncertainty principle to real experiments is also provided by the *probability interpretation*. If one prepares many quantum objects in the same quantum state and performs the measurements of one observable for half of the ensemble, the probability of obtaining a certain eigenvalue is determined by the squared amplitude of the wavefunction in the corresponding representation. The same is true for the measurements of the conjugate observable performed for the remaining half of the ensemble. The product of the variances of the two measurement results obeys the Heisenberg uncertainty relations. The two independent and ideal (noise-free) measurements for each observable form the basis for the probability interpretation.

However, when Heisenberg mentioned the uncertainty principle for the first time via his famous γ-ray microscope and when von Neumann discussed it again in reference to a Doppler speed meter, the uncertainty principle was certainly the statement about the relation between the measurement error of one observable and the inevitable disturbance (back action noise) imposed on the conjugate observable. That is, the uncertainty principle was a statement about the properties of the measuring apparatus, not the measured system.

The Heisenberg uncertainty principle imposes independent constraints on the processes of preparation and measurement of a quantum state. The lack of causality in a single ideal measurement is due to the quantum nature of the measured system and is governed by the uncertainty relation for the preparation of a system. The uncontrollable disturbance imposed on the measured object is due to the quantum nature of the measuring apparatus and is governed by the uncertainty relation for the measurement of a system. These two uncertainty relations are different and independent. This fact leads to a somewhat surprising result when two conjugate observables are measured simultaneously for a single object. The uncertainty product one can optimally achieve in such a case is *four times* larger than the minimum uncertainty product usually anticipated.

We will discuss the uncertainty relation for the preparation of a quantum system in this section. In the next section we will discuss the uncertainty relation for the measurement of a quantum system.

1.2.1. Commutator Bracket and the Minimum Uncertainty State

The observable of a measured system must carry its own uncertainty, independent of how a measuring apparatus is prepared and how the measurement is performed. In fact, the standard formulation of the Heisenberg uncertainty principle describes this aspect of the uncertainty principle.

Consider an ensemble of identical quantum systems prepared in state $|\psi\rangle$. We measure an observable $\hat{A}$ for half of the ensemble and the other observable $\hat{B}$ for the remaining half. Each half of the ensemble practically has an infinite number of identical systems. We assume the measurement is perfect in the sense that the measurement error is identically zero and so the back action noise is infinite. The measurement of $\hat{A}$ gives one of the eigenvalues of $\hat{A}$, say a, with the probability $|\langle a|\psi\rangle|^2$. Similarly, the measurement of $\hat{B}$ gives one of the eigenvalues of $\hat{B}$, say b, with the probability $|\langle b|\psi\rangle|^2$.

The ensemble-averaged values for the measurements of $\hat{A}$ and $\hat{B}$ are

$$\langle \hat{A} \rangle = \langle \psi | \hat{A} | \psi \rangle \quad \langle \hat{B} \rangle = \langle \psi | \hat{B} | \psi \rangle. \tag{1.10}$$

The mean-square deviations for the measurements of $\hat{A}$ and $\hat{B}$ are defined by

$$\langle \Delta \hat{A}^2 \rangle = \langle \hat{A}^2 \rangle - \langle \hat{A} \rangle^2 \quad \langle \Delta \hat{B}^2 \rangle = \langle \hat{B}^2 \rangle - \langle \hat{B} \rangle^2. \tag{1.11}$$

These fluctuations will be zero if and only if the state $|\psi\rangle$ is an eigenstate of either $\hat{A}$ or $\hat{B}$ or both. In such a case, every ideal measurement for that observable gives the identical eigenvalue with no fluctuations.

Suppose that the observables $\hat{A}$ and $\hat{B}$ are conjugate observables; that is, they do not commute but, rather, satisfy the following commutator bracket:

$$[\hat{A}, \hat{B}] = \hat{A}\hat{B} - \hat{B}\hat{A} = i\hat{C}, \tag{1.12}$$

where $\hat{C}$ is either a c-number or an observable (Hermitian operator). In order to calculate the minimum uncertainty product $\langle \Delta \hat{A}^2 \rangle \langle \Delta \hat{B}^2 \rangle$ for such a case, one defines two fluctuation operators:

$$\hat{\alpha} = \hat{A} - \langle \hat{A} \rangle \quad \hat{\beta} = \hat{B} - \langle \hat{B} \rangle. \tag{1.13}$$

Since $\langle \hat{A} \rangle$ and $\langle \hat{B} \rangle$ are real numbers, $\hat{\alpha}$ and $\hat{\beta}$ satisfy the commutator bracket

$$[\hat{\alpha}, \hat{\beta}] = i\hat{C}. \tag{1.14}$$

Since $\langle \hat{\alpha} \rangle = \langle \hat{\beta} \rangle = 0$, the uncertainty product is now rewritten as

$$\langle \Delta \hat{A}^2 \rangle \langle \Delta \hat{B}^2 \rangle = \langle \Delta \hat{\alpha}^2 \rangle \langle \Delta \hat{\beta}^2 \rangle$$
$$= \langle \hat{\alpha}^2 \rangle \langle \hat{\beta}^2 \rangle. \tag{1.15}$$

If we let $|\varphi\rangle = \hat{\alpha}|\psi\rangle$ and $|\chi\rangle = \hat{\beta}|\psi\rangle$, we may use the Schwarz inequality to obtain

$$\langle \varphi | \varphi \rangle \langle \chi | \chi \rangle \geq |\langle \varphi | \chi \rangle|^2. \tag{1.16}$$

Here the equality holds if and only if

$$|\varphi\rangle = c_1 |\chi\rangle, \tag{1.17}$$

where c_1 is a c-number. Since $\hat{\alpha}$ and $\hat{\beta}$ are observables (Hermitian operators), it follows that $\hat{\alpha} = \hat{\alpha}^\dagger$ and $\hat{\beta} = \hat{\beta}^\dagger$. The inequality (1.16) is now written as

$$\langle \hat{\alpha}^2 \rangle \langle \hat{\beta}^2 \rangle \geq |\langle \psi | \hat{\alpha} \hat{\beta} | \psi \rangle|^2. \tag{1.18}$$

Using the identity $\hat{\alpha}\hat{\beta} = \frac{1}{2}(\hat{\alpha}\hat{\beta} + \hat{\beta}\hat{\alpha}) + \frac{1}{2}(\hat{\alpha}\hat{\beta} - \hat{\beta}\hat{\alpha}) = \frac{1}{2}(\hat{\alpha}\hat{\beta} + \hat{\beta}\hat{\alpha}) + \frac{i}{2}\hat{C}$, (1.18) is expressed as

$$\langle \hat{\alpha}^2 \rangle \langle \hat{\beta}^2 \rangle \geq \frac{1}{4} |\langle \psi | \hat{\alpha}\hat{\beta} + \hat{\beta}\hat{\alpha} | \psi \rangle + i \langle \psi | \hat{C} | \psi \rangle |^2. \tag{1.19}$$

Since $\hat{\alpha}\hat{\beta} + \hat{\beta}\hat{\alpha}$ and $\hat{C}$ are Hermitian operators, $\langle \psi | \hat{\alpha}\hat{\beta} + \hat{\beta}\hat{\alpha} | \psi \rangle$ and $\langle \psi | \hat{C} | \psi \rangle$ are real numbers. Accordingly, (1.19) may be written as

$$\langle \hat{\alpha}^2 \rangle \langle \hat{\beta}^2 \rangle = \langle \Delta \hat{A}^2 \rangle \langle \Delta \hat{B}^2 \rangle \geq \frac{1}{4} |\langle \hat{C} \rangle|^2. \tag{1.20}$$

This is the Heisenberg uncertainty principle. The origin of the Heisenberg uncertainty principle can be traced back to the commutator bracket (1.12) for the conjugate observables $\hat{A}$ and $\hat{B}$. This commutator bracket is a *postulate* of the quantum theory and thus we cannot prove it mathematically. Justification of the postulate is the remarkable agreement between the quantum theory based on this postulate and various experimental results. This could arguably be the most profound assumption of the quantum theory.

For the equality to hold in (1.20), one requires

$$\hat{\alpha}|\psi\rangle = c_1 \hat{\beta}|\psi\rangle, \tag{1.21}$$

$$\langle \psi | \hat{\alpha}\hat{\beta} + \hat{\beta}\hat{\alpha} | \psi \rangle = 0. \tag{1.22}$$

The quantum state $|\psi\rangle$, which satisfies (1.21) and (1.22) simultaneously, is called a minimum uncertainty state.

As a special case, we take the operators $\hat{A} = \hat{q}$ (position) and $\hat{B} = \hat{p}$ (momentum) for a free particle. The commutator bracket and the Heisenberg uncertainty relation for this case are

$$[\hat{q}, \hat{p}] = i\hbar \longrightarrow \langle \Delta \hat{q}^2 \rangle \langle \Delta \hat{p}^2 \rangle \geq \frac{\hbar^2}{4}. \tag{1.23}$$

In order to find the minimum uncertainty state, we must solve (1.21) and (1.22). If we use (1.21) and its Hermitian adjoint in (1.22), we obtain

$$(c_1 + c_1^*)\langle \psi | \hat{\beta}^2 | \psi \rangle = 0. \tag{1.24}$$

Since $\langle \hat{\beta}^2 \rangle \neq 0$ (if we exclude the case that $|\psi\rangle$ is an eigenstate of $\hat{p}$), c_1 must be purely imaginary. If we let $c_1 = -ic_2$, where c_2 is real, (1.21) is written as

$$(\hat{q} - \langle \hat{q} \rangle)|\psi\rangle = -ic_2(\hat{p} - \langle \hat{p} \rangle)|\psi\rangle. \tag{1.25}$$

Now we take the scalar product of (1.25) with $\langle q'|$, an eigenbra of $\hat{q}$, and use the identity $\langle q'|\hat{p}|\psi\rangle = \frac{\hbar}{i}\frac{\partial}{\partial q'}\psi(q')$ to obtain [2]

$$\frac{i}{c_2}(q' - \langle \hat{q} \rangle)\psi(q') = \left(\frac{\hbar}{i}\frac{\partial}{\partial q'} - \langle \hat{p} \rangle \right) \psi(q'), \tag{1.26}$$

where $\psi(q') = \langle q'|\psi\rangle$ is the Schrödinger wavefunction in q'-representation. The solution of (1.26) is

$$\psi(q') = c_3 \exp\left[\frac{i}{\hbar}\langle \hat{p} \rangle q' - \frac{1}{2\hbar c_2}(q' - \langle \hat{q} \rangle)^2 \right], \tag{1.27}$$

where c_3 is a constant of integration. To determine c_2 and c_3, we use the following requirements:

$$\langle \psi | \psi \rangle = \int_{-\infty}^{\infty} |\psi(q')|^2 dq' = 1, \tag{1.28}$$

$$\langle \Delta \hat{q}^2 \rangle = \int_{-\infty}^{\infty} (q' - \langle \hat{q} \rangle)^2 |\psi(q')|^2 dq'. \tag{1.29}$$

Using (1.27) in (1.28) and (1.29), we find that $c_2 = \frac{2\langle \Delta \hat{q}^2 \rangle}{\hbar}$ and $|c_3|^2 = \frac{1}{\sqrt{2\pi \langle \Delta \hat{q}^2 \rangle}}$. We can choose c_3 real without loss of generality, and (1.27) becomes

$$\psi(q') = \frac{1}{(2\pi \langle \Delta \hat{q}^2 \rangle)^{\frac{1}{4}}} \exp\left[\frac{i\langle \hat{p} \rangle q'}{\hbar} - \frac{(q' - \langle \hat{q} \rangle)^2}{4\langle \Delta \hat{q}^2 \rangle} \right]. \tag{1.30}$$

8 BASIC CONCEPTS

This is the minimum uncertainty state of a free particle in the position ($q'-$) representation and $|\psi(q')|^2$ is the familiar Gaussian distribution function centered at $q' = \langle \hat{q} \rangle$ with variance $\langle \Delta \hat{q}^2 \rangle$.

The same minimum uncertainty state in the momentum ($p'-$) representation is obtained by the Fourier transform of (1.30) [2]:

$$\varphi(p') = \langle p' | \psi \rangle = \frac{1}{\sqrt{2\pi\hbar}} \int_{-\infty}^{\infty} \exp\left(-\frac{ip'q'}{\hbar}\right) \langle q' | \psi \rangle dq'$$

$$= \frac{1}{[2\pi \langle \Delta \hat{p}^2 \rangle]^{\frac{1}{4}}} \exp\left[-\frac{i}{\hbar} \langle \hat{q} \rangle (p' - \langle \hat{p} \rangle) - \frac{(p' - \langle \hat{p} \rangle)^2}{4 \langle \Delta \hat{p}^2 \rangle}\right], \quad (1.31)$$

where $\langle \Delta \hat{p}^2 \rangle = \hbar^2 / (4 \langle \Delta \hat{q}^2 \rangle)$.

Equation (1.25) can be rewritten as

$$(e^r \hat{q} + ie^{-r} \hat{p}) | \psi \rangle = (e^r \langle \hat{q} \rangle + ie^{-r} \langle \hat{p} \rangle) | \psi \rangle, \quad (1.32)$$

where $c_2 = e^{-2r}$. The preceding relation means that the minimum uncertainty state is defined as an *eigenstate* of a non-Hermitian operator $e^r \hat{q} + ie^{-r} \hat{p}$ with a c-number eigenvalue $e^r \langle \hat{q} \rangle + ie^{-r} \langle \hat{p} \rangle$. The variances of $\hat{q}$ and $\hat{p}$ are respectively given by

$$\langle \Delta \hat{q}^2 \rangle = \frac{\hbar}{2} e^{-2r}, \quad (1.33)$$

$$\langle \Delta \hat{p}^2 \rangle = \frac{\hbar}{2} e^{2r}. \quad (1.34)$$

Here r is often referred to as the *squeezing parameter* and determines the noise distributions between the conjugate observables. When $r = 0$, the minimum uncertainty state has equal noise in the two conjugate observables. When $r \to \pm\infty$, the minimum uncertainty state approaches an eigenstate of either $\hat{q}$ or $\hat{p}$.

1.2.2. Quantum Interference and Wave-Particle Duality

The Schrödinger wavefunction $\psi(\mathbf{q})$ consists of linear superposition of plane waves (1.31) with a *de Broglie wavelength* $\lambda = \frac{h}{p'}$. The amplitude distribution of the de Broglie waves obeys the Gaussian distribution with the variance $\langle \Delta \hat{p}^2 \rangle$ centered at $\langle \hat{p} \rangle$, as shown in Fig. 1-1. Each de Broglie wave interferes with others. In the region close to the average position ($|q' - \langle \hat{q} \rangle| \lesssim \langle \Delta \hat{q}^2 \rangle^{1/2}$), the phase difference between adjacent de Broglie waves is small. On the other hand, in the region away from the average position ($|q' - \langle \hat{q} \rangle| > \langle \Delta \hat{q}^2 \rangle^{1/2}$), the phase difference is large. Hence the de Broglie waves constructively and destructively interfere in respective regions, resulting in localization of the probability of finding the particle in the region close to the average position.

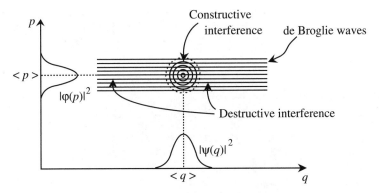

FIGURE 1-1: Constructive and destructive interference of the de Broglie waves of a particle in a minimum uncertainty state.

If the momentum uncertainty increases, the strong constructive/destructive interference played by many de Broglie waves can localize the particle position more tightly and the system features a particle-like behavior. If the momentum uncertainty decreases, the weak constructive/destructive interference played by a few de Broglie waves cannot localize the particle position and the system features a wavelike behavior.

The wave-particle duality in quantum mechanics is a quantum interference effect as mentioned previously and originates from the fact that the probability amplitudes $\psi(\mathbf{q})$ or $\varphi(\mathbf{p})$ are c-numbers (i.e., they carry a phase as well as an amplitude).

1.2.3. Time Evolution of a Minimum Uncertainty State

A free particle prepared in a minimum uncertainty state defined by either (1.30) or (1.31) satisfies the Heisenberg uncertainty relation with equality (i.e., $\langle \Delta \hat{q}^2 \rangle \langle \Delta \hat{p}^2 \rangle = \frac{\hbar^2}{4}$). However, this may not be true at a later time. The unitary evolution operator for a free particle is given by

$$\hat{\mathcal{H}} = \frac{\hat{p}^2}{2m} \longrightarrow \hat{U} = \exp\left(-\frac{i}{\hbar}\frac{\hat{p}^2}{2m}t\right). \tag{1.35}$$

The Schrödinger wavefunction in the coordinate representation is calculated by

$$\psi(q,t) \equiv \langle q|\hat{U}|\psi(0)\rangle = \langle q|\exp\left(-\frac{i}{\hbar}\frac{\hat{p}^2}{2m}t\right)|\psi(0)\rangle$$

$$= \int_{-\infty}^{\infty} \exp\left(-\frac{i}{\hbar}\frac{p^2}{2m}t\right) \langle q|p\rangle \varphi(p,0)\,dp$$

$$= \frac{1}{\sqrt{2\pi\hbar}} \int_{-\infty}^{\infty} \exp\left(-\frac{i}{\hbar}\frac{p^2}{2m}t\right) \exp\left(\frac{ipq}{\hbar}\right) \varphi(p,0)\, dp. \tag{1.36}$$

Here the completeness relation for $|p\rangle$,

$$\int_{-\infty}^{\infty} |p\rangle\langle p| dp = \hat{I}, \tag{1.37}$$

and

$$\langle q|p\rangle = \frac{1}{\sqrt{2\pi\hbar}} \exp\left(\frac{ipq}{\hbar}\right) \tag{1.38}$$

are used to obtain the second and third lines of (1.36), respectively [2]. $\varphi(p,0) = \langle p|\psi(0)\rangle$ is the initial minimum uncertainty state in the momentum representation and is given by (1.31). Since the unitary evolution operator (1.35) commutes with the momentum operator $\hat{p}$, the variance $\langle \Delta \hat{p}^2 \rangle$ is preserved at all time. Using (1.31) in (1.36), one obtains the Schrödinger wavefunction at time t:

$$\psi(q,t) = \frac{1}{\sqrt{2\pi\hbar}} \frac{1}{(2\pi\langle\Delta\hat{p}^2\rangle)^{1/4}}$$

$$\times \int_{-\infty}^{\infty} \exp\left[-\frac{i}{\hbar}\frac{p^2}{2m}t + i\frac{pq}{\hbar} - \frac{i}{\hbar}\langle\hat{q}\rangle(p-\langle\hat{p}\rangle) - \frac{(p-\langle\hat{p}\rangle)^2}{4\langle\Delta\hat{p}^2\rangle}\right] dp \tag{1.39}$$

We can assume $\langle\hat{p}\rangle = \langle\hat{q}\rangle = 0$ at $t=0$ for simplicity. The integral of (1.39) is easily evaluated and we obtain

$$\psi(q,t) = \frac{1}{(2\pi)^{1/4}(\Delta q + \frac{i\hbar t}{2m\Delta q})^{1/2}} \exp\left(-\frac{q^2}{4(\Delta q)^2 + \frac{2i\hbar t}{m}}\right), \tag{1.40}$$

where $\Delta q = \frac{\hbar}{2\langle\Delta\hat{p}^2\rangle^{1/2}}$ is the initial position uncertainty. Even though the momentum uncertainty $\langle\Delta\hat{p}^2\rangle$ is preserved, the position uncertainty increases as time develops,

$$\langle\Delta\hat{q}^2(t)\rangle = (\Delta q)^2 + \frac{\hbar^2 t^2}{4m^2(\Delta q)^2}. \tag{1.41}$$

The wavepacket spreads in width as shown in Fig. 1-2. The physical origin for this quantum (position) diffusion is that the particle simultaneously exists in different momentum eigenstates (de Broglie waves), which propagate with different veloci-

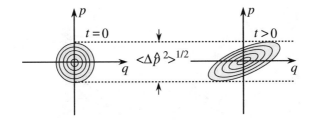

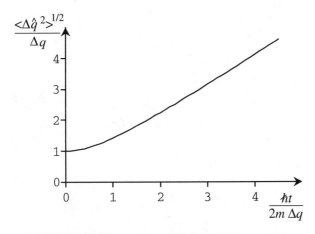

FIGURE 1-2: The quantum diffusion of a free particle.

ties. Therefore, the minimum uncertainty state ceases to be a minimum uncertainty state as time goes on.

1.2.4. Coherent State and Squeezed State

In order to preserve a minimum uncertainty product at all times, we need to suppress the quantum (position) diffusion by means of some contractive forces. A simple example of such a contractive force is the harmonic potential confining a particle

$$\hat{\mathcal{H}} = \frac{\hat{p}^2}{2m} + \frac{1}{2}k\hat{q}^2. \tag{1.42}$$

The harmonic potential (the second term of the preceding equation) counteracts the quantum (position) diffusion and preserves the wavepacket width at all times as shown in Fig. 1-3(a), or allows the periodic oscillation in the wavepacket width as shown in Figs. 1-3(b) and (c), depending on the initial wavepacket width. The (former) stationary wavepacket is called a coherent state and the (latter) pulsating wavepacket is referred to as a squeezed state.

Quantization of a single-mode electromagnetic field is mathematically identical to that of a single (mechanical) harmonic oscillator, as will be discussed in the next chapter. The electric field of a single-mode electromagnetic field is described by two

12   BASIC CONCEPTS

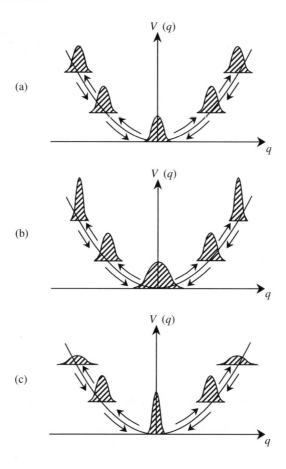

FIGURE 1-3: The minimum uncertain wavepackets in a harmonic potential $V(q) = \frac{1}{2}kq^2$. (a) Coherent state and (b)(c) quadrature amplitude squeezed states.

quadrature amplitudes a_1 and a_2:

$$E = c(a_1 \cos \omega t + a_2 \sin \omega t)u(r), \tag{1.43}$$

where c is a constant, ω an angular oscillation frequency, and $u(r)$ a modal wavefunction determined by boundary conditions. As we shall see, quantization of an electromagnetic field requires that the two quadrature amplitudes a_1 and a_2 are considered as Hermitian operators rather than c-numbers. If the constant c is chosen as

$$c = \sqrt{\frac{\hbar \omega}{2\varepsilon_0 V}}, \tag{1.44}$$

where V is the quantization (cavity) volume and $u(r)$ satisfies the normalization condition,

$$\int |u(r)|^2 dr^3 = V, \tag{1.45}$$

the operators $\hat{a}_1$ and $\hat{a}_2$ satisfy the following commutation relation and the Heisenberg uncertainty relation:

$$[\hat{a}_1, \hat{a}_2] = \frac{i}{2} \longrightarrow \langle \Delta \hat{a}_1^2 \rangle \langle \Delta \hat{a}_2^2 \rangle \geq \frac{1}{16}. \tag{1.46}$$

In analogy with a harmonic oscillator, $\hat{a} = \hat{a}_1 + i\hat{a}_2$ and $\hat{a}^\dagger = \hat{a}_1 - i\hat{a}_2$ are termed photon annihilation and creation operators.

We can construct a minimum uncertainty state of the electromagnetic field that satisfies the Heisenberg uncertainty relation (1.46) with equality, as an eigenstate of the non-Hermitian operator:

$$(e^r \hat{a}_1 + ie^{-r} \hat{a}_2)|\psi\rangle = (e^r \langle \hat{a}_1 \rangle + ie^{-r} \langle \hat{a}_2 \rangle)|\psi\rangle. \tag{1.47}$$

Note that the position and momentum operators in (1.32) are replaced by the two quadrature amplitude operators in (1.47). We can immediately obtain the variance of the two quadrature amplitudes and the uncertainty product:

$$\begin{matrix} \langle \Delta \hat{a}_1^2 \rangle = \frac{1}{4} e^{-2r} & \searrow \\ & & \langle \Delta \hat{a}_1^2 \rangle \langle \Delta \hat{a}_2^2 \rangle = \frac{1}{16}. \\ \langle \Delta \hat{a}_2^2 \rangle = \frac{1}{4} e^{2r} & \nearrow \end{matrix} \tag{1.48}$$

When $r = 0$, the two quadrature amplitudes have identical variances $\langle \Delta \hat{a}_1^2 \rangle = \langle \Delta \hat{a}_2^2 \rangle = \frac{1}{4}$, so the electric field has stationary noise distribution as shown in Fig. 1-4(a). Here it is assumed that $\langle \hat{a}_1 \rangle \neq 0$ and $\langle \hat{a}_2 \rangle = 0$. This specific state is termed a coherent state of the electromagnetic field [3]. A coherent state $|\alpha\rangle$ is an eigenstate of the (non-Hermitian) annihilation operator

$$\hat{a}|\alpha\rangle = \alpha|\alpha\rangle. \tag{1.49}$$

When $r > 0$, one quadrature amplitude $\hat{a}_1$ has smaller variance than the other quadrature amplitude $\hat{a}_2$, so the electric field has pulsating noise distribution as shown in Fig. 1-4(b). This state is called an (amplitude) squeezed state because the measurement of the amplitude has smaller noise as compared to the coherent state [4]–[6]. When $r < 0$, the squeezing between the two quadrature amplitudes is reversed, so the electric field features opposite pulsating noise distribution as shown in Fig. 1-4(c). This state is termed a (phase) squeezed state because the measurement of the phase yields smaller noise as compared to the coherent state [4]–[6].

14 BASIC CONCEPTS

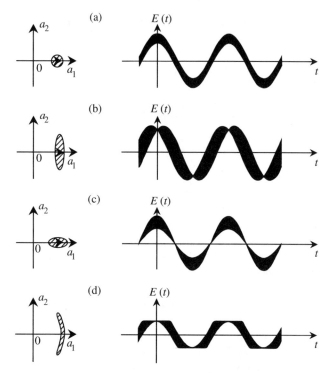

FIGURE 1-4: (a) A coherent state of light, (b) (amplitude) squeezed state of light, (c) (phase) squeezed state of light, and (d) number-phase squeezed state of light.

Figure 1-4(d) shows another type of squeezed state, for which the measurement of photon number has smaller noise as compared to the coherent state [7], [8]. This state is referred to as a number-phase squeezed state. We will discuss the properties and generation schemes of these squeezed states of light in greater detail in Chapter 3.

1.3. VON NEUMANN'S PROJECTION POSTULATE

The standard quantum theory based on the Schrödinger equation cannot describe the process of measurement. The evolution of the wavefunction described by the Schrödinger equation has two key features: it is a *reversible* process, so one can always undo the process in order to return to the initial state from the final state, and it is a *deterministic* evolution in which the final state is uniquely determined by the initial state if the Hamiltonian is known. On the other hand, the process of measurement is *irreversible* and *nondeterministic*. Once the measurement is completed and the information is extracted, it is impossible to return to the initial state. Before the measurement is completed, one cannot predict what the measurement result and the postmeasurement state will be.

Interest among experimental physicists shifted approximately 20 years ago from one-time destructive measurements on ensemble of quantum objects to repeated nondestructive measurements on a single quantum object. For example, an experimentalist prepares the quantum object carefully in an optimum eigenstate: If there exists an unknown external force that acts on the object, it will change the eigenvalue in an unknown manner. The experimentalist may then perform repeated measurements in order to determine the time evolution of eigenvalues and thus learn about the external force. This trend in experimental quantum optics, atomic physics, and scanning microscopy requires more than the standard formulation based on Schrödinger equation and the probability interpretation.

The rigorous description of what happens to a quantum object when it is measured was first formulated by von Neumann and is known as the von Neumann postulate of *wavepacket reduction* [9]. This reduction postulate is one of the two most important and fundamental assumptions in the quantum theory along with the Heisenberg uncertainty principle. The von Neumann postulate says the following:

1. The result of a measurement of an observable $\hat{q}$, for a system prepared in an initial state $\hat{\rho}_{in}$, will be one of the eigenvalues q_n of the observable $\hat{q}$ and the probability of obtaining the specific measurement result q_n is given by

$$P(q_n) = \mathrm{Tr}\left(|q_n\rangle\langle q_n|\hat{\rho}_{in}\right). \tag{1.50}$$

2. After the measurement, the system density operator jumps into a new state given by

$$\hat{\rho}_f = \frac{1}{P(q_n)}|q_n\rangle\langle q_n|\hat{\rho}_{in}|q_n\rangle\langle q_n|. \tag{1.51}$$

Here the eigenstates $|q_n\rangle$ of the observable $\hat{q}$ form a complete orthonormal set (i.e., $\sum_{q_n}|q_n\rangle\langle q_n| = \hat{I}$). Tr stands for the trace operation with respect to the measured system.

This original statement made by von Neumann is valid for an ideal (noise-free and nondestructive) measurement. It is assumed that there is no measurement error. Such a measurement is termed an exact measurement or sharp measurement. However, in order to predict real experimental results, we need to develop a formalism that handles more realistic measurements with a finite measurement error.

1.3.1. Measurement Error and Back Action Noise

The uncertainty relation for measurement of a quantum system should be discussed in the context of the quantum theory of measurement; namely, in the language of the *collapse* (reduction) *of the wavefunction* by a measurement action. The implication of the Heisenberg uncertainty principle on the quantum measurement process did not attract much attention until recently. The quantum theory of measurement has

16 BASIC CONCEPTS

been widely regarded as a problematic and dubious subject of little relevance to real physics. The reason for this irrelevance of the quantum theory of measurement was technological. Experimentalists could not simply perform repetitive nondestructive measurements at the quantum limit on a single quantum object until very recently. In 1980s, experimentalists in quantum optics and atomic physics finally began to catch up with the measurement concepts put forward by the inventors of quantum mechanics. It is now becoming possible to study experimentally the effect of a first measurement on an object (collapse of the wavefunction) at the time of a second measurement. To illustrate this second uncertainty relation, we will consider the following simple experiments.

We can measure the speed v of an object (or the momentum $p = mv$ if mass m is known) by measuring the Doppler frequency shift of a reflected optical pulse in a setup shown in Fig. 1-5:

$$\frac{\delta\omega}{\omega} = -\frac{2v}{c}. \tag{1.52}$$

Suppose the optical pulse (photon wavepacket) has a finite time duration τ. The Fourier-transform-limited spectral width of the wavepacket is $\Delta\omega \sim \frac{1}{\tau}$. We can then measure the velocity (or the momentum) by a single photon only within a measurement error of

$$\Delta v_{\text{meas. error}} = \frac{c}{2}\frac{\Delta\omega}{\omega} = \frac{c}{2\omega\tau} \rightarrow \Delta p_{\text{meas. error}} = \frac{mc}{2\omega\tau}. \tag{1.53}$$

The reflected photon gives the object a momentum (photon recoil) $\frac{2\hbar\omega}{c}$, but the exact moment of time when the momentum is transferred from the photon to the object is uncertain due to finite pulse duration τ. This uncertainty in the collision time between the object and the photon results in the position uncertainty of the object, which is the back action noise of this measurement:

$$\Delta x_{\text{back action}} \simeq \frac{2\hbar\omega}{cm} \times \frac{\tau}{2} \simeq \frac{\hbar\omega\tau}{cm}. \tag{1.54}$$

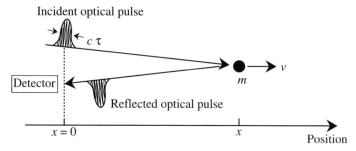

FIGURE 1-5: A setup for measuring the momentum or the position of an object by a single-photon wavepacket.

The product of (1.53) and (1.54) is

$$\Delta p_{\text{meas. error}} \, \Delta x_{\text{back action}} \simeq \frac{\hbar}{2}. \tag{1.55}$$

In order to decrease the measurement error of the momentum, we can use a photon with a shorter wavelength or longer pulse duration or many photons; then we must accept larger back action noise imposed on the position.

We can alternatively measure the position x of the object by measuring the delay time T of the reflected optical pulse in the same setup shown in Fig. 1-5:

$$T = \frac{2x}{c}, \tag{1.56}$$

where the optical pulse is emitted and detected at the same position $x = 0$. We can measure the position by a single photon within a measurement error of

$$\Delta x_{\text{meas. error}} = \frac{c\tau}{4}, \tag{1.57}$$

due to the finite pulse duration. The momentum uncertainty is imposed on the object by the finite spectral width of the photon, which is the back action noise of this measurement:

$$\Delta p_{\text{back action}} = \frac{2\hbar \Delta \omega}{c} = \frac{2\hbar}{c\tau}. \tag{1.58}$$

The product of (1.57) and (1.58) is

$$\Delta x_{\text{meas. error}} \Delta p_{\text{back action}} = \frac{\hbar}{2}. \tag{1.59}$$

We can see that the roles of the position and the momentum are interchanged. The preceding two examples highlight the main elements of a quantum measurement process:

1. The readout of information about the observable of a measured system has a finite measurement error due to the uncertainty of the readout observable of the measuring apparatus. In the preceding examples the photon's time uncertainty determines the measurement error of the object position and the photon's energy uncertainty determines the measurement error of the object momentum.
2. The inevitable back action noise (uncontrollable disturbance) is imposed on the conjugate observable of the measured system by the uncertainty of the conjugate observable of the measuring apparatus. In the preceding examples the photon's energy uncertainty determines the back action noise on the object momentum and the photon's time uncertainty determines the back action noise on the object position.
3. The uncertainty relation between the measurement error and the back action noise originates from the uncertainty relation of the two conjugate observ-

ables of the measuring apparatus (the single photon wavepacket in the preceding examples). This uncertainty relation is independent of the initial state of the measured system; that is, *how the measured system is prepared before the measurement* has nothing to do with this uncertainty relation.

4. An inevitable and irreversible process happens for the measuring apparatus (the death of the photon and birth of the photoelectron in the preceding examples). This process makes the system state jump into a completely new state that depends on the measurement result and the state of the measuring apparatus.

5. The measurement result for each measurement is, in general, unpredictable. As discussed earlier, this fact is often referred to as *lack of causality* in the quantum measurement process. Only statistics for many measurement results performed for identical systems are determined by the system state before measurement.

1.3.2. Pauli's First-Kind and Second-Kind Measurements

In many real measurements, the final state of a measured system is substantially different from the one computed by (1.51). A simple example is the measurement of the photon number of an electromagnetic field by a photodetector. In this example, the final state is always a vacuum state irrespective of the measurement result. In fact, in most cases real measuring devices often produce an excess back action effect on the measured system, above and beyond the fundamental limit that is demanded by the Heisenberg uncertainty principle and computed by the von Neumann projection postulate. Such excess back action is not fundamental, and one can indeed measure the photon number without destruction of the photons. A quantum measurement, in which all such spurious back actions are eliminated and the von Neumann projection postulate can be used to calculate the postmeasurement state, is called Pauli's first-kind measurement [10]. A quantum measurement for which the von Neumann projection postulate does not apply is called Pauli's second-kind measurement [10].

A quantum measurement in which the measured system interacts with the macroscopic measuring device (meter) is called a *direct measurement*. In a direct measurement, considerable randomness exists in the interaction between the quantum system and the classical meter with many degrees of freedom. Consequently, the classical meter perturbs the quantum system far more strongly than the minimum perturbation required by the Heisenberg uncertainty principle. In order to realize Pauli's first-kind measurement, we can employ an *indirect measurement* strategy. An indirect measurement has two steps, as shown in Fig. 1-6. In the first step, the quantum system interacts with the quantum probe, which is optimally prepared for copying the information of the measured observable of the quantum system. This interaction is governed by the standard quantum theory (unitary time evolution) (i.e., reduction of the state vector is not present in the first step). However, quantum correlation is established between the measured observable $\hat{q}$ of the quantum system, and the readout observable $\hat{P}$ of the quantum probe and the (virtual) back action noise is imposed

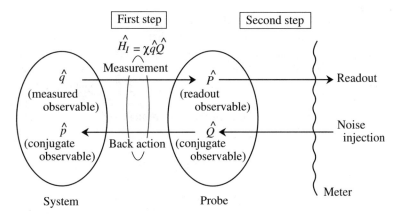

FIGURE 1-6: An indirect quantum measurement process.

on the conjugate observable $\hat{p}$ of the measured system by the quantum noise of the conjugate observable $\hat{Q}$ of the probe.

The second step is the direct measurement of the readout observable of the quantum probe by the classical meter. Prior to the second step, the first step is terminated; that is, the interaction between the quantum system and the quantum probe is switched off and, therefore, the quantum system is never directly perturbed by the classical meter. The readout observable of the quantum probe is exactly measured by the direct measurement. In the second step, a corresponding reduction of the state of the quantum system is enforced since quantum correlation exists between the quantum system and the quantum probe. Since information pertaining to the quantum noise of the conjugate observable $\hat{Q}$ of the quantum probe is lost in this second step, the (virtual) back action noise imposed on the conjugate observable $\hat{p}$ of the system becomes *real* and *permanent*. One cannot counteract the back action nor even obtain knowledge of the back action. In this manner, the measurement produces an irreversible and nondeterministic change in the quantum system. The measurement error of the measured observable and the back action noise imposed on the conjugate observable satisfy the Heisenberg uncertainty relation.

1.3.3. Density Operator and Liouville–von Neumann Equation

If we have complete information about a system, the quantum state of the system is described by a single ket vector $|\psi\rangle$. However, in many cases, it is not possible to obtain complete information about the system; it is only known with a finite probability P_ψ that the system is in a specific state $|\psi\rangle$. Such a system is in a *mixed state* and its density operator is given by

$$\hat{\rho} = \sum_\psi P_\psi |\psi\rangle\langle\psi|, \qquad (1.60)$$

where $P_\psi > 0$ and $\sum_\psi P_\psi = 1$. This probability expressed by P_ψ should not be confused with the intrinsic quantum uncertainty that was discussed in the previous section. Quantum uncertainty implemented by the Heisenberg uncertainty principle exists even if complete information about the system is available. On the other hand, statistical uncertainty implemented by P_ψ is extrinsic (i.e., due to lack of information about the system).

The expectation value for the ensemble measurements of an observable $\hat{O}$ for such a mixed state is calculated by

$$\begin{aligned}\langle \hat{O} \rangle &= \sum_\psi P_\psi \langle \psi | \hat{O} | \psi \rangle \\ &= \sum_\psi P_\psi \mathrm{Tr}(|\psi\rangle\langle\psi|\hat{O}) \\ &= \mathrm{Tr}\left[\sum_\psi P_\psi |\psi\rangle\langle\psi|\hat{O}\right] \\ &= \mathrm{Tr}(\hat{\rho}\hat{O}). \end{aligned} \qquad (1.61)$$

When $P_{\psi_0} = 1$, $\hat{\rho}$ reduces to the density operator of a pure state, $\hat{\rho} = |\psi_0\rangle\langle\psi_0|$, and (1.61) is reduced to the simple form $\langle \hat{O} \rangle = \langle \psi_0 | \hat{O} | \psi_0 \rangle$.

The time-dependent Schrödinger equation for the ket vector $|\psi\rangle$ is

$$i\hbar \frac{\partial}{\partial t}|\psi\rangle = \hat{H}|\psi\rangle, \qquad (1.62)$$

where $\hat{H}$ is the system Hamiltonian. The Hermitian adjoint of (1.62) is

$$-i\hbar \frac{\partial}{\partial t}\langle\psi| = \langle\psi|\hat{H}. \qquad (1.63)$$

Using (1.62) and (1.63), one has the equation of motion for the time evolution of the density operator:

$$\begin{aligned} i\hbar\frac{\partial}{\partial t}\hat{\rho} &\equiv i\hbar\frac{\partial}{\partial t}(|\psi\rangle\langle\psi|) \\ &= i\hbar\left\{\frac{\partial|\psi\rangle}{\partial t}\langle\psi| + |\psi\rangle\frac{\partial\langle\psi|}{\partial t}\right\} \\ &= \hat{H}|\psi\rangle\langle\psi| - |\psi\rangle\langle\psi|\hat{H} \\ &= [\hat{H}, \hat{\rho}]. \end{aligned} \qquad (1.64)$$

It is easily shown that (1.64) is also valid for a mixed state. This is the Liouville–von Neumann equation.

1.3.4. General Description of Indirect Quantum Measurements

Consider an indirect quantum measurement in which the quantum system is initially prepared in a state $\hat{\rho}_s$ and the quantum probe in a state $\hat{\rho}_p$ (Fig. 1-6). An interaction Hamiltonian $\hat{\mathcal{H}}_I$ introduces a unitary evolution $\hat{U}$ for both system and probe. The appropriate interaction Hamiltonian that copies the information of the measured observable $\hat{q}$ of the system onto the readout observable $\hat{P}$ of the probe is given by

$$\hat{\mathcal{H}}_I = \chi \hat{q} \hat{Q}, \tag{1.65}$$

where χ is a coupling constant and $\hat{Q}$ is the conjugate observable to $\hat{P}$. The Heisenberg equations of motion for the observables $\hat{P}$ and $\hat{p}$ are

$$\frac{d}{dt}\hat{P} = \frac{1}{i\hbar}[\hat{P}, \hat{\mathcal{H}}_I] = -\chi \hat{q}, \tag{1.66}$$

$$\frac{d}{dt}\hat{p} = \frac{1}{i\hbar}[\hat{p}, \hat{\mathcal{H}}_I] = -\chi \hat{Q}, \tag{1.67}$$

where the time evolutions by the system and probe Hamiltonians are neglected and the commutation relations, $[\hat{q}, \hat{p}] = i\hbar$ and $[\hat{Q}, \hat{P}] = i\hbar$, are assumed. Equation (1.66) indicates that the time evolution of the readout observable $\hat{P}$ of the probe is governed by the measured observable $\hat{q}$ of the system. This is the required process for the measurement of $\hat{q}$ in terms of $\hat{P}$. Equation (1.67) indicates that the time evolution of the conjugate observable $\hat{p}$ of the system is driven by the conjugate observable $\hat{Q}$ of the probe. This results in the inevitable back action noise imposed on $\hat{p}$ by the quantum uncertainty of $\hat{Q}$.

The interaction $\hat{\mathcal{H}}_I$ of (1.65) brings the system and probe into a correlated quantum state:

$$\hat{\rho}_i = \hat{\rho}_s \otimes \hat{\rho}_p \longrightarrow \hat{\rho}_c = \hat{U} \hat{\rho}_s \otimes \hat{\rho}_p \hat{U}^\dagger. \tag{1.68}$$

The corresponding reduced density operators of the system and the probe alone are

$$\hat{\rho}'_s = \text{Tr}_p(\hat{U} \hat{\rho}_s \otimes \hat{\rho}_p \hat{U}^\dagger), \tag{1.69}$$

$$\hat{\rho}'_p = \text{Tr}_s(\hat{U} \hat{\rho}_s \otimes \hat{\rho}_p \hat{U}^\dagger), \tag{1.70}$$

where Tr_p and Tr_s denote the trace operation over the probe and system coordinates, respectively. The density operator (1.69) corresponds to the system state when we keep all the ensemble members (wavepackets) regardless of the measurement result. This means that even though the probe observable is measured by an experimentist, he or she does not select the wavepackets based on the measurement result and simply keep all the wavepackets for the next measurement. Since this is equivalent to the case that the available information is simply discarded (dissipation rather than measurement), the density operator is generally left in a mixed state even if $\hat{\rho}_s$ is initially in a pure state.

The direct and exact measurement of the readout observable $\hat{P}$ of the probe reports one of the eigenvalues P_n of $\hat{P}$. Since this measurement can be performed without a measurement error, there is a one-to-one correspondence between the outcome P_n of the direct measurement on the probe and the inferred eigenvalue $\tilde{q}_n$ of the measured observable $\hat{q}$ of the system. Due to this one-to-one correspondence, we can use $\tilde{q}$ not only as the inferred value of the system observable $\hat{q}$ but also as the replacement of the readout value P_n of the probe observable $\hat{P}$. The projection operator for the second step is thus written as $|\tilde{q}\rangle\langle\tilde{q}|$, where $|\tilde{q}\rangle$ is an eigenstate of $\hat{P}$. The probability of obtaining the specific result $\tilde{q}$ is given by

$$P(\tilde{q}) = \text{Tr}_p \left(|\tilde{q}\rangle\langle\tilde{q}|\hat{\rho}'_p \right). \tag{1.71}$$

Using (1.70) and (1.71), one obtains [11]

$$P(\tilde{q}) = \text{Tr}_s \left[\hat{X}(\tilde{q})\hat{\rho}_s \right], \tag{1.72}$$

where

$$\hat{X}(\tilde{q}) = \text{Tr}_p \left[\hat{U}^\dagger |\tilde{q}\rangle\langle\tilde{q}|\hat{U}\hat{\rho}_p \right]. \tag{1.73}$$

The operator $\hat{X}(\tilde{q})$ characterizes the three measurement processes completely; that is, the preparation of the initial probe state in terms of $\hat{\rho}_p$, the unitary time evolution in terms of $\hat{U}$ for establishing the quantum correlation between the system and the probe, and the readout of the measurement result $\tilde{q}$ in terms of the projection operator $|\tilde{q}\rangle\langle\tilde{q}|$. The operator $\hat{X}(\tilde{q})$ is termed the *generalized projection operator* and satisfies the completeness relation [11]

$$\int_{-\infty}^{\infty} \hat{X}(\tilde{q})\, d\tilde{q} = \hat{I}. \tag{1.74}$$

The difference between the generalized projection operator $\hat{X}(\tilde{q})$ and the von Neumann projection operator $|q\rangle\langle q|$ that appeared in (1.50) and (1.51) is that $\hat{X}(\tilde{q})$ takes into account the finite measurement error due to the finite quantum noise of the readout observable $\hat{P}$ that is embodied in the initial density operator $\hat{\rho}_p$ of the probe and the finite coupling strength between the system and the probe. In fact, $\hat{X}(\tilde{q})$ may be written as [11]

$$\hat{X}(\tilde{q}) = \int_{-\infty}^{\infty} |q\rangle x(\tilde{q}, q)\langle q|dq, \tag{1.75}$$

where $x(\tilde{q}, q) = \sum_j P_j |_p\langle\tilde{q}|\hat{U}(q)|\psi_j\rangle_p|^2$ is the conditional probability that the classical meter will report a value $\tilde{q}$ when the quantum system is in an eigenstate $|q\rangle$.

Here, we assumed that the initial density operator of the probe is

$$\hat{\rho}_p = \sum_j P_j |\psi_j\rangle\langle\psi_j| \tag{1.76}$$

The density operator of the quantum system after this measurement with the result $\tilde{q}$ is

$$\hat{\rho}_s(\tilde{q}) = \frac{1}{P(\tilde{q})} \langle \tilde{q} | \hat{U} \hat{\rho}_s \otimes \hat{\rho}_p \hat{U}^\dagger | \tilde{q} \rangle. \tag{1.77}$$

Substituting $\hat{\rho}_p$ into this expression we obtain

$$\hat{\rho}_s(\tilde{q}) = \frac{1}{P(\tilde{q})} \sum_j P_j \hat{Y}_j(\tilde{q}) \hat{\rho}_s \hat{Y}_j^\dagger(\tilde{q}), \tag{1.78}$$

where

$$\hat{Y}_j(\tilde{q}) = \langle \tilde{q} | \hat{U} | \psi_j \rangle. \tag{1.79}$$

Equation (1.78) is the generalization of (1.51), and it takes into account finite measurement error. The final state (1.78) is a mixed state. This additional uncertainty is not fundamental, but it stems from the fact that the quantum probe is initially in a mixed state (1.76). If the quantum probe is prepared in a pure state, the final state of the quantum system is left in a pure state (provided, of course, that $\hat{\rho}_s$ is initially in a pure state):

$$\hat{\rho}_s(\tilde{q}) = \frac{1}{P(\tilde{q})} \hat{Y}(\tilde{q}) \hat{\rho}_s \hat{Y}^\dagger(\tilde{q}), \tag{1.80}$$

where $\hat{Y}(\tilde{q}) = \langle \tilde{q} | \hat{U} | \psi \rangle$ is the operator amplitude for the probe to evolve from $|\psi\rangle$ to $|\tilde{q}\rangle$ by the unitary evolution operator $\hat{U}$. In this case, only the intrinsic back action noise dictated by the Heisenberg uncertainty principle is embodied in the final system state. The generalized projection operator (1.73) is written in terms of $\hat{Y}(\tilde{q})$:

$$\hat{X}(\tilde{q}) = \hat{Y}^\dagger(\tilde{q}) \hat{Y}(\tilde{q}). \tag{1.81}$$

The difference between the system state $\hat{\rho}'_s$ (1.69) and $\hat{\rho}_s(\tilde{q})$ (1.78) is that the former represents the statistical properties of all the wavepackets after the measurement if we keep every wavepacket irrespective of the readout, while the latter describes the statistical properties of the specific wavepackets that reported a certain readout $\tilde{q}$. The (mathematical) projection procedure in (1.78) gives this selection of wavepackets depending on the measurement result. On the other hand, the (mathematical) trace operation in (1.69) implies that there is no selection of wavepackets based on the measurement result. This difference of selection and no selection of wavepackets based on the measurement result is the basis for understanding how a quantum measurement is different from simple dissipation. Nonunitary state reduction induced by a projection operator is a natural consequence of the abrupt change of the ensemble members.

1.3.5. Quantum Nondemolition Measurement

In an indirect quantum measurement shown in Fig. 1-6, the measured observable $\hat{q}$ of the system and the readout observable $\hat{P}$ of the probe obey the Heisenberg equations of motion:

$$-i\hbar \frac{d}{dt}\hat{q} = [\hat{H}_s, \hat{q}] + [\hat{H}_I, \hat{q}], \tag{1.82}$$

$$-i\hbar \frac{d}{dt}\hat{P} = [\hat{H}_p, \hat{P}] + [\hat{H}_I, \hat{P}], \tag{1.83}$$

where $\hat{H}_s$ and $\hat{H}_p$ are the unperturbed Hamiltonian of the system and the probe, respectively. $\hat{H}_I$ is the interaction Hamiltonian between the two. The first commutators in (1.82) and (1.83) represent the free evolution of $\hat{q}$ and $\hat{P}$, while the second commutators represent the contribution of the interaction.

In order to measure $\hat{q}$ using $\hat{P}$ as the readout observable, the commutator $[\hat{H}_I, \hat{P}]$ should be nonzero and $\hat{H}_I$ should be a function of $\hat{q}$. The two requirements are satisfied, for instance, if the interaction Hamiltonian is given by (1.65). In general, a measurement of $\hat{q}$ affects the free evolution of $\hat{q}$ in the following two ways. One way is the direct change of $\hat{q}$ driven by the commutator $[\hat{H}_I, \hat{q}]$ in (1.82). Such a direct back action of the measurement is eliminated if the interaction Hamiltonian $\hat{H}_I$ commutes with $\hat{q}$:

$$[\hat{H}_I, \hat{q}] = 0. \tag{1.84}$$

An interaction Hamiltonian satisfying this condition is termed the *back action evading* type.

The other way of changing $\hat{q}$ is through the quantum uncertainty introduced on the conjugate observable $\hat{p}$ by the measurement. The Heisenberg uncertainty principle dictates that when $\hat{q}$ is measured with a finite measurement error, the inevitable uncertainty should be introduced on the conjugate observable $\hat{p}$. If the unperturbed system Hamiltonian $\hat{\mathcal{H}}_s$ contains $\hat{p}$, the free evolution of $\hat{q}$ is affected by the uncertainty introduced on $\hat{p}$ by the measurement. Such an indirect back action of the measurement is eliminated if the unperturbed system Hamiltonian $\hat{\mathcal{H}}_s$ commutes with $\hat{q}$:

$$[\hat{\mathcal{H}}_s, \hat{q}] = 0. \tag{1.85}$$

The observable $\hat{q}$ satisfying this condition is called a *QND observable*.

If (1.84) and (1.85) are simultaneously satisfied, the observable $\hat{q}$ can be measured repeatedly with an infinitesimally small measurement error without introducing the back action noise on $\hat{q}$. Such a measurement is called a quantum nondemolition (QND) measurement.

One example for QND observables is the momentum $\hat{p}$ of a free particle [12]. The unperturbed system Hamiltonian $\hat{\mathcal{H}}_s = \frac{\hat{p}^2}{2m}$ commutes with $\hat{p}$. The measurement of $\hat{p}$ inevitably perturbs the position $\hat{q}$ of the particle but the position uncertainty never couples to the momentum because the system Hamiltonian does not contain $\hat{q}$. This is not the case for the measurement of the position $\hat{q}$ of a free particle. The measurement of $\hat{q}$ inevitably perturbs the momentum $\hat{p}$ of the particle and the momentum uncertainty couples to the position uncertainty at a later time:

$$\hat{q}(t) = \hat{q}(0) + \frac{\hat{p}(0)}{2m}t, \tag{1.86}$$

where $\hat{q}(0)$ and $\hat{p}(0)$ are the position and the momentum operators immediately after the first measurement of the position. The position uncertainty changes due to the back action noise imposed on the momentum as time elapses,

$$\langle \Delta \hat{q}(t)^2 \rangle = \langle \Delta \hat{q}(0)^2 \rangle + \frac{\langle \Delta \hat{p}(0)^2 \rangle}{4m^2}t^2 + \frac{t}{m}\langle \Delta \hat{q}(0)\Delta \hat{p}(0) + \Delta \hat{p}(0)\Delta \hat{q}(0)\rangle. \tag{1.87}$$

If the position and momentum immediately after the measurement are uncorrelated, the third term of (1.87) is zero and the position uncertainty monotonically increases as time elapses. This means that if one measures the position of a free particle very accurately at $t = 0$, the particle position becomes very uncertain at the time of the second measurement due to the back action noise [the second term of (1.87)]: This is due to the large momentum uncertainty in such a case.

Another example for QND observables is the quantum number $\hat{n}$ of a harmonic oscillator [11]. The unperturbed system Hamiltonian $\hat{\mathcal{H}}_s = \hbar\omega(\hat{n} + \frac{1}{2})$ commutes with $\hat{n}$. The measurement of $\hat{n}$ inevitably perturbs the phase $\hat{\phi}$, but the phase uncertainty never couples to the quantum number because the system Hamiltonian does not depend on $\hat{\phi}$.

1.4. SIMULTANEOUS MEASUREMENT OF TWO CONJUGATE OBSERVABLES

As mentioned earlier, the standard formulation of the Heisenberg uncertainty principle is a statement about a measured system. The connection of this formulation to actual experiments requires two ideal and noise-free measurements of each observable for an ensemble of identical systems. If we consider a simultaneous measurement of the position $\hat{q}$ and the momentum $\hat{p}$ of the same system, the second Heisenberg uncertainty principle concerning the measuring apparatus will emerge

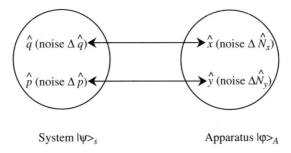

FIGURE 1-7: A model for simultaneous measurements of conjugate observables $\hat{q}$ and $\hat{p}$.

[13]. The model for such a simultaneous measurement with a measured system and measuring apparatus is shown in Fig. 1-7.

The coupled system and apparatus are described by a *tensor product Hilbert space* $H = H_1 \otimes H_2$. If the system and the apparatus are uncorrelated before the measurement interaction, the overall system is described by the state vector

$$|\psi\rangle = |\psi\rangle_S |\varphi\rangle_A, \tag{1.88}$$

where $|\psi\rangle_S$ and $|\varphi\rangle_A$ are the initial system and apparatus states. The conjugate observables of the system and those of the apparatus are described by the product H-space operators:

$$\hat{q} = \hat{q}_1 \otimes \hat{I}_2, \quad \hat{p} = \hat{p}_1 \otimes \hat{I}_2, \tag{1.89}$$

$$\hat{x} = \hat{I}_1 \otimes \hat{x}_2, \quad \hat{y} = \hat{I}_1 \otimes \hat{y}_2, \tag{1.90}$$

where $\hat{I}_1$ and $\hat{I}_2$ are the identity operators for the H_1 and H_2 subspaces. The interaction between the system and the apparatus establishes a quantum correlation simultaneously between $\hat{q}$ and $\hat{x}$ and between $\hat{p}$ and $\hat{y}$. After the interaction described by the unitary operator $\hat{U}$, the expectation values of the apparatus Heisenberg operators, $\hat{x} = \hat{U}^\dagger \hat{x}(0) \hat{U}$ and $\hat{y} = \hat{U}^\dagger \hat{y}(0) \hat{U}$, are given by

$$\langle \hat{x} \rangle \equiv {}_A\langle \varphi |_S\langle \psi | \hat{x} | \psi \rangle_S | \varphi \rangle_A$$
$$= c_x \langle \hat{q} \rangle, \tag{1.91}$$

$$\langle \hat{y} \rangle \equiv {}_A\langle \varphi |_S\langle \psi | \hat{y} | \psi \rangle_S | \varphi \rangle_A$$
$$= c_y \langle \hat{p} \rangle, \tag{1.92}$$

where it is assumed that there are no offsets in the apparatus readout observables, and c_x and c_y are the coupling factors. The noise operators are defined by

SIMULTANEOUS MEASUREMENT OF TWO CONJUGATE OBSERVABLES 27

$$\hat{N}_x = \hat{x} - c_x \hat{q}, \tag{1.93}$$

$$\hat{N}_y = \hat{y} - c_y \hat{p}. \tag{1.94}$$

The average values of the noise operators are zero (i.e., $\langle \hat{N}_x \rangle = \langle \hat{N}_y \rangle = 0$).

We further assume that $\hat{x}$ and $\hat{y}$ can be measured simultaneously. Mathematically, this requires that $\hat{x}$ and $\hat{y}$ commute:

$$[\hat{x}, \hat{y}] = 0. \tag{1.95}$$

Substituting (1.93) and (1.94) into (1.95), we have

$$[c_x \hat{q} + \hat{N}_x, c_y \hat{p} + \hat{N}_y] = c_x c_y [\hat{q}, \hat{p}] + [\hat{N}_x, \hat{N}_y] + c_x [\hat{q}, \hat{N}_y] + c_y [\hat{N}_x, \hat{p}]$$
$$= 0. \tag{1.96}$$

Since $\hat{N}_x$ and $\hat{N}_y$ are the (internal) noise operators associated with the measuring apparatus and are thus uncorrelated with the observables $\hat{q}$ and $\hat{p}$ of the measured system, we have

$$\langle [\hat{q}, \hat{N}_y] \rangle = \langle [\hat{N}_x, \hat{p}] \rangle = 0. \tag{1.97}$$

From (1.96) and (1.97), we have

$$\langle [\hat{N}_x, \hat{N}_y] \rangle = -c_x c_y \langle [\hat{q}, \hat{p}] \rangle = -i\hbar c_x c_y. \tag{1.98}$$

The uncertainty product of the noise operators is therefore constrained by

$$\langle \Delta \hat{N}_x^2 \rangle \langle \Delta \hat{N}_y^2 \rangle \geq \frac{\hbar^2}{4} c_x^2 c_y^2. \tag{1.99}$$

Now we define the normalized readout operators for $\hat{q}$ and $\hat{p}$ by

$$\hat{q}_{\text{obs}} \equiv \frac{\hat{x}}{c_x} = \hat{q} + \frac{\hat{N}_x}{c_x}, \tag{1.100}$$

$$\hat{p}_{\text{obs}} \equiv \frac{\hat{y}}{c_y} = \hat{p} + \frac{\hat{N}_y}{c_y}. \tag{1.101}$$

Since $\hat{q}$ and $\hat{N}_x$ and $\hat{p}$ and $\hat{N}_y$ are uncorrelated, the uncertainty of $\hat{q}_{\text{obs}}$ and $\hat{p}_{\text{obs}}$ is independently contributed by the system and the apparatus:

$$\langle \Delta \hat{q}_{\text{obs}}^2 \rangle = \langle \Delta \hat{q}^2 \rangle + \frac{1}{c_x^2} \langle \Delta \hat{N}_x^2 \rangle, \tag{1.102}$$

$$\langle \Delta \hat{p}_{\text{obs}}^2 \rangle = \langle \Delta \hat{p}^2 \rangle + \frac{1}{c_y^2} \langle \Delta \hat{N}_y^2 \rangle. \tag{1.103}$$

The uncertainty product can therefore be calculated as

$$\langle \Delta \hat{q}_{\text{obs}}^2 \rangle \langle \Delta \hat{p}_{\text{obs}}^2 \rangle = \langle \Delta \hat{q}^2 \rangle \langle \Delta \hat{p}^2 \rangle + \frac{1}{c_x^2 c_y^2} \langle \Delta \hat{N}_x^2 \rangle \langle \Delta \hat{N}_y^2 \rangle$$

$$+ \frac{1}{c_y^2} \langle \Delta \hat{q}^2 \rangle \langle \Delta \hat{N}_y^2 \rangle + \frac{1}{c_x^2} \langle \Delta \hat{p}^2 \rangle \langle \Delta \hat{N}_x^2 \rangle$$

$$\geq \hbar^2. \tag{1.104}$$

The equality holds when $\langle \Delta \hat{q}^2 \rangle = \frac{1}{c_x^2} \langle \Delta \hat{N}_x^2 \rangle$ and $\langle \Delta \hat{p}^2 \rangle = \frac{1}{c_y^2} \langle \Delta \hat{N}_y^2 \rangle$ and when the quantum state of the particle is prepared in a minimum uncertainty state. The minimum uncertainty product (1.104) is *four times* larger than the standard uncertainty product (1.23) and yields an inherent and irreducible lower bound for the simultaneous measurement of conjugate observables [13].

In summary, the role of the Heisenberg uncertainty principle is twofold. The two conjugate observables $\hat{q}$ and $\hat{p}$ cannot be measured precisely due to the commutator bracket $[\hat{q}, \hat{p}] = i\hbar$. If $\hat{q}$ and $\hat{p}$ of such a system are to be simultaneously measured, the *effective* commutator bracket must be eliminated because the commutator bracket itself inhibits the simultaneous measurement. For this purpose, one needs to add new quantum noise operators onto $\hat{q}$ and $\hat{p}$ [see (1.100) and (1.101)]. The new operators $\hat{q}_{\text{obs}} \equiv \hat{q} + \hat{N}_x/c_x$ and $\hat{p}_{\text{obs}} \equiv \hat{p} + \hat{N}_y/c_y$ commute and thus can be simultaneously measured, but only at the expense of additional noise.

REFERENCES

[1] P. A. M. Dirac, *The Principles of Quantum Mechanics* (Clarendon Press, Oxford, 1958).
[2] W. H. Louisell, *Quantum Statistical Properties of Radiation* (Wiley, New York, 1973).
[3] R. Glauber, Phys. Rev. **130**, 2529 (1963); ibid., **131**, 2766 (1963).
[4] H. Takahasi, in *Adv. Commun. Syst.*, ed. A. V. Balakrishnan (Academic Press, New York, 1965), p. 277.
[5] D. Stoler, Phys. Rev. **D1**, 3217 (1970).
[6] H. P. Yuen, Phys. Rev. **A13**, 2226 (1976).
[7] R. Jackiw, J. Math. Phys. **9**, 339 (1968).
[8] M. Kitagawa and Y. Yamamoto, Phys. Rev. A **34**, 3974 (1986).
[9] J. von Neumann, *Mathematical Foundations of Quantum Mechanics* (Princeton University Press, Princeton, 1955).
[10] W. Pauli, Phys. Rev. **58**, 716 (1940).
[11] V. B. Braginsky and F. Y. Khalili, *Quantum Measurement* (Cambridge University Press, Cambridge, 1992).
[12] L. Landau and R. Peierls, Z. Phys. **69**, 56 (1931).
[13] E. Arthurs and J. L. Kelly Jr., Bell Syst. Tech. J. **44**, 725 (1965).

2 Canonical Quantization

The goal of this chapter is to present an overview of the canonical nonrelativistic field quantization procedure based on the minimum action principle and the Lagrangian formalism. After a brief review of the Lagrangian formalism for point particles and fields, the quantization of the electromagnetic field in the Coulomb gauge is considered. The text by C. Cohen-Tannoudji, J. Dupont-Roc, and G. Grynberg [1] gives an excellent detailed treatment of the material presented in Sections 2.1 and 2.2. Section 2.3 is devoted to the *second quantization* of the Schrödinger matter field, which provides a basis for discussions of excitons and many-atom effects.

2.1. MINIMUM ACTION PRINCIPLE AND LAGRANGIAN FORMALISM

We start by considering a system of point particles: If N is the number of degrees of freedom in the system, the Lagrangian formalism can be written as a function of N-coordinates, N-velocity components, and time [$L(x_1, \ldots, x_N; \dot{x}_1, \ldots, \dot{x}_N; t)$]. In its simplest form, the Lagrangian is the difference between the kinetic and potential energies ($L = T - V$). The time-dependent coordinates $x_i(t)$ that the Lagrangian formalism depends on correspond to *paths* that the system could follow. The Lagrangian and its definite time integral

$$S = \int_{t_1}^{t_2} L(x_1, \ldots, x_N; \dot{x}_1, \ldots, \dot{x}_N; t)\, dt, \qquad (2.1)$$

referred to as the *action*, depend on the particular *path* the system follows. The *principle of least action* states that the actual path that the system would follow is the one that makes the action an extremum. Therefore, if we let $x_i(t)$, $i = 1, \ldots, N$ denote the actual path, then small deviations from this path should leave the action unchanged. Mathematically, if we denote the deviated path by

$$x'_i(t) = x_i(t) + \delta x_i(t),$$
$$\dot{x}'_i(t) = \dot{x}_i(t) + \delta \dot{x}_i(t),$$

with $\delta x_i(t_1) = \delta x_i(t_2) = 0$, we should obtain

$$\delta S = \int_{t_1}^{t_2} \delta L \, dt$$
$$= \int_{t_1}^{t_2} \left[\sum_{i=1}^{N} \frac{\partial L}{\partial x_i} \delta x_i + \frac{\partial L}{\partial \dot{x}_i} \delta \dot{x}_i \right] dt. \tag{2.2}$$

We can integrate the second term in (2.2) using integration by parts. Since $\delta x_i(t_1) = \delta x_i(t_2) = 0$, we obtain

$$\delta S = \int_{t_1}^{t_2} dt \sum_{i=1}^{N} \left[-\frac{d}{dt}\left(\frac{\partial L}{\partial \dot{x}_i}\right) + \frac{\partial L}{\partial x_i} \right] \delta x_i + \sum_{i=1}^{N} \left[\frac{\partial L}{\partial \dot{x}_i} \delta x_i \right]_{t_1}^{t_2}$$
$$= \int_{t_1}^{t_2} dt \sum_{i=1}^{N} \left[-\frac{d}{dt}\left(\frac{\partial L}{\partial \dot{x}_i}\right) + \frac{\partial L}{\partial x_i} \right] \delta x_i. \tag{2.3}$$

In order to have $\delta S = 0$ for all δx_i, the expression (in the brackets) of (2.3) should be equal to zero; therefore, for each coordinate x_i

$$\frac{d}{dt}\left(\frac{\partial L}{\partial \dot{x}_i}\right) = \frac{\partial L}{\partial x_i}. \tag{2.4}$$

The set of N second-order differential equations given by (2.4) are referred to as *Lagrange equations* and they directly give the equations of motion for the dynamical system. A detailed discussion of the variational principle and the derivation of the Lagrange equations can be found in the classical mechanics text by Landau and Lifshitz [2].

The Hamiltonian $H(x_1, \ldots, x_N; \dot{x}_1, \ldots, \dot{x}_N; t)$ of the system can be obtained from its Lagrangian only after identifying the *conjugate momentum* p_i associated with each coordinate (i.e., degree of freedom) by

$$p_i = \frac{\partial L}{\partial \dot{x}_i}, \tag{2.5}$$

and then setting

$$H = \sum_i \dot{x}_i p_i - L. \tag{2.6}$$

Starting from the total differential of the Lagrangian and using the Lagrange equation $\partial L / \partial x_i = \dot{p}_i$ [which follows (2.4) and (2.5)], we obtain

MINIMUM ACTION PRINCIPLE AND LAGRANGIAN FORMALISM

$$dH \equiv \sum_i \left[\frac{\partial H}{\partial p_i} dp_i + \frac{\partial H}{\partial x_i} dx_i \right]$$
$$= -\sum_i \dot{p}_i \, dx_i + \sum_i \dot{x}_i \, dp_i. \quad (2.7)$$

Equation (2.7) shows that the independent variables of the Hamiltonian are the coordinates and their associated conjugate momenta [2]. We can use (2.7) to obtain a set of two equations for each coordinate:

$$\dot{x}_i = \frac{\partial H}{\partial p_i}, \qquad \dot{p}_i = -\frac{\partial H}{\partial x_i}. \quad (2.8)$$

These are the *Hamilton–Jacobi equations*, which give the equations of motion for the independent variables x_i and p_i and describe the system dynamics completely. Naturally, the set of $2N$ first-order differential equations given by (2.8) are equivalent to the Lagrange equations of (2.4).

Simple extension of (2.7) gives the total time derivative of the Hamiltonian function $H(x_1, \ldots, x_N; p_1, \ldots, p_N; t)$:

$$\frac{dH}{dt} = \frac{\partial H}{\partial t} + \sum_i \frac{\partial H}{\partial x_i} \dot{x}_i + \sum_i \frac{\partial H}{\partial p_i} \dot{p}_i. \quad (2.9)$$

Using the Hamilton–Jacobi equations of (2.8), we find that $dH/dt = \partial H/\partial t$; that is, in the absence of an explicit dependence on time, the Hamiltonian is a constant of motion ($dH/dt = 0$). This very important property holds for any closed system and is a statement of the conservation of total system energy.

Now consider an arbitrary function u that corresponds to a dynamical variable: Since the coordinates and the momenta are independent variables of the system, we can write any system function as $u(x_i, p_i, t)$. The total time derivative of u is then

$$\frac{du}{dt} = \frac{\partial u}{\partial t} + \sum_i \left[\frac{\partial u}{\partial x_i} \dot{x}_i + \frac{\partial u}{\partial p_i} \dot{p}_i \right]$$
$$= \frac{\partial u}{\partial t} + \sum_i \left[\frac{\partial u}{\partial x_i} \frac{\partial H}{\partial p_i} - \frac{\partial u}{\partial p_i} \frac{\partial H}{\partial x_i} \right], \quad (2.10)$$

where we have used the Hamilton–Jacobi equations to replace the time derivatives of coordinates and momenta with partial derivatives of the Hamiltonian. We can define the *Poisson bracket* of u and v as

$$[u, v]_c \triangleq \sum_i \left[\frac{\partial u}{\partial x_i} \frac{\partial v}{\partial p_i} - \frac{\partial u}{\partial p_i} \frac{\partial v}{\partial x_i} \right]. \quad (2.11)$$

Discarding any explicit time dependence, we can then rewrite (2.10) as

$$\frac{du}{dt} = [u, H]_c, \qquad (2.12)$$

which simply states that the equation of motion of any observable u is determined by its Poisson bracket with the Hamiltonian. Clearly, if this Poisson bracket is zero, the dynamical variable u is a *constant of motion*. We can also consider the Poisson brackets of the independent variables x_i and p_i and find

$$[x_i, x_j]_c = 0 = [p_i, p_j]_c \qquad \forall i, j$$
$$[x_i, p_j]_c = \delta_{ij}. \qquad (2.13)$$

As is well known, classical mechanics fails to describe the dynamics of atoms and molecules, despite its undisputed success with macroscopic particles. In the 1920s it was realized that the commutative law of multiplication for dynamical variables had to be abandoned in order to explain the observed physical properties (i.e., stability of atoms and molecules); thus dynamical variables had to be represented by matrices or operators that, in general, do not commute with each other. An important consideration in the development of the new *quantum mechanics*, however, was the provision that the new equations of motion for dynamical variables be extensions of classical equations (which they reduce to in appropriate limits) [3].

As described previously, the equation of motion for any dynamical variable in classical mechanics is given by its Poisson bracket with the Hamiltonian function. The starting point for the *matrix mechanics* was to find a *quantum Poisson bracket* that had the same basic properties as the classical bracket. These properties include $[u, v]_c = -[v, u]_c$, $[u_1 + u_2, v]_c = [u_1, v]_c + [u_2, v]_c$ and $[u, v_1 v_2]_c = [u, v_1]_c v_2 + v_1[u, v_2]_c$ [3]. Consider a new Poisson bracket that satisfies

$$i\hbar[u, v]_q = uv - vu, \qquad (2.14)$$

where $\hbar$ is a real number. Such a Poisson bracket has the same properties (shown previously) as the classical bracket for any real $\hbar$ and, in the limit as $\hbar \to 0$, reproduces the commutative property of classical mechanics. We can then write the quantum equation of motion for a given observable (i.e., dynamical variable) in analogy with (2.12):

$$\frac{d\hat{u}}{dt} = [\hat{u}, \hat{H}]_q = \frac{1}{i\hbar}(\hat{u}\hat{H} - \hat{H}\hat{u}), \qquad (2.15)$$

where $\hat{u}$ and $\hat{H}$ denote the Hilbert space operators corresponding to the dynamical variable u and the Hamiltonian H, respectively. Equation (2.15) is the Heisenberg equation of motion that is the quantum analog of Hamilton–Jacobi equations.

Since quantum theory is an extension of classical mechanics, it is desirable that the fundamental independent variables (x_i and p_j) have the same Poisson bracket in both theories. More specifically, we require $[x_i, p_j]_c = [\hat{x}_i, \hat{p}_j]_q = \delta_{ij}$, where $\hat{x}_i$ and $\hat{p}_j$ denote the quantum operators for the coordinate and momentum. This requirement, along with the relation between the quantum Poisson bracket and the commutators in (2.14), gives the quantum conditions. Denoting the commutator of two operators (rather confusingly) with $\hat{u}\hat{v} - \hat{v}\hat{u} \equiv [\hat{u}, \hat{v}]$, we obtain the canonical commutation relations of quantum mechanics:

$$[\hat{x}_i, \hat{x}_j] = 0$$
$$[\hat{p}_i, \hat{p}_j] = 0$$
$$[\hat{x}_i, \hat{p}_j] = i\hbar \delta_{ij}. \qquad (2.16)$$

Introduction of these commutation relations between particle coordinates and momenta is generally referred to as *canonical first quantization*. As indicated previously, the dynamics of the new quantized system are determined by the Hamiltonian operator via the Heisenberg equation of motion (2.15).

Due to the central role of the commutation relation between the coordinates and their conjugate momenta, the canonical first-quantization procedure for a general system (i.e., particles and/or fields) must start by identifying a *conjugate momentum* for each *coordinate* (which may be different from mechanical momentum). In this respect, the Lagrangian approach is essential, as it allows for the determination of both the canonical momenta and the Hamiltonian function. The Hamiltonian operator, in turn, is obtained from the Hamiltonian function of (2.6) by replacing the coordinates and the momenta with their corresponding Hilbert space operators.

In contrast to point particles, fields are described by their values at all space-time points. Correspondingly, the Lagrangian depends on a continuum of coordinates, referred to as *generalized coordinates*, and their space and time derivatives [1]. The equations of motion of the field are obtained from a *Lagrangian density* $\mathcal{L}$, which can be defined by the relation

$$S = \int_{t_1}^{t_2} dt\, L = \int_{t_1}^{t_2} dt \int d^3r\, \mathcal{L}(\mathcal{A}_i; \dot{\mathcal{A}}_i; \partial_j \mathcal{A}_i). \qquad (2.17)$$

This complication is a result of the fact that we are now dealing with a system with uncountably infinite degrees of freedom. The particular *paths* that the system could follow are determined by the field variables (generalized coordinates) $\mathcal{A}_i(\mathbf{r}, t)$. The application of the principle of minimum action gives

$$\frac{d}{dt}\left(\frac{\partial \mathcal{L}}{\partial \dot{\mathcal{A}}_i}\right) = \frac{\partial \mathcal{L}}{\partial \mathcal{A}_i} - \sum_j \frac{\partial_j}{\partial r_j} \frac{\partial \mathcal{L}}{\partial \frac{\partial \mathcal{A}_i}{\partial r_j}}. \qquad (2.18)$$

It is important to note that the Lagrangian or Lagrangian density that results in (2.4) or (2.18) is not unique. If we add the time derivative of an analytic function to the Lagrangian, the action is changed by a constant; this new action has the same minimum as the original action and, therefore, Lagrange equations remain unchanged. We shall discuss this issue in detail when we consider the Power-Zienau-Wooley transformation and the long-wavelength approximation in Chapter 6.

Finally, the canonical generalized momentum $\Pi_i(\mathbf{r}, t)$, corresponding to $\mathcal{A}_i(\mathbf{r}, t)$, the Hamiltonian H, and the Hamiltonian density $\mathcal{H}$ are defined as

$$\Pi_i = \frac{\partial \mathcal{L}}{\partial \dot{\mathcal{A}}_i} \qquad (2.19)$$

and

$$H = \int d^3 r \mathcal{H} = \int d^3 r \left(\sum_i \Pi_i \dot{\mathcal{A}}_i - \mathcal{L} \right). \qquad (2.20)$$

The canonical quantization of the field is achieved by replacing the generalized coordinates and momenta by operators (denoted by $\hat{\mathcal{A}}$ and $\hat{\Pi}$, respectively) that satisfy the equal-time commutation relations analogous to (2.16):

$$\left[\hat{\mathcal{A}}_i(\mathbf{r}, t), \hat{\mathcal{A}}_j(\mathbf{r}', t) \right] = 0$$

$$\left[\hat{\Pi}_i(\mathbf{r}, t), \hat{\Pi}_j(\mathbf{r}', t) \right] = 0$$

$$\left[\hat{\mathcal{A}}_i(\mathbf{r}, t), \hat{\Pi}_j(\mathbf{r}', t) \right] = i\hbar \delta_{ij} \delta(\mathbf{r} - \mathbf{r}'). \qquad (2.21)$$

Provided that we are given its Lagrangian density, we are now ready to quantize a physical field and obtain the corresponding quantum field theory. We shall first quantize the electromagnetic field, which is of primary importance in the context of quantum optics.

2.2. QUANTIZATION OF THE ELECTROMAGNETIC FIELD

The fundamental system in standard quantum optics consists of a system of nonrelativistic particles interacting with the electromagnetic field. The Lagrangian for this combined system that results in the correct Maxwell-Lorentz equations as the Lagrange equations is

$$L = L_{\text{particles}} + \frac{\epsilon_o}{2} \int d^3 r \left(\mathbf{E}^2(\mathbf{r}) - c^2 \mathbf{B}^2(\mathbf{r}) \right) + \int d^3 r \left(\mathbf{j}(\mathbf{r}) \cdot \mathbf{A}(\mathbf{r}) - \rho(\mathbf{r}) U(\mathbf{r}) \right), \qquad (2.22)$$

where $L_{\text{particles}}$, $\mathbf{j}(\mathbf{r})$, and $\rho(\mathbf{r})$ denote the Lagrangian, current density, and charge density for the system of nonrelativistic particles. The electric and magnetic fields, denoted by $\mathbf{E}(\mathbf{r})$ and $\mathbf{B}(\mathbf{r})$, are related to the vector $\mathbf{A}(\mathbf{r})$ and scalar $U(\mathbf{r})$ potentials by

$$\mathbf{E}(\mathbf{r}) = -\nabla U(\mathbf{r}) - \dot{\mathbf{A}}(\mathbf{r})$$
$$\mathbf{B}(\mathbf{r}) = \nabla \times \mathbf{A}(\mathbf{r}). \tag{2.23}$$

We can see from (2.22) that the field Lagrangian depends on the electric and magnetic fields and the interaction Lagrangian depends on the vector and scalar potentials. As we have seen in the previous section, the Lagrangian should be expressed in terms of a certain set of suitable generalized coordinates before the canonical quantization can be carried out. The appropriate generalized coordinates in this case are the potentials, since expressing the electric and magnetic fields in terms of these potentials will provide a Lagrangian that depends only on the coordinates and their time and space derivatives.

It is well known that the choices for vector and scalar potentials are not unique. The gauge transformations that allow one to go from one set of potentials to another result in a different interaction Lagrangian [the third term in (2.22)] but leave the equations of motion (i.e., the Maxwell-Lorentz equations) unchanged. We shall use the Coulomb gauge, which is characterized by a transverse vector potential

$$\nabla \cdot \mathbf{A}(\mathbf{r}) = 0. \tag{2.24}$$

Having identified the generalized coordinates and the gauge, we can turn back to the free electromagnetic field that we want to quantize. By omitting the first and last terms in the Lagrangian of (2.22), we can define the Lagrangian density of the free field as

$$\begin{aligned}\mathcal{L} &= \frac{\epsilon_o}{2}\left(\mathbf{E}^2(\mathbf{r}) - c^2\mathbf{B}^2(\mathbf{r})\right) \\ &= \frac{\epsilon_o}{2}\left(\dot{\mathbf{A}}(\mathbf{r})^2 - c^2[\nabla \times \mathbf{A}(\mathbf{r})]^2\right).\end{aligned} \tag{2.25}$$

The conjugate momenta associated with $\mathbf{A}(\mathbf{r})$ are

$$\Pi_i(\mathbf{r}) = \epsilon_o \dot{A}_i(\mathbf{r}). \tag{2.26}$$

We therefore find that the conjugate momenta for a free electromagnetic field is simply related to the electric displacement vector

$$\Pi(\mathbf{r}) = -\mathbf{D}(\mathbf{r}) = -\epsilon_o \mathbf{E}(\mathbf{r}). \tag{2.27}$$

Note that (2.27) is modified in the presence of sources and, in this case, the conjugate momenta are determined by the transverse component of the electric field [1]. It may be shown that the scalar potential $U(\mathbf{r})$ and the longitudinal component of the electric field are determined by the *source coordinates* and hence are not independent field variables. To proceed with the free-field quantization, we first obtain the Hamiltonian

$$H = \frac{1}{2}\int d^3r\left[\frac{1}{\epsilon_o}\Pi^2 + \epsilon_o c^2(\nabla \times \mathbf{A})^2\right]$$

$$= \int d^3k\left[\frac{1}{\epsilon_o}\pi^* \cdot \pi + \epsilon_o c^2 k^2 \mathcal{A}^* \cdot \mathcal{A}\right]. \tag{2.28}$$

The quantization is straightforward in reciprocal space. We first note that, for each given $\mathbf{k}$, there are two independent *generalized coordinates* $\mathcal{A}_\sigma(\mathbf{k})$, $\mathcal{A}_{\sigma'}(\mathbf{k})$ corresponding to the two transverse polarizations. The conjugate momenta corresponding to these independent dynamical variables are $\pi_\sigma(\mathbf{k})$ and $\pi_{\sigma'}(\mathbf{k})$, respectively. As we have seen in the previous section, the canonical quantization principle tells us that the operators (denoted by $\wedge$) corresponding to each independent coordinate and its corresponding conjugate momentum must satisfy the commutation relation:

$$\left[\hat{\mathcal{A}}_\sigma(\mathbf{k}), \hat{\pi}_{\sigma'}(\mathbf{k}')\right] = i\hbar\delta_{\sigma\sigma'}\delta(\mathbf{k}-\mathbf{k}'). \tag{2.29}$$

It is important to point out that the commutation relation in (2.29) is for the two polarization components.

The commutation relation satisfied by the three cartesian components of the vector potential and its conjugate momenta (i.e., the electric field) in free space is more complicated due to the fact that these three components are not linearly independent due to the transversality of the electromagnetic field. The commutation relation satisfied by the components in real space is

$$\left[\hat{A}_i(\mathbf{r}), \hat{\Pi}_j(\mathbf{r}')\right] = i\hbar\delta^\perp_{ij}(\mathbf{r}-\mathbf{r}'), \tag{2.30}$$

where the *transverse delta function* $\delta^\perp_{ij}(\rho)$ is given by [1]

$$\delta^\perp_{ij}(\mathbf{r}) = \frac{2}{3}\delta_{ij}\delta(\mathbf{r}) - \frac{1}{4\pi r^3}\left(\delta_{ij} - \frac{3r_i r_j}{r^2}\right).$$

Finally, we note that the commutation relation introduced for the vector potential and its conjugate momentum in (2.30) implies that the parallel components of the electric and magnetic fields commute, so they can be measured simultaneously. On the other hand, the perpendicular components of the electric and magnetic fields are two incompatible quantities that cannot be determined with arbitrary precision. We shall return to this issue when we discuss the vacuum fluctuations that essentially arise from the postulated commutation relations of (2.29) and (2.30).

The analogy between the Hamiltonian of (2.28) and that of a simple harmonic oscillator is evident if we replace $\mathcal{A}(\mathbf{k})$ with $\mathbf{r}$ and $\pi(\mathbf{k})$ with $\mathbf{p}$. We can therefore define *normal variables* of the electromagnetic field as

$$\alpha_\sigma(\mathbf{k}) = \sqrt{\frac{\epsilon_o}{2\hbar\omega}} \left[\omega \mathcal{A}_\sigma(\mathbf{k}) + \frac{i}{\epsilon_o} \pi_\sigma(\mathbf{k}) \right]. \tag{2.31}$$

The significance of these normal variables (the two components given by σ) is that they have a trivial time dependence ($e^{i\omega_k t}$) in the absence of sources. After quantization, the normal variables become the *annihilation operators* for the two transverse polarization components

$$\hat{a}_\sigma(\mathbf{k}) = \sqrt{\frac{\epsilon_o}{2\hbar\omega}} \left[\omega \hat{\mathcal{A}}_\sigma(\mathbf{k}) + \frac{i}{\epsilon_o} \hat{\pi}_\sigma(\mathbf{k}) \right]. \tag{2.32}$$

The annihilation operators and their *Hermitian conjugates*, which are referred to as *creation* operators, satisfy the well-known commutation relations

$$\left[\hat{a}_\sigma(\mathbf{k}), \hat{a}_{\sigma'}(\mathbf{k}') \right] = 0,$$

$$\left[\hat{a}^\dagger_\sigma(\mathbf{k}), \hat{a}^\dagger_{\sigma'}(\mathbf{k}') \right] = 0,$$

$$\left[\hat{a}_\sigma(\mathbf{k}), \hat{a}^\dagger_{\sigma'}(\mathbf{k}') \right] = \delta_{\sigma\sigma'} \delta(\mathbf{k} - \mathbf{k}'). \tag{2.33}$$

As we will see in latter chapters, application of the annihilation (creation) operators results in the destruction (generation) of a single *photon*, which is the term used to describe the elementary excitation of the electromagnetic field.

The next step is to express the fundamental electromagnetic quantities (i.e., operators) in terms of these annihilation and creation operators and determine the dynamical equations that field parameters obey. We assume that the field is contained in a finite *cube* of size L and satisfies periodic boundary conditions. The allowed electromagnetic field modes in this case form a countably infinite set with wavevectors given by $k_i = n_i 2\pi/L$, where $i = x, y, z$ and $n_i = 0, \pm 1, \pm 2, \ldots$. The correct expressions are obtained by replacing integrals over the reciprocal space by a sum over the allowed discrete modes:

$$\int d^3k \leftrightarrow \sum_{\mathbf{k}} \left(\frac{2\pi}{L} \right)^3, \tag{2.34}$$

where $\mathbf{k} = (k_x, k_y, k_z)$ forms a discrete set. We can analyze any given problem in such a finite volume formalism; the correct results for free space could be obtained by taking the limit as $L \to \infty$.

The operator expressions for the vector potential, the electric and magnetic fields, the Hamiltonian, and the momentum of the free electromagnetic field in terms of traveling plane waves are given by

$$\hat{\mathbf{A}}(\mathbf{r}) = \sum_{\mathbf{k}\sigma_k} \left(\frac{\hbar}{2\epsilon_o \omega_k L^3}\right)^{1/2} \sigma_{\mathbf{k}} \left[\hat{a}_{\mathbf{k}\sigma_k} e^{i\mathbf{k}\cdot\mathbf{r}} + \hat{a}^{\dagger}_{\mathbf{k}\sigma_k} e^{-i\mathbf{k}\cdot\mathbf{r}}\right], \quad (2.35)$$

$$\hat{\mathbf{E}}_{\perp}(\mathbf{r}) = \sum_{\mathbf{k}\sigma_k} i\left(\frac{\hbar \omega_k}{2\epsilon_o L^3}\right)^{1/2} \sigma_{\mathbf{k}} \left[\hat{a}_{\mathbf{k}\sigma_k} e^{i\mathbf{k}\cdot\mathbf{r}} - \hat{a}^{\dagger}_{\mathbf{k}\sigma_k} e^{-i\mathbf{k}\cdot\mathbf{r}}\right], \quad (2.36)$$

$$\hat{\mathbf{B}}(\mathbf{r}) = \sum_{\mathbf{k}\sigma_k} i\left(\frac{\hbar \omega_k}{2\epsilon_o c^2 L^3}\right)^{1/2} \left(\frac{\mathbf{k}}{|\mathbf{k}|} \times \sigma_{\mathbf{k}}\right) \left[\hat{a}_{\mathbf{k}\sigma_k} e^{i\mathbf{k}\cdot\mathbf{r}} - \hat{a}^{\dagger}_{\mathbf{k}\sigma_k} e^{-i\mathbf{k}\cdot\mathbf{r}}\right], \quad (2.37)$$

$$\hat{H} = \frac{\epsilon_o}{2} \int d^3 r \left[\hat{\mathbf{E}}_{\perp}^2(\mathbf{r}) + c^2 \hat{\mathbf{B}}^2(\mathbf{r})\right]$$

$$= \sum_{\mathbf{k}\sigma_k} \hbar \omega_k \left[\hat{a}^{\dagger}_{\mathbf{k}\sigma_k} \hat{a}_{\mathbf{k}\sigma_k} + \frac{1}{2}\right], \quad (2.38)$$

$$\hat{\mathbf{G}} = \frac{1}{c^2} \int d^2 r \left[\hat{\mathbf{E}}(\mathbf{r}) \times \hat{\mathbf{H}}(\mathbf{r})\right]$$

$$= \sum_{\mathbf{k}\sigma_k} \hbar \mathbf{k} \left[\hat{a}^{\dagger}_{\mathbf{k}\sigma_k} \hat{a}_{\mathbf{k}\sigma_k} + \frac{1}{2}\right], \quad (2.39)$$

where we have used the commutation relation $[\hat{a}_{\mathbf{k}\sigma_k}, \hat{a}^{\dagger}_{\mathbf{k}'\sigma_{k'}}] = \delta_{\sigma_k \sigma_{k'}} \delta_{\mathbf{k}\mathbf{k}'}$. The dynamics of the electromagnetic field can be analyzed either in the Heisenberg or Schrödinger pictures. The equations of motion for the operators corresponding to the field variables in the Heisenberg picture are identical to those given by classical Maxwell's equations. Using the Hamiltonian of (2.38), we can also write a *Schrödinger equation* for the time-dependent *wavefunction* of the electromagnetic field. A fundamentally new concept that arises in the quantization is that of the wavefunction of the field. This wavefunction describes the *state* of the field and, naturally, does not have a classical analog. We shall consider some of the important states of the electromagnetic field in the next chapter.

2.3. SECOND QUANTIZATION OF THE SCHRÖDINGER FIELD

In quantum optics, we are interested in interactions between the quantized electromagnetic field and particles, which may be atoms or carriers in semiconductors. When the thermal de Broglie wavelength λ_T of these particles is much smaller than their average separation ($N^{-1/3}$, where N is the volume density of the atoms or the carriers), it is a good approximation to assume that these particles are independent; in fact, practically all traditional treatments of atom-field interactions make this assumption. At low temperatures, the situation is reversed and the quantum statistics of the particles come into play: The interacting particles can no longer be assumed to be independent and a quantum field theoretical approach is necessary. In this section, we present the quantization of the nonrelativistic Schrödinger field, which can be used to describe the dynamics of a cold dense ensemble of atoms.

Before proceeding, we note a cooperative effect between atoms that does not depend on the thermal de Broglie wavelength. When the atomic density is high enough (i.e., when $N^{-1/3} < \lambda_L$) (where λ_L is the optical wavelength), the fluorescence from the atoms is collective, essentially creating a *coherent spontaneous emission* with a magnitude proportional to N^2. This phenomenon, commonly referred to as *superradiance*, has nothing to do with the quantum statistics of the Schrödinger field and can be analyzed using the Dicke formalism (collective angular momentum operator).

The goal of the second quantization of the Schrödinger field is to describe the dynamics of an ensemble of identical (interacting) particles by means of a field operator. Such an operator obeys the equation of motion, which is identical to that of single-particle wavefunctions. In our case, this equation is Schrödinger's equation for a single particle:

$$i\hbar \dot{\psi}(\mathbf{r}, t) = \left(-\frac{\hbar^2}{2m} \Delta + V(\mathbf{r}) \right) \psi(\mathbf{r}, t). \tag{2.40}$$

We assume that the Hamiltonian $[-\frac{\hbar^2}{2m}\Delta + V(\mathbf{r})]$ acting on a *single-particle Hilbert-space* $\mathcal{H}(1)$ has a discrete spectrum and denote its eigenstates by $\varphi_n(\mathbf{r})$. Clearly, these eigenstates form an orthonormal basis for $\mathcal{H}(1)$.

We now define *annihilation and creation* operators for the particles obeying (2.40) as a mapping between Hilbert spaces of different particle numbers [4]:

$$\hat{c}_n : \mathcal{H}(N) \to \mathcal{H}(N-1)$$
$$\hat{c}_n^\dagger : \mathcal{H}(N) \to \mathcal{H}(N+1), \tag{2.41}$$

where $\hat{c}_n$ and $\hat{c}_n^\dagger$ denote the annihilation and creation operators for the single-particle eigenstate denoted by n, respectively, and N denotes the number of initial particles. We note that the definition in (2.41) allows us to write the Hamiltonian for a *noninteracting*, many-body system in a very simple form:

$$\hat{H} = \sum_n E_n \hat{c}_n^\dagger \hat{c}_n, \tag{2.42}$$

where E_n is the energy eigenvalue corresponding to the eigenstate $\varphi_n(\mathbf{r})$. One of the objectives of this section is to arrive at this Hamiltonian using the canonical quantization techniques for the matter fields, starting from an appropriate Lagrangian.

The field operator $\hat{\Psi}(\mathbf{r})$ ($\hat{\Psi}^\dagger(\mathbf{r})$) is defined as a superposition of the annihilation (creation) operators

$$\hat{\Psi}(\mathbf{r}) = \sum_n \varphi_n(\mathbf{r}) \hat{c}_n, \tag{2.43}$$

$$\hat{\Psi}^\dagger(\mathbf{r}) = \sum_n \varphi_n^*(\mathbf{r}) \hat{c}_n^\dagger. \tag{2.44}$$

Physically, the field operators defined by (2.43) and (2.44) annihilate and create a particle at position **r**, respectively. The inverse relations follow (2.43) and (2.44):

$$\hat{c}_n = \int d^3 r \varphi_n^*(\mathbf{r}) \hat{\Psi}(\mathbf{r}), \tag{2.45}$$

$$\hat{c}_n^\dagger = \int d^3 r \varphi_n(\mathbf{r}) \hat{\Psi}^\dagger(\mathbf{r}). \tag{2.46}$$

As we shall prove a posteriori, the classical Lagrangian density describing the correct dynamics for the Schrödinger field operator $\hat{\Psi}(\mathbf{r})$ is given by [1]

$$\mathcal{L} = \frac{i\hbar}{2}(\psi^* \dot{\psi} - \dot{\psi}^* \psi) - \frac{\hbar^2}{2m} \nabla \psi^* \cdot \nabla \psi - V(\mathbf{r}) \psi^* \psi, \tag{2.47}$$

where $\psi(\mathbf{r}, t)$ denotes the classical complex Schrödinger field and $V(\mathbf{r})$ is as defined in (2.40). First, we show (using Lagrange equations) that $\psi(\mathbf{r})$ satisfies Schrödinger's equation [(2.40)]. Since (2.40) is first order in time, specifying $\psi(\mathbf{r}, t_o)$ is sufficient to determine $\psi(\mathbf{r}, t)$ at all subsequent times. This fact implies that $\psi(\mathbf{r}, t)$ and $\psi^*(\mathbf{r}, t)$ are not independent variables and, therefore, an independent generalized coordinate needs to be identified in order to obtain the correct Hamiltonian. In the case of a classical Schrödinger field, it may be shown that the real part of the complex field $\psi(\mathbf{r}, t) = \psi_r(\mathbf{r}, t) + i\psi_i(\mathbf{r}, t)$ acts as the independent field variable, under an appropriate transformation of the Lagrangian [1] that removes the $\dot{\psi}_i(\mathbf{r}, t)$ dependence of $\mathcal{L}$. The conjugate momentum to the generalized coordinate $\psi_r(\mathbf{r}, t)$ is

$$\Pi_r(\mathbf{r}, t) = \frac{\partial \mathcal{L}}{\partial \dot{\psi}_r(\mathbf{r}, t)} = 2\hbar \psi_i(\mathbf{r}, t), \tag{2.48}$$

which implies that $\psi(\mathbf{r}, t) = \psi_r(\mathbf{r}, t) + (i/2\hbar)\Pi_r(\mathbf{r}, t)$.

We now proceed with the quantization procedure by introducing the canonical commutation relations. We shall see, however, that there are two consistent ways to quantize the Schrödinger field, either by introducing commutation or anticommutation relations. As is well known, this choice determines the quantum statistics of the quantized matter field and is of fundamental importance. The quantized Hamiltonian in both cases is

$$H = \int d^3 r \hat{\Psi}^\dagger(\mathbf{r}) \left(-\frac{\hbar^2}{2m} \Delta + V(\mathbf{r}) \right) \hat{\Psi}(\mathbf{r}). \tag{2.49}$$

Bosonic Field: If we choose to proceed by introducing the commutation relation

$$\left[\hat{\Psi}_r(\mathbf{r}), \hat{\Pi}_r(\mathbf{r}') \right] = i\hbar \delta(\mathbf{r} - \mathbf{r}'), \tag{2.50}$$

we obtain

$$\left[\hat{\Psi}(\mathbf{r}), \hat{\Psi}(\mathbf{r}')\right] = \left[\hat{\Psi}^\dagger(\mathbf{r}), \hat{\Psi}^\dagger(\mathbf{r}')\right] = 0,$$
$$\left[\hat{\Psi}(\mathbf{r}), \hat{\Psi}^\dagger(\mathbf{r}')\right] = \delta(\mathbf{r} - \mathbf{r}'). \quad (2.51)$$

The many-body wavefunctions generated by successive applications of $\hat{\Psi}^\dagger(\mathbf{r})$ on the vacuum state [i.e., the unique eigenstate of the Hilbert space $\mathcal{H}(0)$] are symmetric under the interchange of the coordinates of any two particles.

Fermionic Field: If instead, we introduce an anticommutation relation for the field operators

$$\left[\hat{\Psi}(\mathbf{r}), \hat{\Psi}(\mathbf{r}')\right]_+ = \hat{\Psi}(\mathbf{r})\hat{\Psi}(\mathbf{r}') + \hat{\Psi}(\mathbf{r}')\hat{\Psi}(\mathbf{r}) = 0,$$
$$\left[\hat{\Psi}(\mathbf{r}), \hat{\Psi}^\dagger(\mathbf{r}')\right]_+ = \delta(\mathbf{r} - \mathbf{r}'), \quad (2.52)$$

we find that the opposite is true: The wavefunction is now antisymmetric under the interchange of the coordinates of two particles.

In both bosonic and fermionic cases, the Heisenberg equation of motion for the quantized field is given by $i\hbar\dot{\hat{\Psi}}(\mathbf{r}) = [\hat{\Psi}(\mathbf{r}), \hat{H}]$. Using the Hamiltonian of (2.49), we can show that the equations of motion for both the bosonic and fermionic fields are given by the Schrödinger equation. The two quantization procedures are then compatible with the underlying dynamical equation that describes the time translations [1].

The quantum statistical differences between the two quantization procedures become more apparent after transformation into the new set of operators given by (2.45). Using the transformation of (2.43) and (2.44), we can rewrite the Hamiltonian as

$$\hat{H} = \int d^3r \left(\sum_n \varphi_n^*(\mathbf{r})\hat{c}_n^\dagger\right)\left(-\frac{\hbar^2}{2m}\Delta + V(\mathbf{r})\right)\left(\sum_m \varphi_m(\mathbf{r})\hat{c}_m\right)$$
$$= \sum_{n,m} \hat{c}_n^\dagger \hat{c}_m \int d^3r \varphi_n^*(\mathbf{r}) E_m \varphi_m(\mathbf{r})$$
$$= \sum_{n,m} \hat{c}_n^\dagger \hat{c}_m E_m \delta_{n,m}$$
$$= \sum_n E_n \hat{c}_n^\dagger \hat{c}_n. \quad (2.53)$$

This is the result obtained using the definition of the creation and annihilation operators as mappings between Hilbert spaces of different particle numbers. We have therefore shown that the Schrödinger field operator $\hat{\Psi}(\mathbf{r})$ obtained following the quantization of the classical Lagrangian of (2.47) is the correct field operator that annihilates an excitation of the field (i.e., particle) at position $\mathbf{r}$. Moreover, this theory allows for the existence of two different fields that obey identical dynamical equations for different commutation relations. The relevant commutation relation is determined by the *spin-statistics theorem* [5]; for example, if the total spin of the composite particle is half-integer, anticommutation relations need to be introduced in the quantization procedure.

The commutation relations for the field operators introduced in (2.51) and (2.52) imply the following relations for the corresponding annihilation and creation operators:

$$\left[\hat{b}_n, \hat{b}_m^\dagger\right] = \delta_{nm}, \quad \left[\hat{f}_n, \hat{f}_m^\dagger\right]_+ = \delta_{nm}.$$

$$\left[\hat{b}_n, \hat{b}_m\right] = 0, \quad [f_n, f_m]_+ = 0 \tag{2.54}$$

Here, we have used $\hat{c}_n \to \hat{b}_n$ for a bosonic field and $\hat{c}_n \to \hat{f}_n$ for a fermionic field. In the case of field theory of bosonic atoms, the operator $\hat{b}_n^\dagger \to \hat{b}_{i,\mathbf{k}}^\dagger$ creates an atom in the internal atomic state $|i>$ with a center-of-mass momentum $\hbar\mathbf{k}$. In the case of fermionic atoms, (2.54) tells us that we cannot have more than 1 fermionic atom in a single mode.

Throughout this chapter we have analyzed the quantization procedures for free Maxwell and Schrödinger fields. The fundamental paradigm of quantum optics, however, consists of interacting Maxwell and Schrödinger fields. The second quantized Hamiltonian for this interacting system is given by

$$\hat{H} = \int d^3 r \hat{\Psi}^\dagger(\mathbf{r}) \left[-\frac{1}{2m}\left(\frac{\hbar}{i}\nabla - q\hat{\mathbf{A}}\right)^2 + V(\mathbf{r}) + qU(\mathbf{r}) \right] \hat{\Psi}(\mathbf{r})$$

$$+ \int d^3 r \frac{\epsilon_o}{2} \left[\hat{\mathbf{E}}^2(\mathbf{r}) + c^2 \hat{\mathbf{B}}^2(\mathbf{r}) \right]. \tag{2.55}$$

The Hamiltonian of (2.55) is, unfortunately, too general to be of practical use: before being able to use it in our calculations, we first apply a transformation that allows us to make the so-called *long-wavelength approximation* and then rewrite the resulting Hamiltonian in terms of the creation and annihilation operators. This procedure will be carried out in Chapter 6. In most cases, we simplify the source Hamiltonian by assuming that only two or three internal atomic states are relevant in determining the system dynamics.

REFERENCES

[1] C. Cohen-Tannoudji, J. Dupont-Roc, and G. Grynberg, *Photons and Atoms, Introduction to Quantum Electrodynamics* (Wiley-Interscience, New York, 1989).

[2] L. D. Landau and E. M. Lifshitz, *Mechanics* (Pergamon, New York, 1976).
[3] P. A. M. Dirac, *The Principles of Quantum Mechanics* (Oxford University Press, Oxford, UK, 1958).
[4] E. K. U. Gross, E. Runge, and O. Heinonen, *Many-Particle Theory* (IOP Publishing, Bristol, UK, 1991).
[5] F. Mandel and G. Shaw, *Quantum Field Theory* (Wiley, New York, 1984).

3 Quantum States of the Electromagnetic Fields

When a quantum system is not perturbed by measurements, it evolves in a completely causal manner. The state vector for the system $|\psi(t)\rangle$ changes in accordance with the Schrödinger equation. As discussed in Chapter 1, a free particle prepared in a minimum uncertainty wavepacket ceases to be a minimum uncertainty wavepacket as time elapses; instead, the wavepacket spreads due to the initial momentum uncertainty. However, certain special wavepackets keep the minimum uncertainty product for all time when they are subject to a simple harmonic potential. The wavefunction of the displaced ground state of a simple harmonic oscillator is stationary and always preserves the minimum uncertainty product. This stationary minimum uncertainty state is termed a coherent state $|\alpha\rangle$.

As we have seen in Chapter 2, the quantized electromagnetic field is mathematically identical to an ensemble of quantum harmonic oscillators. The corresponding stationary minimum uncertainty wavepacket of the electromagnetic field is called a coherent state of light and plays an important role in laser physics and quantum optics. The coherent state of light has the same uncertainties for the electric and magnetic fields as a vacuum state $|0\rangle$. The coherent states are often used as the basis set for various probability functions that describe the quantum statistical properties of radiation. The book by Klauder and Skagerstam [1] provides an excellent collection of original papers on coherent states, and the reader is referred to references contained therein for a detailed discussion of the subject.

It is often stated that coherent states of light resemble the classical field as closely as quantum mechanics permits. An ideal laser produces light that is not very far from a coherent state. A coherent state remains a coherent state even when it is attenuated. Due to these reasons, many optical experiments are truly constrained by the quantum noise of coherent states.

The quantum theory of light was developed in connection with the invention of laser in the 1960s. However, the fluctuations of laser light, which are well approximated by a coherent state, can be described by the classical photodetection theory. An experimental foundation for the quantum theory of light was firmly laid only after squeezed states of light were generated in the 1980s.

The basic concept of the squeezed state dates back to the work by Schrödinger in the 1930s [2]. Generation and application of a squeezed state of light were first studied by Takahasi in 1965 [3]. His idea of using a degenerate parametric amplifier turned out to be the most efficient way of generating squeezed states.

The squeezed state of light offers the promise of achieving quantum noise reduction beyond the standard shot noise limit. However, as will be discussed later, losses introduce uncorrelated vacuum fluctuations into a squeezed state and degrade the degree of quantum noise reduction. This universal character of dissipation of a quantum system places serious limitations on the generation and application of squeezed states.

The goal of this chapter is to present the properties of the coherent states, the photon number eigenstates, and the squeezed states. The latter two are often referred to as nonclassical light because some of the measurement results performed for the photon number states and the squeezed states cannot be explained by the semiclassical photodetection theory. Properties of the photon number eigenstates are described in Sec. 3.1. Properties of the coherent and squeezed states are discussed in Secs. 3.2 and 3.3. Section 3.4 describes properties of correlated photon twins that are generated by a parametric down-conversion process. Section 3.5 discusses various statistical distribution functions for a density operator associated with coherent states [i.e., the diagonal $P(\alpha)$ distribution, the quasi-probability density $Q(\alpha)$, and the Wigner distribution function $W(\alpha, \alpha^*)$].

3.1. PHOTON NUMBER EIGENSTATES

As we have already seen in Chapter 2, the quantized electromagnetic fields are mathematically equivalent to an ensemble of harmonic oscillators. In this chapter, we are interested in the quantum states of a single-mode electromagnetic field. The electric field operator (with a polarization in the x-direction) and magnetic field operator (with a polarization in the y-direction) are expressed in terms of the annihilation and creation operators $\hat{a}$ and $\hat{a}^\dagger$:

$$\hat{E}_x = \mathcal{E}\left[\hat{a}(t) + \hat{a}^\dagger(t)\right]\sin kz, \tag{3.1}$$

$$\hat{H}_y = i\sqrt{\frac{\epsilon_0}{\mu_0}}\mathcal{E}\left[\hat{a}(t) - \hat{a}(t)^\dagger\right]\cos kz, \tag{3.2}$$

where $\mathcal{E} = \sqrt{\frac{\hbar\omega}{2V\epsilon_0}}$ is the electric field created by one photon inside the cavity and V is the mode volume. If we use the trivial time dependence of $\hat{a}(t) = \hat{a}(0)e^{-i\omega t}$ and $\hat{a}^\dagger(t) = \hat{a}^\dagger(0)e^{i\omega t}$ in (3.1), we obtain

$$\hat{E}_x = 2\mathcal{E}\left[\hat{a}_1(0)\cos(\omega t) + \hat{a}_2(0)\sin(\omega t)\right]\sin(kz), \tag{3.3}$$

where $\hat{a}_1(0) \equiv \frac{1}{2}[\hat{a}(0)+\hat{a}^\dagger(0)]$ and $\hat{a}_2(0) = \frac{1}{2i}[\hat{a}(0)-\hat{a}(0)^\dagger]$ are the two quadrature-phase amplitudes. They obey the commutation relation, $[\hat{a}_1, \hat{a}_2] = \frac{i}{2}$.

3.1.1. Properties of Photon Number Eigenstates

The eigenstates of the Hamiltonian $\hat{\mathcal{H}}$ or the photon number operator $\hat{n} \equiv \hat{a}^\dagger \hat{a}$ are called photon number eigenstates (or Fock states). The photon number eigenstates are defined by

$$\hat{\mathcal{H}}|n\rangle \equiv \hbar\omega\left(\hat{n} + \frac{1}{2}\right)|n\rangle = \hbar\omega\left(n + \frac{1}{2}\right)|n\rangle, \tag{3.4}$$

$$\hat{n}|n\rangle = n|n\rangle. \tag{3.5}$$

The lowest energy eigenstate is the vacuum state $|0\rangle$ and is defined by

$$\hat{a}|0\rangle = 0. \tag{3.6}$$

The eigenvalue $\frac{1}{2}\hbar\omega$ for the vacuum state $|0\rangle$ is called the (quantum mechanical) zero-point energy. The eigenvalue relation for $\hat{a}^\dagger|0\rangle$ is

$$\hat{\mathcal{H}}\hat{a}^\dagger|0\rangle = \hbar\omega\left(\hat{a}^\dagger\hat{a} + \frac{1}{2}\right)\hat{a}^\dagger|0\rangle$$

$$= \hbar\omega\left(\hat{a}^{\dagger 2}\hat{a} + \frac{3}{2}\hat{a}^\dagger\right)|0\rangle$$

$$= \frac{3}{2}\hbar\omega\hat{a}^\dagger|0\rangle. \tag{3.7}$$

Therefore, the new state $\hat{a}^\dagger|0\rangle$ is also an eigenstate of the Hamiltonian with an eigenvalue $\frac{3}{2}\hbar\omega$. Similarly, we obtain

$$\hat{\mathcal{H}}(\hat{a}^\dagger)^n|0\rangle = \hbar\omega\left(n + \frac{1}{2}\right)\left(\hat{a}^\dagger\right)^n|0\rangle. \tag{3.8}$$

The non-Hermitian operator $\hat{a}^\dagger$ raises the eigenvalue by one if it operates on the photon number eigenstate,

$$\hat{a}^\dagger|n\rangle = C_{n+1}|n+1\rangle, \tag{3.9}$$

where C_{n+1} is a constant. From (3.9) and its adjoint, we obtain

$$\langle n|\hat{a}\hat{a}^\dagger|n\rangle = |C_{n+1}|^2\langle n+1|n+1\rangle = |C_{n+1}|^2. \tag{3.10}$$

If we use the commutation relation, $\hat{a}\hat{a}^\dagger = \hat{a}^\dagger\hat{a} + 1$, in the preceding equation, we can determine the normalization constant as $C_{n+1} = \sqrt{n+1}$ and thus (3.9) becomes

$$\hat{a}^\dagger|n\rangle = \sqrt{n+1}\,|n+1\rangle. \tag{3.11}$$

Similarly, the non-Hermitian operator $\hat{a}$ lowers the eigenvalue by one if it operates on the photon number eigenstate,

$$\hat{a}|n\rangle = \sqrt{n}\,|n-1\rangle. \tag{3.12}$$

The non-Hermitian operators $\hat{a}^\dagger$ and $\hat{a}$ are called the creation operator and the annihilation operator, respectively.

The photon number eigenstate is mathematically constructed by successive operation of the photon creation operator $\hat{a}^\dagger$ on the vacuum state $|0\rangle$,

$$|n\rangle = \frac{1}{\sqrt{n!}}(\hat{a}^\dagger)^n|0\rangle, \tag{3.13}$$

where the relation (3.11) is used. Similarly, the vacuum state is generated by operation of the photon annihilation operator $\hat{a}$ on the photon number eigenstate $|n\rangle$,

$$|0\rangle = \frac{1}{\sqrt{n!}}\hat{a}^n|n\rangle, \tag{3.14}$$

where the relation (3.12) is used.

The electric field expectation value for the photon number eigenstate is zero irrespective of the eigenvalue n:

$$\langle n|\hat{E}|n\rangle = \mathcal{E}\langle n|\hat{a} + \hat{a}^\dagger|n\rangle \sin kz = 0, \tag{3.15}$$

where we use the fact that the photon number eigenstates are orthogonal with each other (i.e., $\langle m|n\rangle = \delta_{mn}$). However, the electric field intensity has a nonzero value,

$$\langle n|\hat{E}^2|n\rangle = \mathcal{E}^2\langle n|\hat{a}^\dagger\hat{a}^\dagger + \hat{a}\hat{a}^\dagger + \hat{a}^\dagger\hat{a} + \hat{a}\hat{a}|n\rangle \sin^2 kz$$
$$= 2\mathcal{E}^2\left(n + \frac{1}{2}\right)\sin^2 kz. \tag{3.16}$$

That is, if one measures the photon number for an ensemble of quantum systems, all in the same photon number eigenstate $|n\rangle$, the measurement results are always equal to the eigenvalue n; but if one measures the electric field or the magnetic field for the ensemble, the measurement results are scattered around its ensemble-averaged value of zero.

This can be visualized by calculating the q-representation of the photon number eigenstate, $\phi_n(q) \equiv \langle q|n\rangle$, where $\langle q|$ is an eigenbra of the generalized coordinate $\hat{q} = \sqrt{\frac{\hbar}{2m\omega}}\,(\hat{a} + \hat{a}^\dagger)$. Using the relation $\hat{a}|0\rangle = 0$, we obtain [4]

$$\left(m\omega q + \hbar\frac{d}{dq}\right)\phi_0(q) = 0. \tag{3.17}$$

The solution of this differential equation that satisfies the normalization condition,

$$\int_{-\infty}^{\infty} |\phi_0(q)|^2 \, dq = 1,$$

is

$$\phi_0(q) = \left(\frac{m\omega}{\pi\hbar}\right)^{\frac{1}{4}} \exp\left[-\frac{1}{2}\left(\frac{m\omega}{\hbar}\right) q^2\right]. \tag{3.18}$$

For the higher-photon number eigenfunctions, we obtain

$$\phi_n(q) = \langle q | \frac{(\hat{a}^\dagger)^n}{\sqrt{n!}} | 0 \rangle = \frac{1}{\sqrt{(2m\hbar\omega)^n n!}} \left(m\omega q - \hbar \frac{d}{dq}\right)^n \phi_0(q)$$

$$= H_n\left[\sqrt{\frac{m\omega}{\hbar}} q\right] \phi_0(q), \tag{3.19}$$

where H_n is the Hermite polynomial function. Figure 3-1 shows the noise distributions in $a_1 - a_2$ phase space [i.e., the quasi-probability density $Q(\alpha) = |\langle\alpha|n\rangle|^2$, which will be discussed later in this chapter, the *Schrödinger wavefunction in q-representation* $\phi_n(q)$, and the photon number distributions $P(n)$ of the three lowest-energy photon number eigenstates $|0\rangle$, $|1\rangle$ and $|2\rangle$]. The distribution functions $P(n)$, $Q(\alpha)$, and $|\phi(q)|^2$ are experimentally obtained by ensemble measurements with a photon counter, optical heterodyne, and homodyne detectors, respectively. $Q(\alpha)$ represents the distributions of the amplitude and the phase of the electric field. The circular distribution of $Q(\alpha)$ around $a_1 = a_2 = 0$ indicates that the phase of the photon number eigenstate is completely random. $|\phi_n(q)|^2$ is the probability distribution of the electric field, which is obtained by optical homodyne detection for a single quadrature amplitude of the electric field. These plots show that the average value of the electric field is zero for all photon number eigenstates. The oscillatory behavior of $|\phi_n(q)|^2$ is the result of the remarkable quantum interference effect induced by the quantum measurement process. For instance, $\phi_1(q = 0)$ is identically zero because of the destructive interference between the positive p component and the negative p component at $q = 0$.

The photon number eigenstates behave classically for the measurement of photon number. As shown in Fig. 3-1, the measurement result is always the eigenvalue n for the state $|n\rangle$. However, if we measure a single quadrature of the electric field by an optical homodyne detector, the quantum aspect of the photon number eigenstate emerges as shown in Fig. 3-1(c). The closest classical analog to the photon number eigenstate is the ensemble of wavepackets with constant amplitudes and random phases, as shown in Fig. 3-2(a). Such a classical electromagnetic field features the same statistics $P(n)$ for the photon number measurement and $Q(\alpha)$ for the optical heterodyne measurement but does not feature an oscillatory behavior for $|\phi_n(q)|^2$ in the optical homodyne measurement due to the absence of *quantum interference* as

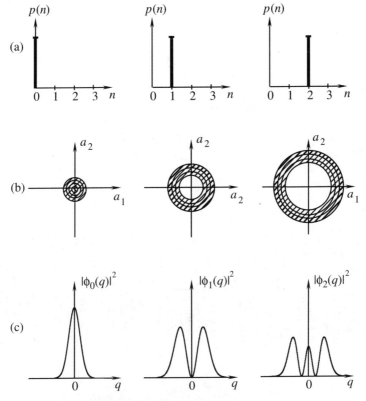

FIGURE 3-1: The photon number distribution $P(n)$, the quasi-probability density $Q(\alpha)$, and the Schrödinger wavefunction $|\phi_n(q)|^2$ of the photon number eigenstates.

shown in Fig. 3-2(b). The oscillatory behavior of $|\phi_n(q)|^2$ shown in Fig. 3-1(c) is the manifestation of the fact that each wavepacket of $|n\rangle$ simultaneously occupies all different phases rather than taking a random hidden but definite phase.

The photon number eigenstates form a complete orthonormal set of basis vectors that describe an arbitrary single-mode quantized electromagnetic field. Pertinent relations for the photon number eigenstates are summarized as follows [4]:

$$\langle n'|n''\rangle = \delta_{n'n''} \quad \text{(orthonormality)}, \tag{3.20}$$

$$\sum_{n=0}^{\infty} |n\rangle\langle n| = \hat{I} \quad \text{(completeness)}, \tag{3.21}$$

$$\hat{a}^\dagger \hat{a} = \sum_{n=0}^{\infty} n|n\rangle\langle n|, \tag{3.22}$$

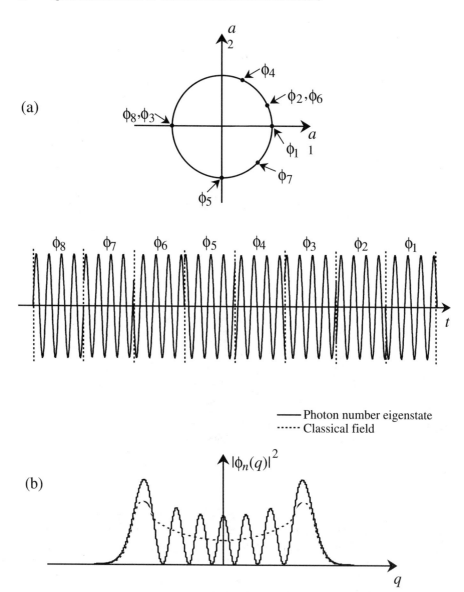

FIGURE 3-2: Analogy and difference between the photon number eigenstate and the classical field with constant amplitude and random phase.

$$\hat{a} = \sum_{n=0}^{\infty} \sqrt{n+1}\, |n\rangle\langle n+1|, \tag{3.23}$$

$$\hat{a}^\dagger = \sum_{n=0}^{\infty} \sqrt{n+1}\, |n+1\rangle\langle n|, \tag{3.24}$$

$$f(\hat{a}^\dagger \hat{a}) = \sum_{n=0}^{\infty} f(n)|n\rangle\langle n|. \tag{3.25}$$

Equation (3.21) indicates that the projection operator $|n\rangle\langle n|$ forms a decomposition of unity (identity operator). Any projection operator that forms a decomposition of unity describes a physically realizable measurement. The projection operator $|n\rangle\langle n|$ describes an ideal QND measurement of photon number, which will be discussed later. When the electromagnetic field interacts with a dissipative or gain medium and one photon is absorbed from the field or emitted to the field via the interaction, the new state after photon absorption/emission process is calculated by the projection operators (3.23) and (3.24). We will discuss these projection operators further in the treatment of system-reservoir coupling and quantum Monte-Carlo wavefunction methods.

3.1.2. Generation of Photon Number Eigenstates

Equation (3.13) indicates that the photon number eigenstate $|n\rangle$ is generated by the successive n photon emission process starting from the vacuum state $|0\rangle$. Injection of an excited atom or N excited atoms into a resonant cavity and appropriate atom-field interaction can produce a photon number eigenstate inside a cavity. This subject will be discussed later in the treatment of cavity quantum electrodynamics. An alternative way to realize this process is the nondegenerate parametric amplifier shown in Fig. 3-3. A pump photon at angular frequency ω_p is split into a signal photon at ω_s and an idler photon at ω_i in a second-order nonlinear crystal, if the energy con-

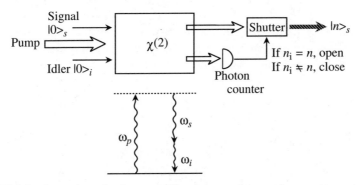

FIGURE 3-3: Generation of a photon number eigenstate in a nondegenerate parametric amplifier with an idler measurement-feedforward loop.

servation ($\omega_s + \omega_i = \omega_p$) and the momentum conservation ($\mathbf{k}_s + \mathbf{k}_i = \mathbf{k}_p$) are simultaneously satisfied. The interaction Hamiltonian for this process is given by

$$\hat{\mathcal{H}}_I = \hbar(\kappa \hat{a}_s^\dagger \hat{a}_i^\dagger + \kappa^* \hat{a}_s \hat{a}_i), \tag{3.26}$$

where the coupling constant κ is proportional to the second-order nonlinear susceptibility $\chi^{(2)}$ and the pump wave amplitude E_p. Equation (3.26) assumes that the pump field is well represented by a classical field. When both input fields at ω_s and ω_i are vacuum states, the output field state of this nondegenerate parametric amplifier is a joint-correlated state given by [5]

$$|\psi\rangle_{\text{out}} = \exp\left(\frac{i}{\hbar}\hat{\mathcal{H}}_I t\right)|0\rangle_s|0\rangle_i$$

$$= N \sum_n \frac{[-ie^{i\theta}\tanh(r)]^n}{\cosh(r)}|n\rangle_s|n\rangle_i, \tag{3.27}$$

where N is a normalization constant, $\kappa = |\kappa|e^{i\theta}$ and $r = |\kappa|t$. The preceding equation suggests that the numbers of emitted photons in the two channels are identical. If the idler output photon number is measured by a photon counter and the result is n, the corresponding signal wavepacket also has n photons due to the perfect correlation of photon number in the two channels. The essence of this generation scheme consists of two parts: creation of the joint-correlated states (3.27) by a unitary evolution process and projection onto the photon number eigenstate by a nonunitary quantum measurement performed on the idler wave.

The preceding generation scheme does not always produce a desired photon number eigenstate, but rather the signal output field is determined by an accidental measurement result for the idler output photon number. It was shown that the following interaction Hamiltonian generates a photon number eigenstate $|n\rangle$ from an input vacuum state $|0\rangle$ [6]:

$$\hat{\mathcal{H}}_I = \chi^{(n)}[\hat{a}^n E_p^* + (\hat{a}^\dagger)^n E_p] + \chi^{(n+2)}[\hat{a}^\dagger \hat{a}^{n+1} E_p^* + (\hat{a}^\dagger)^{n+1}\hat{a} E_p]. \tag{3.28}$$

This interaction Hamiltonian describes the process in which a pump photon at ω_p is converted into n signal photons at $\omega_s = \omega_p/n$ in two ways: $\omega_p \to n\omega_s$ and $\omega_p + \omega_s \to (n+1)\omega_s$ simultaneously. The two nonlinear coefficients must satisfy

$$\chi^{(n+2)} = -\frac{\chi^{(n)}}{n}. \tag{3.29}$$

and the interaction time must be equal to

$$t = \frac{\pi\left(m + \frac{1}{2}\right)}{\chi^{(n)}|E_p|\sqrt{n!}}, \tag{3.30}$$

where m is an integer and $\hbar = 1$. If these two conditions are satisfied, it is straightforward to show

$$|\psi\rangle_{\text{out}} = \exp(i\hat{\mathcal{H}}_I t)|0\rangle = |n\rangle. \qquad (3.31)$$

3.2. COHERENT STATES

The generalized minimum uncertainty state for the two Hermitian (quadrature) operators $\hat{a}_1 = \frac{1}{2}(\hat{a}+\hat{a}^\dagger)$ and $\hat{a}_2 = \frac{1}{2i}(\hat{a}-\hat{a}^\dagger)$ is defined mathematically as an eigenstate of the non-Hermitian operator $e^r \hat{a}_1 + ie^{-r} \hat{a}_2$ with an eigenvalue of $e^r \langle\hat{a}_1\rangle + ie^{-r} \langle\hat{a}_2\rangle$. The eigenvalue equation (1.47) can be rewritten as

$$e^r(\hat{a}_1 - \langle\hat{a}_1\rangle)|\psi\rangle_{MUS} = -ie^{-r}(\hat{a}_2 - \langle\hat{a}_2\rangle)|\psi\rangle_{MUS}, \qquad (3.32)$$

The uncertainties of $\hat{a}_1$ and $\hat{a}_2$ are calculated by taking the adjoint of (3.32) and identifying $\langle\Delta\hat{a}_1^2\rangle = {}_{MUS}\langle\psi|(\hat{a}_1 - \langle\hat{a}_1\rangle)^2|\psi\rangle_{MUS}$ and $\langle\Delta\hat{a}_2^2\rangle = {}_{MUS}\langle\psi|(\hat{a}_2 - \langle\hat{a}_2\rangle)^2|\psi\rangle_{MUS}$. The result was given by (1.48). In the special case of $r = 0$, the two quadrature amplitude operators $\hat{a}_1$ and $\hat{a}_2$ have the same uncertainties. The non-Hermitian operator $e^r \hat{a}_1 + ie^{-r} \hat{a}_2$ for generating this specific minimum uncertainty state is reduced to the photon annihilation operator $\hat{a}$. The eigenstates of $\hat{a}$ are called coherent states for reasons that will become clear in Chapter 4. In the general case of $r \neq 0$, the two quadrature amplitude operators $\hat{a}_1$ and $\hat{a}_2$ have different uncertainties but satisfy the minimum uncertainty product. The non-Hermitian operator $e^r \hat{a}_1 + ie^{-r} \hat{a}_2$ can be rewritten as $\hat{b} \equiv \mu\hat{a} + \nu\hat{a}^\dagger$, where two real numbers $\mu = \cosh(r)$ and $\nu = \sinh(r)$ satisfy $\mu^2 - \nu^2 = 1$. The new non-Hermitian operator $\hat{b}$ satisfies the boson commutation relation $[\hat{b}, \hat{b}^\dagger] = 1$. This is the Bogoliubov transformation of the photon annihilation and creation operators. The eigenstates of $\hat{b} = \mu\hat{a} + \nu\hat{a}^\dagger$ are called squeezed states.

3.2.1. Properties of Coherent States

The coherent state is defined by (1.49). By using (3.13) and (3.21) in (1.49), we obtain

$$\begin{aligned}
|\alpha\rangle &= \sum_n |n\rangle\langle n|\alpha\rangle \\
&= \sum_n |n\rangle\langle 0|\frac{\hat{a}^n}{\sqrt{n!}}|\alpha\rangle \\
&= \sum_n \frac{\alpha^n}{\sqrt{n!}} \langle 0|\alpha\rangle |n\rangle.
\end{aligned} \qquad (3.33)$$

54 QUANTUM STATES OF THE ELECTROMAGNETIC FIELDS

Imposing the normalization condition, $\langle \alpha | \alpha \rangle = 1$, we obtain

$$\begin{aligned} 1 = \langle \alpha | \alpha \rangle &= \sum_n \sum_m \langle m | n \rangle \frac{(\alpha^*)^m \alpha^n}{\sqrt{m!}\sqrt{n!}} |\langle 0 | \alpha \rangle|^2 \\ &= \sum_n \frac{|\alpha|^{2n}}{n!} |\langle 0 | \alpha \rangle|^2 \\ &= e^{|\alpha|^2} |\langle 0 | \alpha \rangle|^2. \end{aligned} \tag{3.34}$$

Therefore, we can determine the expansion coefficient in (3.33) without loss of generality as

$$|\langle 0 | \alpha \rangle|^2 = e^{-|\alpha|^2} \longrightarrow \langle 0 | \alpha \rangle = e^{-\frac{|\alpha|^2}{2}}. \tag{3.35}$$

Using (3.35) in (3.33), the coherent state can be expressed using the photon number eigenstates:

$$|\alpha\rangle = e^{-\frac{1}{2}|\alpha|^2} \sum_{n=0}^{\infty} \frac{\alpha^n}{\sqrt{n!}} |n\rangle. \tag{3.36}$$

The probability of finding the photon number n for the coherent state obeys the Poisson distribution

$$P(n) \equiv |\langle n | \alpha \rangle|^2 = \frac{e^{-|\alpha|^2} |\alpha|^{2n}}{n!}. \tag{3.37}$$

The mean, mean square, and variance of the photon number for the coherent state $|\alpha\rangle$ are

$$\langle \hat{n} \rangle \equiv \sum_n n \, P(n) = |\alpha|^2, \tag{3.38}$$

$$\langle \hat{n}^2 \rangle \equiv \sum_n n^2 \, P(n) = |\alpha|^2(|\alpha|^2 + 1) = \langle \hat{n} \rangle(\langle \hat{n} \rangle + 1), \tag{3.39}$$

$$\langle \Delta \hat{n}^2 \rangle \equiv \langle \hat{n}^2 \rangle - \langle \hat{n} \rangle^2 = |\alpha|^2 = \langle \hat{n} \rangle. \tag{3.40}$$

The coherent states $|\alpha\rangle$ should not be confused with a statistical mixture of photon number eigenstates with Poisson distributions. The density operator of such a mixed state does not have off-diagonal terms and is simply given by

$$\hat{\rho}_{\text{mix}} = \sum_n \frac{e^{-|\alpha|^2} |\alpha|^{2n}}{n!} |n\rangle\langle n|. \tag{3.41}$$

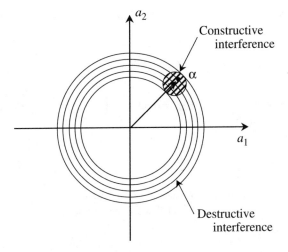

FIGURE 3-4: The noise distributions of a coherent state.

The density operator of the coherent state, on the other hand, has off-diagonal terms (simultaneous occupation of different photon number eigenstates) as well as the diagonal terms:

$$\hat{\rho}_{\text{pure}} = \sum_n \sum_m \frac{e^{-|\alpha|^2} \alpha^n \alpha^{*m}}{\sqrt{m!n!}} |n\rangle\langle m|. \tag{3.42}$$

As illustrated in Fig. 3-4, the coherent state has a well-defined phase due to the constructive and destructive interference between different photon number eigenstates. This quantum interference effect is a unique property of the linear superposition state and is absent in the mixed state.

Coherent states have many fascinating and useful properties. Coherent states are generated by translating the vacuum state $|0\rangle$ to have a finite excitation amplitude α. To see this, we use (3.36) and (3.13) to obtain

$$|\alpha\rangle = e^{-\frac{|\alpha|^2}{2}} \sum_n \frac{\alpha^n}{\sqrt{n!}} |n\rangle$$

$$= e^{-\frac{|\alpha|^2}{2}} \sum_n \frac{(\alpha \hat{a}^\dagger)^n}{n!} |0\rangle$$

$$= e^{-\frac{|\alpha|^2}{2}} e^{\alpha \hat{a}^\dagger} |0\rangle. \tag{3.43}$$

Since $\hat{a}|0\rangle = 0$, we have $e^{-\alpha^* \hat{a}}|0\rangle = [1 - (\alpha^*\hat{a}) + \frac{1}{2}(\alpha^*\hat{a})^2 + \cdots]|0\rangle = |0\rangle$. Therefore, (3.43) may be rewritten as

$$|\alpha\rangle = e^{-\frac{|\alpha|^2}{2}} e^{\alpha \hat{a}^\dagger} e^{-\alpha^* \hat{a}} |0\rangle. \tag{3.44}$$

56 QUANTUM STATES OF THE ELECTROMAGNETIC FIELDS

Any two noncommuting operators $\hat{A}$ and $\hat{B}$ satisfy the following Baker-Hausdorff relation [4]:

$$e^{\hat{A}+\hat{B}} = e^{\hat{A}}e^{\hat{B}}e^{-\frac{1}{2}[\hat{A},\hat{B}]} = e^{\hat{B}}e^{\hat{A}}e^{\frac{1}{2}[\hat{A},\hat{B}]}, \tag{3.45}$$

provided $[\hat{A}, [\hat{A}, \hat{B}]] = [\hat{B}, [\hat{A}, \hat{B}]] = 0$. Using $\hat{A} = \alpha\hat{a}^\dagger$, $\hat{B} = -\alpha^*\hat{a}$, and $[\hat{A}, \hat{B}] = |\alpha|^2$ in (3.44), we can show that

$$|\alpha\rangle = \hat{D}(\alpha)|0\rangle = e^{\alpha\hat{a}^\dagger - \alpha^*\hat{a}}|0\rangle, \tag{3.46}$$

which shows that $|\alpha\rangle$ is obtained by acting on $|0\rangle$ by the *displacement operator* $\hat{D}(\alpha)$. In the next section we will show that the displacement operator $\hat{D}(\alpha)$ is physically realized by a classical oscillating current.

Coherent states form a complete basis set. The mathematical proof is as follows:

$$\int \frac{d^2\alpha}{\pi}|\alpha\rangle\langle\alpha| = \frac{1}{\pi}\int_0^\infty d|\alpha|\int_0^{2\pi}|\alpha|d\phi e^{-|\alpha|^2}\sum_m\sum_n \frac{|\alpha|^m e^{im\phi}}{\sqrt{m!}} \cdot \frac{|\alpha|^n e^{-in\phi}}{\sqrt{n!}}|m\rangle\langle n|$$

$$\left(\longleftarrow \int_0^{2\pi} d\phi e^{i(m-n)\phi} = 2\pi\delta_{mn}\right)$$

$$= \frac{1}{\pi}\int_0^\infty d|\alpha| \cdot |\alpha|2\pi e^{-|\alpha|^2}\sum_n \frac{|\alpha|^{2n}}{n!}|n\rangle\langle n|$$

$$\left(\longleftarrow x = |\alpha|^2,\ dx = 2|\alpha|d|\alpha|\right)$$

$$= \sum_n \frac{1}{n!}\int_0^\infty dx\, e^{-x}x^n|n\rangle\langle n|\ \left(\longleftarrow n! = \int_0^\infty dx\, e^{-x}x^n\right)$$

$$= \sum_n |n\rangle\langle n|$$

$$= \hat{I}. \tag{3.47}$$

As we discussed in Sec. 3.1, a projection operator that is a decomposition of unity describes a physically realizable measurement. The projection operator $|\alpha\rangle\langle\alpha|$ represents the simultaneous measurement of the two conjugate observables $\hat{a}_1$ and $\hat{a}_2$ with equal strengths. Such a simultaneous measurement of $\hat{a}_1$ and $\hat{a}_2$ is physically realized by a 50%-50% beam splitter followed by two optical homodyne detectors or an optical heterodyne detector [7].

Coherent states are not orthogonal to each other:

$$\langle\beta|\alpha\rangle = e^{-\frac{1}{2}(|\beta|^2+|\alpha|^2)}\sum_m\sum_n \frac{(\beta^*)^m}{\sqrt{m!}}\frac{\alpha^n}{\sqrt{n!}}\langle m|n\rangle$$

$$= e^{-\frac{1}{2}(|\beta|^2+|\alpha|^2)} \sum_n \frac{(\beta^*\alpha)^n}{n!}$$

$$= e^{-\frac{1}{2}(|\beta|^2+|\alpha|^2)+\beta^*\alpha}$$

$$= e^{-\frac{1}{2}|\beta-\alpha|^2}. \tag{3.48}$$

Coherent states are approximately orthogonal only in the limit of large separation of the two eigenvalues, $|\beta - \alpha| \to \infty$. Therefore, any coherent state can be expanded using other coherent states,

$$|\alpha\rangle = \int \frac{d^2\beta}{\pi} |\beta\rangle\langle\beta|\alpha\rangle = \int \frac{d^2\beta}{\pi} e^{-\frac{1}{2}|\beta-\alpha|^2} |\beta\rangle. \tag{3.49}$$

This means that a coherent state forms an *overcomplete set* and that the simultaneous measurement of $\hat{a}_1$ and $\hat{a}_2$, represented by the projection operator $|\alpha\rangle\langle\alpha|$, is not an exact measurement but instead an approximate measurement with a finite measurement error.

We next obtain the q-representation of the coherent state. If we multiply the left-hand side of (1.49) by $\langle q|$ and use $\hat{a} = \frac{1}{\sqrt{2\hbar\omega}}(\omega\hat{q} + i\hat{p})$, we obtain

$$\left(\omega q + \hbar \frac{d}{dq}\right) \langle q|\alpha\rangle = \sqrt{2\hbar\omega}\,\alpha \langle q|\alpha\rangle. \tag{3.50}$$

By integrating (3.50) and requiring the normalization condition,

$$\int_{-\infty}^{\infty} |\langle q|\alpha\rangle|^2 \, dq = 1, \tag{3.51}$$

we obtain

$$\langle q|\alpha\rangle = \left(\frac{\omega}{\pi\hbar}\right)^{\frac{1}{4}} \exp\left[-\frac{\omega}{2\hbar}[q - \langle q\rangle]^2 + i\frac{\langle p\rangle}{\hbar}q + i\theta\right], \tag{3.52}$$

where θ is an arbitrary real phase. This is indeed the minimum uncertainty wavepacket with stationary quantum uncertainty $\langle \Delta\hat{q}^2\rangle = \frac{\hbar}{2\omega}$ and $\langle \Delta\hat{p}^2\rangle = \frac{\hbar\omega}{2}$.

The time-varying electric field of the coherent state is obtained by the expectation value of the electric field operator (3.1),

$$\langle\alpha|\hat{E}(t)|\alpha\rangle = \mathcal{E}\left[\alpha e^{-i\omega t} + \alpha^* e^{i\omega t}\right] \sin kz$$

$$= 2\mathcal{E}|\alpha| \cos(\omega t + \phi) \sin kz. \tag{3.53}$$

Similarly,

$$\langle\alpha|\hat{E}^2(t)|\alpha\rangle = \mathcal{E}^2\left[4|\alpha|^2\cos^2(\omega t+\phi)+1\right]\sin^2 kz. \tag{3.54}$$

The root-mean-square deviation in the electric field is therefore given by

$$\langle\Delta\hat{E}(t)^2\rangle^{1/2} = \sqrt{\frac{\hbar\omega}{2\varepsilon_0 V}}\,|\sin kz|. \tag{3.55}$$

Note that $\langle\Delta\hat{E}(t)^2\rangle^{1/2}$ is independent of the field strength $|\alpha|$. Since $\langle\Delta\hat{E}(t)^2\rangle^{1/2}$ is independent of $|\alpha|^2$, the quantum noise becomes less important as $|\alpha|^2$ increases. The same argument also applies to the magnetic field, and it is the reason why a highly excited coherent state ($|\alpha|\gg 1$) can be treated as a classical electromagnetic field.

3.2.2. Generation of Coherent States

Equation (3.46) indicates that the coherent state $|\alpha\rangle$ is generated by translating the vacuum state $|0\rangle$ to a point with finite excitation amplitude α without changing its quantum statistical properties. This suggests that the harmonic oscillator originally in a ground state $|0\rangle$ can be driven to a coherent state $|\alpha\rangle$ if a classical external force $f(t)$ couples linearly to the generalized coordinate of the harmonic oscillator. The Hamiltonian of such a driven harmonic oscillator is given by

$$\hat{\mathcal{H}} = \hbar\omega\left(\hat{a}^\dagger\hat{a}+\frac{1}{2}\right)+\hbar[f(t)\hat{a}+f^*(t)\hat{a}^\dagger]. \tag{3.56}$$

The solution of the Schrödinger equation, $i\hbar\frac{\partial}{\partial t}|\psi(t)\rangle = \hat{\mathcal{H}}|\psi(t)\rangle$, for the initial state $|\psi(0)\rangle = |0\rangle$ is obtained as [4]

$$|\psi(t)\rangle = \exp[A(t)+C(t)\hat{a}^\dagger]|0\rangle, \tag{3.57}$$

where

$$A(t) = -\int_0^t dt''\, f(t'')\int_0^{t''} e^{i\omega(t'-t'')}f(t')\,dt', \tag{3.58}$$

$$C(t) = -i\int_0^t e^{i\omega(t'-t)}f^*(t')\,dt'. \tag{3.59}$$

When the classical driving force $f(t)$ is resonant with the harmonic oscillator, that is,

$$f(t) = f_0\, e^{i\omega t} \quad (f_0;\text{ constant real number}) \tag{3.60}$$

(3.57), (3.58), and (3.59) are reduced to

$$C(t) = -i\, e^{-i\omega t}(f_0 t) \equiv \alpha, \tag{3.61}$$

$$A(t) = -\frac{1}{2}(f_0 t)^2 = -\frac{|\alpha|^2}{2}, \tag{3.62}$$

$$|\psi(t)\rangle = |\alpha\rangle \quad \text{(coherent state)}. \tag{3.63}$$

The preceding argument for an external-force-driven harmonic oscillator indicates that to generate a coherent state of light, all we need is a classical oscillating current that is resonantly coupled to the electromagnetic field. Glauber showed that a classical oscillating current in free space produces a multimode coherent state of light [8]. In a more realistic context, an oscillating atomic dipole moment (current) in a highly saturated laser oscillator produces a state that is not very far from a coherent state of light. We will show in Chapter 15 that the quantum noise of a laser operating at far above threshold is close to that of a coherent state.

A coherent state does not change its quantum statistical properties if it is attenuated by scattering and/or absorption. A simple model that demonstrates this remarkable feature is illustrated in Fig. 3-5, in which the input coherent state $|\alpha\rangle$ and the vacuum state $|0\rangle$ are combined by a beam splitter. The interaction Hamiltonian and the unitary evolution operator describing the beam splitter are given by

$$\hat{\mathcal{H}}_I = \hbar\kappa(\hat{a}^\dagger \hat{b} + \hat{a}\hat{b}^\dagger), \tag{3.64}$$

$$\hat{U} = \exp\left[i\kappa t(\hat{a}^\dagger \hat{b} + \hat{a}\hat{b}^\dagger)\right], \tag{3.65}$$

where κ is a coupling constant between two modes and t is an interaction time. The output state $|\psi\rangle_{\text{out}}$ is given by [4]

$$|\psi\rangle_{\text{out}} = \hat{U}|\alpha\rangle_a |0\rangle_b = |\beta\rangle_a |\gamma\rangle_b, \tag{3.66}$$

where $\beta = \sqrt{T}\alpha$ and $\gamma = \sqrt{1-T}\alpha$. Equation (3.66) indicates that the two output states of the beam splitter are uncorrelated coherent states with the eigenvalues given by the corresponding classical field amplitudes. A coherent state is unique in this

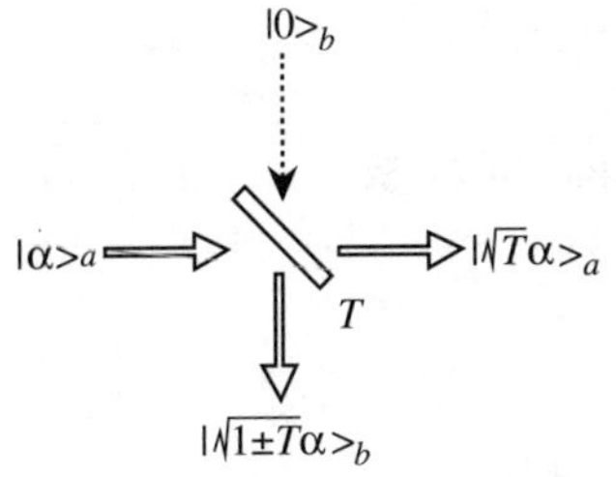

FIGURE 3-5: A simple beam splitter driven by a coherent state $|\alpha\rangle$ and a vacuum state $|0\rangle$.

respect in that it remains as a coherent state after attenuation. The reservoirs consisting of ground state harmonic oscillators inject the vacuum fluctuation and partially replace the original quantum noise of the coherent state; however, since the vacuum state is also a coherent state, the overall noise distributions are unchanged. This fact lies at the heart of coherent state generation in a laser oscillator, which is an open dissipative system.

3.3. SQUEEZED STATES

3.3.1. Properties of Quadrature Amplitude Squeezed States

A quadrature amplitude squeezed state is mathematically constructed from a vacuum state by the following unitary evolution operator [9], [10]:

$$|\beta, r\rangle = \hat{D}(\beta)\hat{S}(r)|0\rangle, \tag{3.67}$$

where

$$\hat{S}(r) = \exp\left\{\frac{r}{2}\left[\hat{a}^2 - (\hat{a}^\dagger)^2\right]\right\}, \tag{3.68}$$

$$\hat{D}(\beta) = \exp(\beta \hat{a}^\dagger - \beta^* \hat{a}). \tag{3.69}$$

The unitary operator $\hat{S}(r)$ produces a squeezed vacuum state if it acts directly on a vacuum state and if r is a real number. This is easily understood in the Heisenberg picture. The Heisenberg operator evolves as

$$\hat{a}_{\text{out}} \equiv \hat{S}^\dagger(r)\hat{a}_{\text{in}}\hat{S}(r) = \cosh(r)\hat{a}_{\text{in}} - \sinh(r)\hat{a}_{\text{in}}^\dagger$$
$$= e^{-r}\hat{a}_{\text{in},1} + ie^r \hat{a}_{\text{in},2}. \tag{3.70}$$

If the input state is the vacuum state $|0\rangle$, we have

$$\langle \hat{a}_{\text{out}} \rangle = 0, \tag{3.71}$$

$$\langle \Delta \hat{a}_{\text{out},1}^2 \rangle = e^{-2r} \langle \Delta \hat{a}_{\text{in},1}^2 \rangle = \frac{1}{4}e^{-2r}, \tag{3.72}$$

$$\langle \Delta \hat{a}_{\text{out},2}^2 \rangle = e^{2r} \langle \Delta \hat{a}_{\text{in},2}^2 \rangle = \frac{1}{4}e^{2r}. \tag{3.73}$$

The corresponding output state in the Schrödinger picture is referred to as a squeezed vacuum state $|0, r\rangle$. It may be expressed as a linear superposition of photon number eigenstates:

$$|0, r\rangle = \left[1 - \left(\frac{\nu}{\mu}\right)^2\right]^{1/4} \sum_{n=0}^{\infty} \frac{[(2n)!]^{\frac{1}{2}}}{2^n n!} \left(\frac{\nu}{\mu}\right)^n |2n\rangle, \tag{3.74}$$

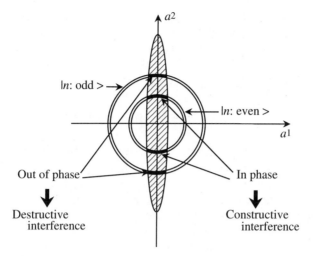

FIGURE 3-6: Constructive and destructive interference between the two intersections of $|0, r\rangle$ and $|n\rangle$ states.

where $\nu = \cosh(r)$ and $\mu = \sinh(r)$. The squeezed vacuum state has finite probabilities only for even photon number eigenstates. This is again the result of the quantum interference between the two intersections between the squeezed vacuum state and the photon number eigenstate in the phase space as shown in Fig. 3-6.

The displacement operator $\hat{D}(\beta)$ simply translates the coherent excitation of the squeezed vacuum state to a finite value β from zero without changing the quantum uncertainties. Such a squeezed state has the following linear expansion in terms of photon number eigenstates:

$$|\beta, r\rangle = \frac{1}{\sqrt{\mu}} \exp\left[-\frac{\left(1-\frac{\nu}{\mu}\right)}{2}\beta^2\right] \sum_{n=0}^{\infty} \frac{1}{\sqrt{n!}} \left(\frac{\nu}{2\mu}\right)^{\frac{n}{2}} H_n\left(\frac{\beta}{\sqrt{2\mu\nu}}\right) |n\rangle, \quad (3.75)$$

where $H_n(x)$ is the Hermite polynomial. The squeezed states with a positive real β and positive real r and negative real r are depicted in Figs. 1-4(b) and (c), respectively.

3.3.2. Generation of Quadrature Amplitude Squeezed States

Quadrature amplitude squeezed states can be generated by the unitary evolution expressed by (3.68) from a vacuum state. The interaction Hamiltonian required for realizing this unitary evolution is quadratic in photon annihilation and creation operators:

$$H_I = \frac{\hbar}{2}[\chi(\hat{a}^\dagger)^2 + \chi^*\hat{a}^2]. \quad (3.76)$$

The Hamiltonian expresses the simultaneous two-photon generation and absorption processes, which can be realized either by second- or third-order nonlinear processes [11]–[14]. The nonlinear interaction parameter χ for these two cases is

$$\chi \propto \begin{cases} \chi^{(2)} E_p : & \text{degenerate parametric amplification} \\ \chi^{(3)} E_p^2 : & \text{degenerate four-wave mixing,} \end{cases} \quad (3.77)$$

where E_p is the (classical) pump field amplitude and $\chi^{(2)}$ and $\chi^{(3)}$ denote the corresponding nonlinear susceptibility. The phase of the pump wave E_p determines which quadrature amplitude is amplified and which one is deamplified. As long as the pump wave is not depleted by the generated wave, the pump wave fluctuation does not affect the quantum fluctuation of the generated wave. In such a case we can treat the pump wave amplitude as a c-number.

In a degenerate parametric amplifier (DPA) [12], [13], a pump photon of frequency ω_p is converted to two nearly degenerate photons of frequency ω_s and ω_i ($\omega_s \simeq \omega_i$) with $\omega_p = \omega_s + \omega_i$ as shown in Fig. 3-7. The effective nonlinear interaction requires the phase matching condition, $k_p = k_s + k_i$.

The two simultaneously generated signal and idler photons starting from a single pump photon have conjugate phases, so they interfere with each other to suppress one quadrature amplitude noise and enhance the other quadrature amplitude noise as shown in Fig. 3-7.

The pump field is usually generated via second harmonic generation from a single-frequency laser radiation; this is experimentally convenient since the detection of quadrature amplitude squeezed states at $\omega_s = \omega_p/2$ by an optical homodyne detector requires a reference phase local oscillator wave at the same frequency $\omega_{LO} = \omega_p/2$. Most of the experiments carried out so far used a resonant cavity to enhance the effective nonlinear interaction strength. The squeezing bandwidth in this case is limited by the cavity bandwidth.

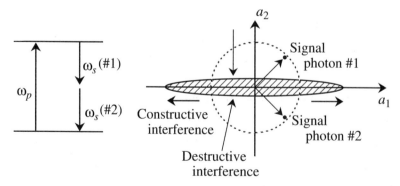

FIGURE 3-7: Generation of a quadrature amplitude squeezed state by a degenerate parametric amplifier.

In degenerate four-wave mixing (DFWM) [14], two pump photons of frequency ω_p are converted to two nearly degenerate signal and idler photons of frequencies ω_s and ω_i ($\omega_s \simeq \omega_i$), where $2\omega_p = \omega_s + \omega_i$. The phase matching condition $2k_p = k_s + k_i$ can be automatically realized for such a degenerate case. The frequency up-conversion required for preparing the local oscillator wave and down-conversion required for generating the quadrature squeezed state are not necessary in this four-wave mixing method. The squeezed state generation experiments by degenerate parametric amplifiers and four-wave mixers are discussed in great detail in the text by D. Walls and G. Milburn and by H. Bachor [15].

3.3.3. Properties of Number-Phase Squeezed States

Unfortunately, the phase cannot be defined easily in quantum mechanics, because there is no simple and experimentally accessible Hermitian phase operator. The standard approach to define the phase in quantum mechanics is based on the following decomposition of annihilation and creation operators:

$$\hat{a} = (\hat{n} + 1)^{1/2} \hat{E}_-, \tag{3.78}$$

$$\hat{a}^\dagger = \hat{E}_+ (\hat{n} + 1)^{1/2}. \tag{3.79}$$

We cannot define the Hermitian phase operator $\hat{\phi}$ simply by $\hat{E}_- = e^{i\hat{\phi}}$ and $\hat{E}_+ = e^{-i\hat{\phi}}$, since $\hat{E}_-$ and $\hat{E}_+$ are not unitary (i.e., $\hat{E}_+\hat{E}_- \neq 1$ even though $\hat{E}_-\hat{E}_+ = 1$). Nevertheless, we can define the cosine and sine operators by

$$\hat{C} \equiv \frac{1}{2}(\hat{E}_- + \hat{E}_+) = \frac{1}{2}\left[(\hat{n}+1)^{-\frac{1}{2}}\hat{a} + \hat{a}^\dagger(\hat{n}+1)^{-\frac{1}{2}}\right], \tag{3.80}$$

$$\hat{S} \equiv \frac{1}{2i}[\hat{E}_- - \hat{E}_+] = \frac{1}{2i}\left[(\hat{n}+1)^{-\frac{1}{2}}\hat{a} - \hat{a}^\dagger(\hat{n}+1)^{-\frac{1}{2}}\right]. \tag{3.81}$$

Using the commutation relation $[a, \hat{a}^\dagger] = 1$, we obtain the following commutation relations for $\hat{C}$, $\hat{S}$, and $\hat{n}$:

$$\left[\hat{n}, \hat{C}\right] = -i\hat{S}, \tag{3.82}$$

$$\left[\hat{n}, \hat{S}\right] = i\hat{C}. \tag{3.83}$$

The Heisenberg uncertainty relations for the cosine, sine, and photon number operators are directly obtained from (3.82) and (3.83):

$$P_{ns} \equiv \frac{\langle \Delta \hat{n}^2 \rangle \langle \Delta \hat{S}^2 \rangle}{\langle \hat{C} \rangle^2} \geq \frac{1}{4}, \tag{3.84}$$

$$P_{nc} \equiv \frac{\langle \Delta \hat{n}^2 \rangle \langle \Delta \hat{C}^2 \rangle}{\langle \hat{S} \rangle^2} \geq \frac{1}{4}. \tag{3.85}$$

When an electromagnetic field mode is highly excited with $\langle \hat{a}_1 \rangle \gg 1$ and $\langle \hat{a}_2 \rangle = 0$, one has $\langle \hat{C} \rangle \simeq 1$ and $\langle \Delta \hat{S}^2 \rangle$ is approximately identified as the phase noise $\langle \Delta \hat{\phi}^2 \rangle$. The alternative definitions of the Hermitian phase operator and the phase eigenstate are given in Sec. 4.5.

The number-phase minimum uncertainty state that precisely satisfies (3.84) with an equality sign is mathematically constructed as an eigenstate of a non-Hermitian operator:

$$(e^{-r}\hat{n} + ie^r \hat{S})|\psi\rangle = (e^{-r}\langle\hat{n}\rangle + ie^r \langle\hat{S}\rangle)|\psi\rangle. \tag{3.86}$$

The squeezing parameter r determines the photon number noise and the sine operator noise under the constraint of the minimum uncertainty product (3.84). The eigenstate of (3.86) is mathematically constructed only for the following special case [16]:

$$I_{-(1+\langle\hat{n}\rangle)}\left[\left(\frac{\langle\Delta\hat{n}^2\rangle}{\langle\Delta\hat{S}^2\rangle}\right)^{1/2}\right] = 0, \tag{3.87}$$

where $I_\mu(x)$ is a modified Bessel function of the first kind of order μ. This constraint requires that $\langle\hat{n}\rangle$ must be chosen as $\langle\hat{n}\rangle \in [2k, 2k+1](k = 0, 1, 2, \ldots)$ and that the squeezing parameter $r = \frac{1}{4}\ln\frac{\langle\Delta\hat{S}^2\rangle}{\langle\Delta\hat{n}^2\rangle}$ cannot be chosen independently of $\langle\hat{n}\rangle$. The eigenstate of (3.86) is given by

$$|\psi\rangle = \xi \sum_{n=0}^{\infty} I_{n-\langle\hat{n}\rangle}\left[\left(\frac{\langle\Delta\hat{n}^2\rangle}{\langle\Delta\hat{S}^2\rangle}\right)^{1/2}\right]|n\rangle, \tag{3.88}$$

where ξ is a normalization factor. The photon number noise $\langle\Delta\hat{n}^2\rangle$ is zero at the boundaries $2k$ and $2k+1$ (photon number eigenstates) and takes a maximum value at k near the midpoint.

The variances of the photon number and the (approximate) phase of this state satisfy

$$\langle\Delta\hat{n}^2\rangle = \frac{1}{2}e^{-2r} \searrow$$
$$\langle\Delta\hat{n}^2\rangle\langle\Delta\hat{S}^2\rangle \simeq \langle\Delta\hat{n}^2\rangle\langle\Delta\hat{\phi}^2\rangle = \frac{1}{4}. \tag{3.89}$$
$$\langle\Delta\hat{S}^2\rangle \simeq \langle\Delta\hat{\phi}^2\rangle = \frac{1}{2}e^{+2r} \nearrow$$

When the squeezing parameter r is equal to $\ln\left(\frac{1}{\sqrt{2\langle\hat{n}\rangle}}\right)$, the number-phase squeezed state is close to a coherent state and the variances are given by $\langle\Delta\hat{n}^2\rangle = \langle\hat{n}\rangle$ and $\langle\Delta\hat{S}^2\rangle \simeq \langle\Delta\hat{\phi}^2\rangle = \frac{1}{4\langle\hat{n}\rangle}$. The electric field of a number-phase squeezed state with $r < \ln\left(\frac{1}{\sqrt{2\langle\hat{n}\rangle}}\right)$ is depicted in Fig. 1-4 (d).

3.3.4. Generation of Number-Phase Squeezed States

The number-phase squeezed state defined by (3.86) is not connected to a coherent state by any unitary operator. No physical process is known to produce this mathematically defined state. However, the self-phase modulation process in a third-order nonlinear medium produces a number-phase squeezed state starting from a coherent state when the average field excitation is properly manipulated after the self-phase modulation process [17].

The interaction Hamiltonian for the self-phase modulation is

$$\hat{\mathcal{H}}_I = \hbar\chi \hat{a}^{\dagger 2}\hat{a}^2 = \hbar\chi\hat{n}(\hat{n}-1), \tag{3.90}$$

where χ is proportional to the third-order Kerr susceptibility. Since $\hat{\mathcal{H}}_I$ commutes with the photon number operator $\hat{n}$, the photon number is preserved in this unitary evolution process. However, a quantum correlation is established between the photon number and the phase. The unitary evolution operator and the input and output Heisenberg operators are

$$\hat{U}_1 = \exp\left[i\frac{\gamma}{2}\hat{n}(\hat{n}-1)\right], \tag{3.91}$$

$$\hat{a}_{\text{out}} = \hat{U}_1^\dagger \hat{a}_{\text{in}} \hat{U}_1 = e^{i\gamma\hat{n}}\hat{a}_{\text{in}}, \tag{3.92}$$

where $\gamma = 2\chi t$ and t is the interaction time. Figure 3-8(a) shows the quasi-probability density $Q(\alpha)(= \langle\alpha|\hat{\rho}|\alpha\rangle)$ representation for such a self-phase modulated state when the input state is a coherent state. If we can translate the average field amplitude by adding a c-number excitation ξ that is $-\frac{\pi}{2}$ out of phase from the original field amplitude $\langle\hat{a}_{\text{out}}\rangle$, the Heisenberg operator is given by

$$\hat{a}'_{\text{out}} = \hat{a}_{\text{out}} + \xi. \tag{3.93}$$

The corresponding output state becomes a number-phase squeezed state as shown by solid lines in Fig. 3-8(b). The unitary operator for such a translation is given by $\hat{U}_2 = \exp(\xi\hat{a}^\dagger - \xi^*\hat{a})$. This is the displacement operator that translates the vacuum state $|0\rangle$ to the coherent state $|\xi\rangle$ or the squeezed vacuum state $|0,r\rangle$ to the squeezed state $|\xi,r\rangle$. This translation operation is physically realized by adding a phase-coherent laser field to the signal field with an extremely small coupling efficiency.

The number-phase squeezed state generated in this overall process is given by

$$|\psi\rangle = \hat{U}_2\hat{U}_1|\alpha\rangle = \exp(\xi\hat{a}^\dagger - \xi^*\hat{a})\exp\left[i\frac{\gamma}{2}\hat{n}(\hat{n}-1)\right]|\alpha\rangle. \tag{3.94}$$

The photon number noise and phase noise are calculated as [15]

$$\langle\Delta\hat{n}^2\rangle = \langle\hat{n}\rangle\left[\sqrt{\gamma^2\langle\hat{n}\rangle^2 + 1} - \gamma\langle\hat{n}\rangle\right]$$

$$\langle\Delta\hat{\phi}^2\rangle = \frac{1}{4\langle\hat{n}\rangle} \cdot \frac{1}{\left[\sqrt{\gamma^2\langle\hat{n}\rangle^2+1} - \gamma\langle\hat{n}\rangle\right]} \qquad \langle\Delta\hat{n}^2\rangle\langle\Delta\hat{\phi}^2\rangle \simeq \frac{1}{4}. \tag{3.95}$$

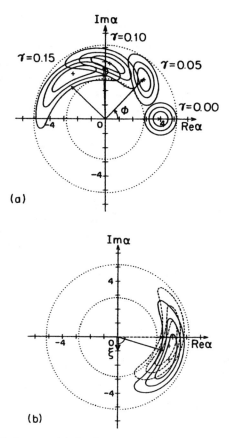

FIGURE 3-8: (a) Self-phase modulation in Kerr medium and (b) interference at high reflectivity mirror; γ is a nonlinear interaction parameter.

Figure 3-9 shows the normalized photon number noise $\frac{\langle \Delta \hat{n}^2 \rangle}{\langle n \rangle}$ and the normalized phase noise $\langle \hat{n} \rangle \langle \Delta \hat{\phi}^2 \rangle$ as a function of the nonlinear interaction parameter γ, where ξ is chosen to minimize $\langle \Delta \hat{n}^2 \rangle$ at each point. When γ is smaller than the critical value $\gamma_1 = \frac{1}{2} \langle \hat{n} \rangle^{-\frac{2}{3}}$, the number-phase uncertainty product is equal to the minimum value. The normalized photon number noise $\langle \Delta \hat{n}^2 \rangle / \langle \hat{n} \rangle$ is approximately given by $\langle \hat{n} \rangle^{-\frac{1}{3}}$.

The experimental realization of this self-phase modulation and displacement operation has been demonstrated by optical solitons in a fiber [18]. A completely different approach for generating a number-phase squeezed state will be discussed in Chapter 15, where we consider a constant current driven semiconductor laser. The text by D. Walls and G. Milburn, and by H. Bachor [15] also gives an excellent detailed treatment of the number-phase squeezed state generation experiments.

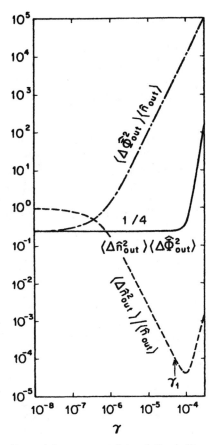

FIGURE 3-9: Number-phase minimum uncertainty relation in Kerr nonlinear interferometer.

3.4. CORRELATED TWIN PHOTONS AND QUANTUM ENTANGLEMENT

The correlated twin photons generated in a nondegenerate parametric down-conversion process can be used for the measurement-induced photon number eigenstate realization as already discussed in Sec. 3.1. A pump photon at frequency ω_p is converted to signal and idler photons at frequencies ω_s and ω_i, respectively, in a second-order nonlinear crystal (see Fig. 3-3). The unitary evolution operator corresponding to the interaction Hamiltonian (3.26) is given by

$$\hat{U} = \exp\left[-i|\kappa|t\left(e^{i\theta}\hat{a}_s^\dagger\hat{a}_i^\dagger + e^{-i\theta}\hat{a}_s\hat{a}_i\right)\right]. \tag{3.96}$$

This unitary operator can be expressed in the normally ordered form [5]:

$$\hat{U} = \frac{1}{\cosh(r)} \exp\left[-ie^{i\theta} \tanh(r) \hat{a}_i^\dagger \hat{a}_s^\dagger\right] \exp\left[-\ln\{\cosh(r)\}\hat{a}_i^\dagger \hat{a}_i\right]$$
$$\times \exp\left[-\ln\{\cosh(r)\}\hat{a}_s^\dagger \hat{a}_s\right] \exp\left[-ie^{-i\theta} \tanh(r) \hat{a}_i \hat{a}_s\right], \quad (3.97)$$

where $r = |\kappa| t$. Such correlated twin photons have interesting quantum properties. In this section we will discuss the two aspects of twin photons, generation of arbitrary quantum states by projective measurement and nonlocal quantum entanglement.

3.4.1. Generation of Arbitrary Quantum States by Projective Measurements

If we input a coherent state $|\alpha\rangle_s$ instead of the vacuum state at the signal frequency ω_s and the vacuum state $|0\rangle_i$ at the idler frequency ω_i, the joint-correlated output signal and idler waves are described by

$$|\psi_{si}\rangle = \hat{U}|\alpha\rangle_s |0\rangle_i$$
$$= N e^{-\frac{|\alpha|^2}{2}} \sum_{k,l} \binom{k+l}{k}^{1/2} \frac{[-ie^{i\theta} \tanh(r)]^l}{\cosh(r)} \left[\frac{\alpha}{\cosh(r)}\right]^k \left(\frac{1}{k!}\right)^{1/2}$$
$$\times |k+l\rangle_s |l\rangle_i. \quad (3.98)$$

To see how strongly the output signal and idler waves are correlated, we can take a trace over the idler coordinates or project onto a specific measurement readout. An unconditional signal output wave is obtained by taking a simple trace over the idler coordinates:

$$\hat{\rho}_s^{(\text{red})} \equiv Tr_i [|\psi_{si}\rangle\langle\psi_{si}|]$$
$$= N' e^{-|\alpha|^2} \sum_{k,l,h} \binom{k+l}{k}^{1/2} \binom{h+l}{h}^{1/2}$$
$$\frac{[\tanh(r)]^{2l}}{\cosh^2(r)} \left[\frac{\alpha}{\cosh(r)}\right]^k \left[\frac{\alpha^*}{\cosh(r)}\right]^h$$
$$\times \left(\frac{1}{k!h!}\right)^{1/2} |k+l\rangle_{ss}\langle h+l|. \quad (3.99)$$

This is a mixed state with large fluctuations in the photon number and in the two quadrature amplitudes. Both input signal and idler waves are amplified and added as the output signal wave, so it is understandable that the unconditional signal output is very *noisy*. This is also seen by the following Heisenberg operator evolution:

$$\hat{b}_s \equiv \hat{U}^\dagger \hat{a}_s \hat{U} = \cosh(r)\hat{a}_s - ie^{i\theta}\sinh(r)\hat{a}_i^\dagger, \quad (3.100)$$

$$\hat{b}_i \equiv \hat{U}^\dagger \hat{a}_i \hat{U} = \cosh(r)\hat{a}_i - ie^{i\theta}\sinh(r)\hat{a}_s^\dagger. \quad (3.101)$$

The signal wave is amplified by a factor of $\sqrt{G} = \cosh(r)$ and the idler wave is amplified by a factor of $\sqrt{G-1} = \sinh(r)$, and both components are added as the output signal wave.

The postmeasurement signal state is calculated by the probability operator amplitude

$$\hat{Y}(\phi) = {}_i\langle\phi|\hat{U}|0\rangle_i, \quad (3.102)$$

where $|\phi\rangle_i$ is the eigenstate corresponding to the measurement result for the idler wave. The postmeasurement signal state is given by [see (1.80)]

$$\hat{\rho}_s(\phi) = \frac{1}{P(\phi)}\hat{Y}(\phi)\hat{\rho}_{si}\hat{Y}^\dagger(\phi). \quad (3.103)$$

$P(\phi) \equiv Tr_s(\hat{Y}^\dagger(\phi)\hat{Y}(\phi)|\alpha\rangle_{ss}\langle\alpha|)$ is the probability of obtaining a specific readout ϕ. When the idler wave is measured by a photon counter and the measurement result is m, the probability operator amplitude is [5]

$$\hat{Y}(m) = {}_i\langle m|\hat{U}|0\rangle_i$$
$$= \frac{1}{\cosh(r)}\frac{[-ie^{i\theta}\tan(r)]^m}{(m!)^{1/2}}(a_s^\dagger)^m \exp[-\ln\{\cosh(r)\}\hat{a}_s^\dagger\hat{a}_s]. \quad (3.104)$$

Using (3.104) in (3.103), we have

$$\hat{\rho}_s(m) = \frac{1}{P(m)}\hat{Y}(m)|\alpha\rangle_{ss}\langle\alpha|\hat{Y}^\dagger(m). \quad (3.105)$$

This is a photon number eigenstate if the signal input state is vacuum state and a number-phase squeezed state if $\alpha \neq 0$. In fact, calculation of the photon number variance $\langle\Delta\hat{n}^2\rangle$ and the phase variance $\langle\Delta\hat{\phi}^2\rangle$ shows that the postmeasurement state satisfies the minimum uncertainty product (3.89) [5]. From (3.100) and (3.101), we have the following (operator) Manley-Rowe relation:

$$\hat{n}_{s,\text{out}} - \hat{n}_{i,\text{out}} = \hat{n}_{s,\text{in}} - \hat{n}_{i,\text{in}}, \quad (3.106)$$

where $\hat{n}_{\text{out}} = \hat{b}^\dagger\hat{b}$ and $\hat{n}_{\text{in}} = \hat{a}^\dagger\hat{a}$ are the output and input photon number operators, respectively. This suggests that the photon number correlation is perfect when both signal and idler input waves are in vacuum states. When the input signal is in a coherent state, there exists a strong quantum correlation in the limit of large gain:

$$\langle\Delta(\hat{n}_{s,\text{out}} - \hat{n}_{i,\text{out}})^2\rangle = \langle\Delta\hat{n}_{s,\text{in}}^2\rangle = \langle\hat{n}_{s,\text{in}}\rangle. \quad (3.107)$$

Therefore, the conditional variance $\langle \Delta(\hat{n}_{s,\text{out}} - \hat{n}_{i,\text{out}})^2 \rangle = \langle n_{s,\text{in}} \rangle$ is smaller than the photon number noise of the coherent state $\langle \hat{n}_{s,\text{out}} \rangle = G \langle \hat{n}_{s,\text{in}} \rangle$ if $G \gg 1$.

When the idler wave is measured by a homodyne detector for a single quadrature amplitude and the measurement result is α_1, the probability operator amplitude is [5]

$$\hat{Y}(\alpha_1) = {}_i\langle \alpha_1 | \hat{U} | 0 \rangle_i$$

$$= \frac{1}{\cosh(r)} \left(\frac{2}{\pi}\right)^{\frac{1}{4}} e^{-\alpha_1^2} \sum_{m=0}^{\infty} \left\{ \frac{[-ie^{i\theta} \tanh(r)]^m}{m!} \frac{H_m(\sqrt{2}\alpha_1^*)}{2^{m/2}} (\hat{a}_s^\dagger)^m \right\}$$

$$\times \exp[-\ln\{\cosh(r)\}\hat{a}_s^\dagger \hat{a}_s]. \tag{3.108}$$

Using (3.108) in (3.103), we observe that we have a quadrature amplitude squeezed state for the signal wave. In fact, the calculation of the two quadrature amplitudes noise $\langle \Delta \hat{a}_1^2 \rangle$ and $\langle \Delta \hat{a}_2^2 \rangle$ shows that the postmeasurement state satisfies the minimum uncertainty product (1.48) [5].

From (3.100) and (3.101), we have the following relation:

$$\hat{b}_{s1} - \hat{b}_{i1} = (\sqrt{G} - \sqrt{G-1})(\hat{a}_{s1} - \hat{a}_{i1}) \longrightarrow 0 \, (G \gg 1), \tag{3.109}$$

$$\hat{b}_{s2} + \hat{b}_{i2} = (\sqrt{G} - \sqrt{G-1})(\hat{a}_{s2} + \hat{a}_{i2}) \longrightarrow 0 \, (G \gg 1), \tag{3.110}$$

where $\theta = \frac{\pi}{2}$ is assumed. Therefore, one set of quadrature amplitudes, $\hat{b}_{s1}$ and $\hat{b}_{i1}$, is positively correlated, and the other quadrature amplitudes, $\hat{b}_{s2}$ and $\hat{b}_{i2}$, are negatively correlated. The conditional variance $\langle \Delta(\hat{b}_{sj} - \hat{b}_{ij})^2 \rangle (j = 1, 2)$ is smaller than the quadrature amplitude noise of the coherent state $\langle \Delta \hat{a}_{sj}^2 \rangle = \frac{1}{4}$ if $G \gg 1$.

The postmeasurement state by a heterodyne detector for simultaneous measurements of two quadrature amplitudes is similarly calculated and found to be the coherent state irrespective of the gain G [5].

3.4.2. Nonlocal Quantum Entanglement

When parametric down-conversion efficiency is small and both signal and idler input states are vacuum states, the joint correlated output state is well approximated by $|\psi_{si}\rangle = c_0|0\rangle_s|0\rangle_i + c_1|1\rangle_s|1\rangle_i$ [see (3.98)]. This joint-correlated linear superposition carries the information about the pump phase, which is absent in the hypothetical mixed state $\hat{\rho}_{\text{mix}} = |c_0|^2 |0\rangle_s |0\rangle_{ii}\langle 0|_s\langle 0| + |c_1|^2 |1\rangle_s |1\rangle_{ii}\langle 1|_s\langle 1|$. Moreover, when the polarizations of the signal and idler photons are arranged to be orthogonal and the two photons are mixed appropriately [19], the output state can be one of the EPR-Bell states:

$$|\chi\rangle = \frac{1}{\sqrt{2}}[|k\rangle_1|q\rangle_2 - |q\rangle_1|k\rangle_2] \otimes [|V\rangle_1|H\rangle_2 - |H\rangle_1|V\rangle_2]. \tag{3.111}$$

where the two spatial modes are designated by $|k\rangle$ and $|q\rangle$. In general, the two photons are generated with different wave vector, so they can be separated far apart.

Therefore, when the measurement is performed for one photon, no influence resulting from this measurement can propagate to the other photon in a given time. However, if the polarization measurement result for one photon is H (horizontal linear polarization), the polarization measurement result for the other is always V (vertical linear polarization), and vice versa. According to Einstein, Podolsky, and Rosen (EPR) [20], such experimental results force us to accept that there exists an element of physical reality corresponding to this physical quantity (local hidden variable). However, the entangled quantum state (3.111) can be easily converted to a canonically conjugate polarization basis:

$$|\chi\rangle = \frac{1}{\sqrt{2}}[|k\rangle_1|q\rangle_2 - |q\rangle_1|k\rangle_2] \otimes [|R\rangle_1|L\rangle_2 - |L\rangle_1|R\rangle_2]. \qquad (3.112)$$

Now, if the polarization measurement result for one photon is L (left circular polarization), the polarization measurement result for the other is always R (right circular polarization), and vice versa. According to the preceding arguments, the photons should also have this value of circular polarization. However, the linear and circular polarizations are canonically conjugate, so they cannot be determined simultaneously. Moreover, the decision whether the linear or circular polarization is to be measured can be made when the two photons are separated far apart and cannot influence each other in a given time. This counterintuitive result is sometimes referred to as the EPR paradox or violation of Einstein locality.

In quantum mechanics a dynamic variable is represented by an operator and produces a real value only when it is projected on the state vector. In a sense, the measurement creates the physical reality. A nonlocal quantum entangled state realizes such a fundamentally quantum mechanical effect, which no realistic local hidden variable theory can explain. The recent text by A. Peres [21] gives an excellent review of this issue.

3.5. PROBABILITY DISTRIBUTION FUNCTIONS FOR A DENSITY OPERATOR

In order to describe the quantum statistical properties of the electromagnetic field, it is useful to define various probability distribution functions for a density operator. The simplest expansion of a density operator is given by the diagonal representation of coherent states:

$$\hat{\rho} = \int d^2\alpha \, P(\alpha)|\alpha\rangle\langle\alpha|. \qquad (3.113)$$

The diagonal $P(\alpha)$ function is calculated by the following procedure:

$$\hat{\rho}(\hat{a}, \hat{a}^\dagger) = \hat{\rho}^{(a)}(\hat{a}, \hat{a}^\dagger) \qquad \longleftarrow \quad \text{antinormal ordering}$$

$$= \sum_{r,s} \rho_{rs}^{(a)} \hat{a}^r \int \frac{d^2\alpha}{\pi} |\alpha\rangle\langle\alpha|(\hat{a}^\dagger)^s$$

$$= \int \frac{d^2\alpha}{\pi} |\alpha\rangle\langle\alpha| \sum_{r,s} \rho_{rs}^{(a)} \alpha^r (\alpha^*)^s \quad \leftarrow \quad \begin{array}{l}\text{antinormal ordered}\\ \text{associated function}\end{array} \quad (3.114)$$

$$= \int d^2\alpha |\alpha\rangle\langle\alpha| P(\alpha).$$

That is,

$$P(\alpha) = \frac{1}{\pi}\overline{\rho}^{(a)}(\alpha, \alpha^*) = \frac{1}{\pi}\sum_{r,s}\rho_{rs}^{(a)}\alpha^r(\alpha^*)^s. \quad (3.115)$$

The diagonal $P(\alpha)$ function is useful for calculating the mean value of the operator expressed in a normal ordered form,

$$\hat{O} = \sum_{r,s} O_{rs}^{(n)}(\hat{a}^\dagger)^r \hat{a}^s, \quad (3.116)$$

$$\langle \hat{O}\rangle = \text{Tr}(\hat{\rho}\hat{O}) = \text{Tr}\left[\int d^2\alpha |\alpha\rangle\langle\alpha| P(\alpha)\hat{O}\right]$$

$$= \int d^2\alpha\, P(\alpha)\langle\alpha|\hat{O}|\alpha\rangle$$

$$= \int d^2\alpha\, P(\alpha)\sum_{r,s} O_{rs}^{(n)}(\alpha^*)^r \alpha^s$$

$$= \int d^2\alpha\, P(\alpha)\overline{O}^{(n)}(\alpha, \alpha^*), \quad (3.117)$$

where $\overline{O}^{(n)}(\alpha, \alpha^*)$ is the normal ordered associated function for the operator $\hat{O}$. A pure coherent state $|\alpha_0\rangle$ has a δ-function $P(\alpha)$, $P(\alpha) = \delta(\alpha - \alpha_0)$. The diagonal $P(\alpha)$ representation is very useful for describing the statistical mixture of coherent states such as the field produced by a laser oscillator. We will derive the quantum mechanical Fokker-Planck equation for $P(\alpha)$ in order to discuss the quantum statistical properties of a laser in Chapter 15.

However, certain nonclassical states, such as photon number eigenstates and squeezed states, do not have well-behaved $P(\alpha)$ functions and we must use alternative probability distribution functions.

The quasi-probability density $Q(\alpha)$ is defined by

$$Q(\alpha) \equiv \langle\alpha|\hat{\rho}(\hat{a}, \hat{a}^\dagger)|\alpha\rangle$$

$$= \langle\alpha|\sum_{r,s}\rho_{rs}^{(n)}(\hat{a}^\dagger)^r \hat{a}^s|\alpha\rangle \quad \leftarrow \quad \text{normal ordering}$$

$$= \sum_{r,s} \rho_{rs}^{(n)} (\alpha^*)^r \alpha^s$$

$$= 1/\pi \overline{\rho}^{(n)}(\alpha, \alpha^*) \quad \leftarrow \quad \text{normal ordered associated function.}$$
(3.118)

The $Q(\alpha)$ representation is therefore the normal ordered associated function of a density operator. All states have well-behaved $Q(\alpha)$ functions. The diagonal $P(\alpha)$ function and the quasi-probability density $Q(\alpha)$ are related by

$$Q(\alpha) \equiv \langle \alpha | \hat{\rho}(\hat{a}, \hat{a}^\dagger) | \alpha \rangle$$

$$= \langle \alpha | \int d^2 \alpha' | \alpha' \rangle \langle \alpha' | P(\alpha') | \alpha \rangle$$

$$= \int d^2 \alpha' P(\alpha') \exp\left(-|\alpha - \alpha'|^2\right).$$
(3.119)

That is, $Q(\alpha)$ is given by the convolution of $P(\alpha)$ and the Gaussian function with a variance of $1/2$. The quasi-probability density $Q(\alpha)$ is useful for calculating the mean value of an operator expressed in an antinormal ordered form,

$$\hat{O} = \sum_{r,s} O_{rs}^{(a)} \hat{a}^r (\hat{a}^\dagger)^s,$$
(3.120)

$$\langle \hat{O} \rangle = \mathrm{Tr}(\hat{\rho}\hat{O}) = \mathrm{Tr}\left[\hat{\rho} \sum_{r,s} O_{rs}^{(a)} \hat{a}^r \int \frac{d^2 \alpha}{\pi} |\alpha\rangle\langle\alpha| (\hat{a}^\dagger)^s \right]$$

$$= \int \frac{d^2 \alpha}{\pi} \langle \alpha | \hat{\rho} | \alpha \rangle \sum_{r,s} O_{rs}^{(a)} \alpha^r (\alpha^*)^s$$

$$= \int \frac{d^2 \alpha}{\pi} Q(\alpha) \overline{O}^{(a)}(\alpha, \alpha^*),$$
(3.121)

where $\overline{O}^{(a)}(\alpha, \alpha^*)$ is an antinormal ordered associated function of the operator $\hat{O}$. The characteristic functions of a density operator are defined by

$$C^{(w)}(\xi) \equiv \mathrm{Tr}\left[\hat{\rho} e^{i(\xi \hat{a} + \xi^* \hat{a}^\dagger)} \right] \quad \text{(Wigner characteristic function)}, \quad (3.122)$$

$$C^{(n)}(\xi) \equiv \mathrm{Tr}\left[\hat{\rho} e^{i\xi^* \hat{a}^\dagger} e^{i\xi \hat{a}} \right] \quad \begin{pmatrix} \text{normal-ordered} \\ \text{characteristic function} \end{pmatrix}, \quad (3.123)$$

$$C^{(a)}(\xi) \equiv \mathrm{Tr}\left[\hat{\rho} e^{i\xi \hat{a}} e^{i\xi^* \hat{a}^\dagger} \right] \quad \begin{pmatrix} \text{anti-normal-ordered} \\ \text{characteristic function} \end{pmatrix}. \quad (3.124)$$

The Wigner characteristic function $C^{(w)}(\xi)$ is useful for calculating the mean value of an operator in the symmetric ordered form: $\hat{O} = (\xi \hat{a} + \xi^* \hat{a}^\dagger)^l$.

Using the Baker-Hausdorff relation [4], the three characteristic functions are related by

$$C^{(w)}(\xi) = e^{-\frac{1}{2}|\xi|^2} C^{(n)}(\xi) = e^{\frac{1}{2}|\xi|^2} C^{(a)}(\xi). \tag{3.125}$$

The Fourier transforms of the three characteristic functions are

$$\mathcal{F}[C^{(n)}(\xi)] \equiv \int e^{-i(\xi\alpha+\xi^*\alpha^*)} C^{(n)}(\xi) \frac{d^2\xi}{\pi^2}$$
$$= P(\alpha) \quad \longleftarrow \quad \text{diagonal } P(\alpha) \text{ function,} \tag{3.126}$$

$$\mathcal{F}[C^{(a)}(\xi)] \equiv \int e^{-i(\xi\alpha+\xi^*\alpha^*)} C^{(a)}(\xi) \frac{d^2\xi}{\pi^2}$$
$$= Q(\alpha)/\pi \quad \longleftarrow \quad \text{quasi-probability density,} \tag{3.127}$$

$$\mathcal{F}[C^{(w)}(\xi)] \equiv \int e^{-i(\xi\alpha+\xi^*\alpha^*)} C^{(w)}(\xi) \frac{d^2\xi}{\pi^2}$$
$$= W(\alpha, \alpha^*)/\pi \quad \longleftarrow \quad \text{Wigner distribution function.} \tag{3.128}$$

The quasi-probability density $Q(\alpha)$ corresponds to the statistics obtained by the simultaneous measurements of the two quadrature amplitudes $\hat{a}_1$ and $\hat{a}_2$ with minimum and equal additional noise. A heterodyne detector provides one physical realization of this measurement. This detector simultaneously "measures" the two conjugate observables $\hat{a}_1$ and $\hat{a}_2$, and additional noise is imposed on the measured results. As indicated in (3.119), $Q(\alpha)$, for a pure coherent state $|\alpha_0\rangle$, has a variance of $1/2$. One-half of the total variance stems from the quantum noise of the measured coherent state $|\alpha_0\rangle$ and the remaining half is due to the quantum noise of the measuring heterodyne detector. On the other hand, a homodyne detector measures a single observable $\hat{a}_1$ and does not suffer from additional noise. The statistics obtained by this measurement are calculated by integrating the Wigner distribution function with respect to the conjugate coordinate α_2.

REFERENCES

[1] J. R. Klauder and B. S. Skagerstam, *Coherent States* (World Scientific, Singapore, 1985).
[2] E. Schrödinger, Naturwiss. **14**, 664 (1926).
[3] H. Takahasi, in *Adv. Commun. System*, ed. A. V. Barakrishnan (Academic Press, New York, 1965), p. 277.
[4] W. H. Louisell, *Quantum Statistical Properties of Radiation* (Wiley, New York, 1973).
[5] K. Watanabe and Y. Yamamoto, Phys. Rev. A **38**, 3556 (1988).
[6] S. Ya. Kilin and D. B. Horoshko, Phys. Rev. Lett. **74**, 5206 (1995).
[7] Y. Yamamoto and H. A. Haus, Rev. Mod. Phys. **58**, 1001 (1985).

REFERENCES

[8] R. J. Glauber, Phys. Rev. **130**, 2529 (1963); **131**, 2766 (1963); R. J. Glauber, in *Quantum Optics and Electronics*, C. deWitt et al., eds. (Gordon and Breach, New York, 1965).

[9] D. Stoler, Phys. Rev. D **1**, 3217 (1970).

[10] H. P. Yuen, Phys. Rev. A **13**, 2226 (1976).

[11] Y. Yamamoto et al., in *Progress in Optics*, vol. **28**, ed. E. Wolf (North-Holland, Amsterdam, 1990), p. 89.

[12] H. J. Kimble, in, eds. J. Dalibard and J. M. Raimond (North-Holland, Amsterdam, 1992).

[13] S. Reynaud, A. Heidmann, E. Giacobino, and C. Fabre, in *Progress in Optics*, vol. 30, ed. E. Wolf (North-Holland, Amsterdam, 1992).

[14] R. E. Slusher, L. W. Hollberg, B. Yurke, J. C. Mertz, and J. F. Valley, Phys. Rev. Lett. **55**, 2409 (1985).

[15] D. F. Walls and G. Milburn, Quantum Optics (Springer-Verlag, Berlin, 1994); H. A. Bachor, *A quide to experiments in quantum optics* (Wiley-VCH, Weinheim, 1998).

[16] R. Jackiw, J. Math. Phys. **9**, 339 (1968).

[17] M. Kitagawa and Y. Yamamoto, Phys. Rev. A **34**, 3974 (1986).

[18] P. Drummond, S. Friberg, R. Shelby, and Y. Yamamoto, *Nature* **365**, 307 (1993).

[19] L. Mandel and E. Wolf, *Optical Coherence and Quantum Optics* (Cambridge University Press, 1995).

[20] A. Einstein, B. Podolsky, and N. Rosen, Phys. Rev. **47**, 777 (1935).

[21] A. Peres, *Quantum Theory: Concepts and Methods* (Kluwer, Dordrecht, 1995).

4 Coherence of the Electromagnetic Fields

The goal of this chapter is to introduce the correlation functions that describe the coherence properties of optical fields. The discussion is centered on the quantum description of *Young's interference* and *Hanbury–Brown–Twiss* experiments, which measure the first- and second-degree coherence, respectively. It is shown that purely quantum phenomena such as nonlocal quantum interference, *photon antibunching*, and *sub-Poisson photon distribution* arise in experiments that depend on the second (or higher) degree coherence functions. The coherence properties of thermal, coherent, and photon number states of light are evaluated and a derivation of photon count distribution is given. The final two sections of the chapter deal with the quantum mechanical representation of the phase of an electromagnetic field and the fundamental limits to optical interferometry.

4.1. PHOTODETECTION

The fundamental process in any optical measurement is the photodetection, where a fraction of photons in the incident light are converted into free carriers in the detector circuit. Most photodetectors are based on the process of stimulated absorption, which is caused by the electric-dipole interaction between an electromagnetic field and *detector atoms* [which are initially in their (bound) ground state]. The current generated in the photodetector circuit is proportional to the stimulated absorption rate W. The detailed quantum theory of this basic interaction will be given in Chapter 6; for our current purposes, we only need to recall the Fermi's golden rule for a bound-free transition rate [1]

$$W = \lim_{T\to\infty} \frac{1}{T}|\langle F|\hat{U}(T)|I\rangle|^2 \simeq \frac{2\pi}{\hbar}\sum_F \left|\langle F|\hat{H}_{int}|I\rangle\right|^2 \cdot \delta(E_F - E_I) \quad (4.1)$$

Here $\hat{H}_{int} = -\mu \cdot \hat{\mathbf{E}}$, μ and $\hat{\mathbf{E}} = \hat{\mathbf{E}}^{(+)} + \hat{\mathbf{E}}^{(-)}$ denote the atom-field interaction Hamiltonian, the dipole moment, and the electric field operators, respectively. $\hat{U}(T)$ is the time-evolution operator in the interaction picture; the last expression in (4.1) is obtained by expanding $\hat{U}(T)$ to first order in $\hat{H}_{int}$. The initial ($|I\rangle = |g, i\rangle$) and final ($|F\rangle = |e, f\rangle$) states are atom-field product states, with $|g\rangle, |e\rangle$ ($|i\rangle, |f\rangle$) denoting the inital and final atom (field) states. Since the atoms are assumed to be in their ground state initally, the negative frequency component of the electric field operator

$\hat{E}^{(-)}$ ($\propto \hat{a}^\dagger$) could be discarded in evaluating the matrix element $\langle F|\hat{H}_{\text{int}}|I\rangle$ in (4.1). Due to the δ-function appearing in the right-hand side, the sum over final states can be extended to all atom-field states. For a monochromatic field (with frequency ω), the δ-function acts exclusively on atomic parameters. The transition rate can then be written as

$$W = \frac{2\epsilon_o c}{\hbar \omega} \cdot \sigma_{\text{abs}} \cdot \langle \hat{I}(\bar{r}, t)\rangle, \qquad (4.2)$$

where

$$\langle \hat{I}(\mathbf{r}, t)\rangle = I(\mathbf{r}, t) = \langle i \left| \hat{E}^{(-)}(\mathbf{r}, t) \hat{E}^{(+)}(\mathbf{r}, t) \right| i\rangle = Tr_{\text{field}} \left\{ \hat{\rho}_{\text{field}} \hat{E}^{(-)} \hat{E}^{(+)} \right\}. \qquad (4.3)$$

σ_{abs} appearing in (4.2) denotes the atomic absorption (photoionization) cross section. It may be shown that $2\epsilon_o c \langle \hat{I}(\mathbf{r}, t)\rangle$ in (4.2) is simply the expectation value of the electromagnetic field intensity operator [2], [3].

We therefore see from this simple quantum mechanical calculation that photoabsorption, which is the basic detection mechanism utilized in all optical measurements, gives us the expectation value of the intensity operator of the incident field. From now on, we shall neglect the particular form of the measurement interaction and assume that the measured quantity is simply $\langle \hat{I}(\mathbf{r}, t)\rangle$. The normal ordering of the field operators in (4.3) has important implications and is a consequence of the fact that detection is based on photon absorption rather than photon emission.

4.2. YOUNG'S INTERFERENCE EXPERIMENT AND FIRST-ORDER COHERENCE

In this section, we shall briefly analyze Young's interference experiment. Our principal goal here is to introduce the quantum formalism for the description of first-order coherence of optical fields. For a detailed classical and quantum description of the experiment, the reader is referred to Ref. [2].

Figure 4-1 shows the simplified (idealized) diagram of the Young's interference experiments: A plane wave is incident on a screen with two pinholes at $\mathbf{r}_1$ and $\mathbf{r}_2$. Classical electromagnetic theory tells us that the two pinholes act as secondary (point) sources of spherical radiation. The experiment is essentially the measurement of the light intensity on a second screen located in the far field of the two secondary sources. Assuming that a photodetector is utilized, we obtain the (position dependent) average intensity given by $\langle \hat{I}(\mathbf{r}, t)\rangle = \langle \hat{E}^{(-)}(\mathbf{r}, t) \hat{E}^{(+)}(\mathbf{r}, t)\rangle$. The total electric field operator ($\hat{E}(\mathbf{r}, t) = \hat{E}^{(+)} + \hat{E}^{(-)}$) is a superposition of the electric fields at the two pinholes at earlier times. If we denote $s_i = |\mathbf{r} - \mathbf{r}_i|$ ($i = 1, 2$) we obtain

$$\hat{E}(\mathbf{r}, t) = u_1 \hat{E}\left(\mathbf{r}_1, t - \frac{s_1}{c}\right) + u_2 \hat{E}\left(\mathbf{r}_2, t - \frac{s_2}{c}\right), \qquad (4.4)$$

where $u_i \propto 1/s_i$ is the spherical mode function [2], [3]. It is assumed in (4.4) that the electric fields at the two pinholes are identical due to the plane-wave excitation.

78 COHERENCE OF THE ELECTROMAGNETIC FIELDS

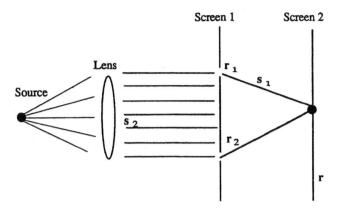

FIGURE 4-1: Young's interference experiment.

Using (4.4), it is straightforward to show that the measured average intensity is given by

$$
\begin{aligned}
I(\mathbf{r},t) = &\ |u_1|^2 \langle \hat{E}^{(-)}\left(\mathbf{r}_1, t - \frac{s_1}{c}\right) \hat{E}^{(+)}\left(\mathbf{r}_1, t - \frac{s_1}{c}\right) \rangle \\
&+ |u_2|^2 \langle \hat{E}^{(-)}\left(\mathbf{r}_2, t - \frac{s_2}{c}\right) \hat{E}^{(+)}\left(\mathbf{r}_2, t - \frac{s_2}{c}\right) \rangle \\
&+ 2\mathrm{Re}\left\{ u_1^* u_2 \langle \hat{E}^{(-)}\left(\mathbf{r}_1, t - \frac{s_1}{c}\right) \hat{E}^{(+)}\left(\mathbf{r}_2, t - \frac{s_2}{c}\right) \rangle \right\}.
\end{aligned}
\tag{4.5}
$$

The first and second terms on the right-hand side of (4.5) correspond to the intensities from the respective pinholes. The last term is a correlation function $G^{(1)}(\mathbf{r}_1 t_1, \mathbf{r}_2 t_2)$ of the electric field operator at two different space-time points: It appears as a result of the interference of the two beams. The magnitude of the correlation function is comparable to the direct intensity terms and its phase varies rapidly as a function of distance along the direction $\mathbf{r}_1 - \mathbf{r}_2$. Since classically coherence is regarded as *the ability of an optical field to produce interference fringes* [2], the correlation function appearing in (4.5) is a natural measure of first-order coherence.

The general definition of the normalized first order correlation function is

$$
\begin{aligned}
g^{(1)}(\mathbf{r}_1 t_1, \mathbf{r}_2 t_2) &\overset{\Delta}{=} \frac{\langle \hat{E}^{(-)}(\mathbf{r}_1, t_1) \hat{E}^{(+)}(\mathbf{r}_2, t_2) \rangle}{\left[\langle \hat{E}^{(-)}(\mathbf{r}_1, t_1) \hat{E}^{(+)}(\mathbf{r}_1, t_1) \rangle \langle \hat{E}^{(-)}(\mathbf{r}_2, t_2) \hat{E}^{(+)}(\mathbf{r}_2, t_2) \rangle\right]^{1/2}} \\
&= \frac{G^{(1)}(\mathbf{r}_1 t_1, \mathbf{r}_2 t_2)}{\sqrt{G^{(1)}(\mathbf{r}_1 t_1, \mathbf{r}_1 t_1) G^{(1)}(\mathbf{r}_2 t_2, \mathbf{r}_2 t_2)}}
\end{aligned}
\tag{4.6}
$$

In most cases, we are interested in experiments where the $\mathbf{r}_1, \mathbf{r}_2$ dependence of the correlation function $G^{(1)}$ is negligible. This is the case in Young's experiment as both pinholes are illuminated by the same plane wave. Moreover, for optical fields that are generated by stationary sources, the correlation function only depends on the time difference $\tau = t_2 - t_1 \;(= (s_2 - s_1)/c)$.

Classically, it may readily be shown that the Fourier transform of the correlation function $G^{(1)}(\tau)$ gives the spectrum of the field. Consequently, all (classical) monochromatic fields have the same first-order correlation function, given by

$$g^{(1)}(\tau) = e^{-i\omega\tau + \phi}, \tag{4.7}$$

where ω denotes the frequency of the monochromatic field and ϕ is a (fixed) initial phase difference. One could easily show that such a monochromatic field gives interference fringes with maximum possible fringe contrast. Similarly, a broadband field with a Lorentzian spectrum (such as that obtained when the field obeys the Wiener-Levy phase-diffusion model [4]) gives

$$g^{(1)}(\tau) = e^{-i\omega\tau - \gamma|\tau|}, \tag{4.8}$$

where γ^{-1} is the coherence time of the field.

We have already seen in Chapter 2 that field quantization requires the introduction of a new concept: the *wavefunction* describing the *quantum state* of the electromagnetic field. For example, a monochromatic field mode may be in a photon number eigenstate, a coherent state, or in any arbitrary linear superposition of the photon number eigenstates. The coherence properties should be calculated for each particular state of interest. We first recall the explicit form of the quantized electric field operator for a single mode

$$\hat{E}_1^{(+)}(\mathbf{r},t) = i\left(\frac{\hbar\omega}{2\epsilon_0 L^3}\right)^{1/2} \cdot e^{i\mathbf{k}\cdot\mathbf{r}} \cdot \hat{a}_1(t) = i\left(\frac{\hbar\omega}{2\epsilon_0 L^3}\right)^{1/2} \cdot e^{i(\mathbf{k}\cdot\mathbf{r}-\omega t)} \cdot \hat{a}_1(0), \tag{4.9}$$

where the subscript 1 refers to the field originating from the pinhole 1, rather than the quadrature amplitude. From (4.9) we obtain an expression for the normalized first-order correlation function in terms of annihilation and creation operators:

$$g^{(1)}(\tau) = \frac{\langle a_1^\dagger(t+\tau) a_2(t)\rangle}{\sqrt{\langle a_1^\dagger(t+\tau) a_1(t+\tau)\rangle \langle a_2^\dagger(t) a_2(t)\rangle}} = \frac{\langle a_1^\dagger a_2\rangle}{\sqrt{\langle a_1^\dagger a_1\rangle \langle a_2^\dagger a_2\rangle}} \cdot e^{i\omega\tau}. \tag{4.10}$$

In (4.9) and (4.10), the rightmost expressions are obtained for a free electromagnetic field. For an interacting field, the general time dependence of the annihilation and creation operators must be kept; in such a case, the Heisenberg picture provides the natural formalism for the calculation of the correlation function. In Young's experiment, we have a free field with $\tau = (s_1 - s_2)/c$. In the limit $\langle \hat{I}_1\rangle = \langle \hat{I}_2\rangle = \langle \hat{I}\rangle$, $u_1 = u_2 = u$, the measured average intensity simplifies to

$$I(\mathbf{r},t) = |u|^2 \langle \hat{I}\rangle \left(1 + \text{Re}\left\{\frac{\langle a_1^\dagger a_2\rangle}{\langle I\rangle} e^{-ik(s_2-s_1)}\right\}\right). \tag{4.11}$$

In most cases, we would consider the (single-mode) illumination of the pinholes by a common light source, in a yet unspecified quantum state. To simplify further the quantum problem, we could assume that the first screen is replaced by a 50%–50% beam splitter and then construct a Mach–Zehnder interferometer with phase adjustable two arms and a second 50%–50% beam splitter: This assumption enables us to relate the state of the two interfering fields to that of the incident state.

The first case that we consider is a driving field that is initially in a single-photon eigenstate $|1\rangle_1$ at input port 1 and a vacuum state $|0\rangle_2$ at the other input port ($|\psi_{in}\rangle = |1, 0\rangle$) [3]. The wavefunction of the field in the Mach–Zehnder interferometer is

$$|\psi\rangle = \frac{1}{\sqrt{2}} (|1, 0\rangle + |0, 1\rangle). \quad (4.12)$$

The superposition appearing in (4.12) is a manifestation that the photon can reach the second beam splitter by going through either one of the two arms. Each path has a nonzero probability amplitude, and the net probability of finding the photon in either one of the two outputs of the second beam splitter is the square of the sum of the two probability amplitudes. For this state, we find $|g^{(1)}(\tau)| = 1$, indicating perfect fringe visibility. Therefore, it appears that the interference pattern observed in Young's experiment or Mach–Zehnder interferometer is a single-photon phenomenon and does not at all depend on the strength of the radiation field [3]. In fact, Dirac has indicated that interference is strictly a single-photon phenomenon and a photon may only interfere with itself [5]. However, we will see later in this chapter that this is not a true statement.

Next we consider the case where the input state of port 1 is in a coherent state $|\alpha\rangle_1$ and the other port is a vacuum state $|0\rangle_2$. The state in the interferometer is an uncorrelated coherent state $|\alpha_1, \alpha_2\rangle$, with $|\alpha_1| = |\alpha_2| = |\alpha|/\sqrt{2}$) [3]: Since the coherent state is an eigenstate of the annihilation operator, calculation of $g^{(1)}(\tau)$ is straigthforward and yields the intensity pattern

$$I(\mathbf{r}, t) = \frac{\langle \hat{I} \rangle}{2} (1 + \cos [k(s_1 - s_2) + \phi]) \quad (4.13)$$

in the case of Young's interference experiment. The extension of these results to other pure or mixed single-mode field states is straightforward. The conclusion is the same as that of the classical approach: All single-mode light states are first-order coherent and yield perfect fringe visibility [2].

An important extension of the preceding discussion is obtained when we consider the illumination of each path by an independent source. Assuming that the two field modes are prepared in coherent states with the identical amplitude $|\alpha_1| = |\alpha_2| = |\alpha|$) and identical frequency, the interference output obtained is the same as that given in Eq. (4.13), implying perfect fringe visibility [6]. Such a result may seem to contradict Dirac's argument as it is photons from two different sources that create the interference output. As pointed out by Glauber [7], the apparent conflict between the two interpretations is resolved when one emphasizes that it is the probability amplitudes (and not the particles) that interfere with each other. In the preceding example, the detector can never tell whether the detected photon originates from source 1 or source 2. This lack of information (= probability amplitudes) interferes to produce the fringe pattern.

Finally, we remark that in a two-beam interference experiment that we described previously, the fringes are not obtained for independent and definite excitation with a photon number eigenstate, which results in $g^{(1)}(\tau) = 0$ for an ensemble measurement. For amplitude squeezed states, one would observe fringes with degraded contrast. Apart from this discrepancy in independent excitation, quantum and classical approaches predict identical results for measurements that depend on the first-order coherence.

4.3. HANBURY BROWN–TWISS EXPERIMENT AND SECOND-ORDER COHERENCE

As indicated earlier, the classical and quantum theories predict drastically different results for the photon correlation experiments of the type we are going to describe here. The fundamental experiment that measures intensity correlations is the Hanbury–Brown–Twiss experiment, detailed in Fig. 4-2: An incoming electric field $\hat{E}(\mathbf{r}, t)$ is split into two using a 50%–50% beam splitter. The intensities of the split beams are measured by two photodetectors. The detector outputs are then (electronically) manipulated to give the correlation of the two measured intensities. The fundamental quantum process is then the (joint) observation of one photoionization event at $\mathbf{r}_1, t_1$ *and* one at $\mathbf{r}_2, t_2$. Quantum mechanically, the joint detection probability is obtained by expanding the time evolution operator (of the combined field and two-detector system) to second order in $\hat{H}_{\text{int}}$. The corresponding probability amplitude

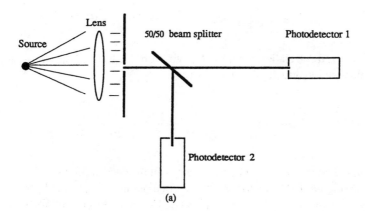

FIGURE 4-2: Hanbury-Brown-Twiss experiment.

M_2 is given by [1], [7]

$$M_2 = \int_0^T dt' \int_0^T dt'' \langle F| \hat{H}_{\text{int}}(\mathbf{r}_2, t'')\hat{H}_{\text{int}}(\mathbf{r}_1, t') |I\rangle. \qquad (4.14)$$

Just as in the single photodetector case, it is only the positive frequency component $\hat{\mathbf{E}}^{(+)}$ of the electric field operator that contributes to the transition amplitude [i.e., the relevant matrix element is $\langle f|\hat{\mathbf{E}}^{(+)}(\mathbf{r}_2, t'')\hat{\mathbf{E}}^{(+)}(\mathbf{r}_1, t')|i\rangle$]. It may be shown that the joint observation rate W_{joint} could be obtained by taking the time derivative of the transition probability ($|M_2|^2$) and is given by

$$W_{\text{joint}} = A\langle i|\hat{\mathbf{E}}^{(-)}(\mathbf{r}_1, t_1)\hat{\mathbf{E}}^{(-)}(\mathbf{r}_2, t_2)\hat{\mathbf{E}}^{(+)}(\mathbf{r}_2, t_2)\hat{\mathbf{E}}^{(+)}(\mathbf{r}_1, t_1)|i\rangle, \qquad (4.15)$$

where A is a constant determined by the atomic dipole correlation function [1]. The significance of the particular form of the transition rate in (4.15) is the appearance of normal ordering of the field operators: A naive quantum extension of the classical intensity correlation function $G_{\text{classical}}(\tau) = \langle I(t+\tau) I(t)\rangle$ (without introducing normal ordering) would not describe the observed experimental results. We reiterate that the normal ordering in (4.15) is a direct result of the fact that we are considering (joint) photodetection based on stimulated absorption.

The intensity correlation function that is measured in a Hanbury–Brown–Twiss (HBT) type of experiment is linearly proportional to the joint detection rate W_{joint} of (4.15). The electric field operators at the two detectors (located at $\mathbf{r}_1$ and $\mathbf{r}_2$) are related to the incident field operator: For a 50%–50% beam-splitter, one obtains $\hat{\mathbf{E}}(\mathbf{r}_1) = \hat{\mathbf{E}}_1 = (\hat{\mathbf{E}} + \hat{\mathbf{E}}_{\text{vac}})/\sqrt{2}$ and $\hat{\mathbf{E}}(\mathbf{r}_2) = \hat{\mathbf{E}}_2 = (\hat{\mathbf{E}} - \hat{\mathbf{E}}_{\text{vac}})/\sqrt{2}$. The vacuum field input ($\hat{\mathbf{E}}_{\text{vac}}$) into the unused port could be neglected as it does not contribute to the expectation value in (4.15). We can then write the measured second-order field correlation function in terms of the incident electric field operator as

$$G^{(2)}(\tau) = \langle \hat{\mathbf{E}}^{(-)}(t)\hat{\mathbf{E}}^{(-)}(t+\tau)\hat{\mathbf{E}}^{(+)}(t+\tau)\hat{\mathbf{E}}^{(+)}(t)\rangle. \qquad (4.16)$$

The normalized correlation function that will form the basis of our discussion in the rest of this section directly follows (4.16):

$$g^{(2)}(\tau) = \frac{\langle \hat{a}^\dagger(t)\hat{a}^\dagger(t+\tau)\hat{a}(t+\tau)\hat{a}(t)\rangle}{\langle \hat{a}^\dagger(t)\hat{a}(t)\rangle \langle \hat{a}^\dagger(t+\tau)\hat{a}(t+\tau)\rangle}. \qquad (4.17)$$

We now evaluate $g^{(2)}(\tau)$ for particular quantum states of light that are of interest. Since the original HBT experiment was carried out using a chaotic light source, the first case that we consider is thermal state. The density operator for single-mode thermal light is

$$\hat{\rho}_t = \left(1 - \exp\left[-\frac{\hbar\omega}{k_B T}\right]\right) \exp\left[-\frac{\hbar\omega\hat{a}^\dagger\hat{a}}{k_B T}\right]$$
$$= \left(1 - \exp\left[-\frac{\hbar\omega}{k_B T}\right]\right) \sum_n \exp\left[-\frac{n\hbar\omega}{k_B T}\right] |n\rangle\langle n|, \quad (4.18)$$

where T is the temperature of the radiation field mode. Since there is no nontrivial time dependence, the second-order correlation function may be easily evaluated to give

$$g^{(2)}(\tau) = g^{(2)}(0) = 2. \quad (4.19)$$

In order to see the implication of (4.19) more clearly, we should consider the generalization of this result to multimode chaotic light. If we assume that the state of the field is described by the Gaussian distribution of the frequencies ω in the density operator (4.18) [8], then we can factorize the numerator in (4.17) as

$$\langle\hat{a}^\dagger(t)\hat{a}^\dagger(t+\tau)\hat{a}(t+\tau)\hat{a}(t)\rangle = \langle\hat{a}^\dagger(t)\hat{a}^\dagger(t+\tau)\rangle\langle\hat{a}(t+\tau)\hat{a}(t)\rangle$$
$$+ \langle\hat{a}^\dagger(t)\hat{a}(t+\tau)\rangle\langle\hat{a}^\dagger(t+\tau)\hat{a}(t)\rangle$$
$$+ \langle\hat{a}^\dagger(t)\hat{a}(t)\rangle\langle\hat{a}^\dagger(t+\tau)\hat{a}(t+\tau)\rangle. \quad (4.20)$$

The first term on the right-hand side is zero if we assume that the mean field is zero and that there are no phase-dependent fluctuations (such as those in a squeezed state) [3]. Using (4.20), we then find

$$g^{(2)}(\tau) = 1 + |g^{(1)}(\tau)|^2. \quad (4.21)$$

Since $|g^{(1)}(\tau)| = 1$ for any single-mode light, (4.19) follows (4.21). If the field has a Lorentzian spectrum (with width γ), then $g^{(2)}(\tau) = 1 + \exp[-\gamma|\tau|]$ (Fig. 4-3). When the light intensities at two time instants are uncorrelated, we obtain $g^{(2)}(\tau) = 1$. For thermal light, we see that within the field correlation time γ^{-1}, the detection of a photon makes a second subsequent detection event very likely: This phenomenon is referred to as *photon bunching* and has been demonstrated experimentally in the original HBT experiment.

Next, we consider a single-mode coherent-state $|\alpha\rangle$: The evaluation of $g^{(2)}(\tau)$ is particularly easy for this case as $|\alpha\rangle$ is an eigenstate of the annihilation operator. We obtain $g^{(2)}(\tau) = 1$. In fact, we can envision a generalization of the correlation function in (4.17) to nth order: For a single-mode field in a coherent state, we obtain

$$g^{(n)}(t_1, \ldots, t_n) = \frac{\langle\hat{a}^\dagger(t_1)\cdots\hat{a}^\dagger(t_n)\hat{a}(t_n)\cdots\hat{a}(t_1)\rangle}{(\langle\hat{a}^\dagger(t)\hat{a}(t)\rangle)^n} = 1. \quad (4.22)$$

Higher-order correlation functions may be important in n-detector setups or in nonlinear optical processes [3]. Equation (4.22) tells us that the coherent state is *all*

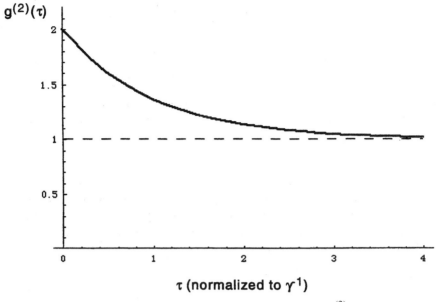

FIGURE 4-3: The normalized second-order correlation function $g^{(2)}(\tau)$ for broadband Gaussian light.

orders coherent. The same result would apply to a classical single-mode light; this is not unexpected since we have already seen that the coherent state is the quantum state of light that most closely resembles a classical field. Photon detection events for a coherent state are completely uncorrelated; detection of a photon at $\tau = 0$ gives us no information regarding the detection time of a second photon. As we shall see in the next section, the photocount distribution obtained for a coherent state is Poissonian, which is another indication of the lack of correlation.

Before proceeding, we discuss some of the inequalities satisfied by the second-order correlation function in both classical and quantum theories. One can use Cauchy's inequality to show that

$$g^{(2)}_{\text{classical}}(0) \geq 1, \tag{4.23}$$

$$g^{(2)}_{\text{classical}}(0) \geq g^{(2)}_{\text{classical}}(\tau). \tag{4.24}$$

The quantum second-order correlation function $g^{(2)}(\tau)$ can violate both of these inequalities. The first inequality is generally used to define the purely quantum mechanical regime

$$0 \leq g^{(2)}(0) < 1. \tag{4.25}$$

We reiterate that there are no classical fluctuations that would give Eq. (4.25). This value of $g^{(2)}(0)$ implies that photons are *antibunched*; that is, detection of a photon makes a subsequent detection event less likely. In most cases of interest, the second inequality is violated along with the first one, as $g^{(2)}(\infty) = 1$ for all finite bandwidth electromagnetic fields.

The prototypical example of a purely quantum state of light is the single-mode photon number eigenstate $|n\rangle$, for which we obtain

$$g^{(2)}(0) = \frac{n-1}{n} < 1. \qquad (4.26)$$

It is easy to understand the origin of photon antibunching in a photon number eigenstate: The total number of photons is known and the detection of one makes it less likely to detect other photons. Particularly, for $|n\rangle = |1\rangle$, $g^{(2)}(0) = 0$, as there are no photons left to detect after the first one.

Using simple operator algebra, one may show that [3]

$$g^{(2)}(0) = 1 + \frac{\langle \Delta \hat{n}^2 \rangle - \langle \hat{n} \rangle}{\langle \hat{n} \rangle^2}, \qquad (4.27)$$

where $\langle \hat{n} \rangle$ and $\langle \Delta \hat{n}^2 \rangle$ denote the mean value and variance of the number operator $\hat{n}$. We therefore see that photon antibunching for a single-mode field implies sub-Poisson statistics $\langle \Delta \hat{n}^2 \rangle < \langle \hat{n} \rangle$. One-to-one correspondance does not in general hold for multimode fields [3]. Finally, we note that in Chapter 12 we shall discuss an interesting multimode field state that exhibits strong antibunching ($g^{(2)}(0) \ll 1$) and sub-Poissonian statistics: The second-order correlation function of these *heralded single-photon states* indicates that one has photon generation (detection) time information.

So far we have only discussed the first- and second-order correlation functions obtained for certain states of the single- and multimode free electromagnetic fields. In many quantum optics experiments, we are interested in the quantum statistical properties of light generated by atoms or excitons, under different excitation conditions. Calculation of the correlation functions for light generated by open dissipative *quantum sources* requires the use of the *quantum regression theorem*: We shall apply this theorem to specific experiments after introducing the master equation formalism in Chapter 7.

4.4. PHOTON COUNTING

An interesting extension of simple photodetection measurements is photon counting, where using appropriate electronics we count the number of photons detected by a photodiode or a photomultiplier in a fixed time interval of length T. If the measurement is repeated many times for identical input states, we obtain a probability distribution $P_n(T)$ for detecting n photons in the time interval T [2], [3]. The results of photon-counting measurements give information about the diagonal elements of the density operator of the electromagnetic field and are therefore of fundamental importance.

4.4.1. Classical Theory of Photon Count Distribution

As we have already seen in (4.1), the photoelectron generation rate is proportional to the intensity of the incident field. The probability $p(t)\,dt$ that a photodetection event occurs in the time interval $(t, t+dt)$ is then given by $p(t)\,dt = \eta I(t)\,dt$, where η is a constant proportional to the detector efficiency. It is assumed that the time interval is small enough that the probability for two photoelectron detection events is negligibly small. The probability $P_n(t_i, t+dt)$ for detecting n photons in the time interval $(t_i, t+dt)$ can be written as

$$P_n(t_i, t+dt) = P_n(t_i, t)(1 - p(t)\,dt) + P_{n-1}(t_i, t)p(t)\,dt. \tag{4.28}$$

The first term in (4.28) refers to the case where all n photons are detected in the interval (t_i, t) and none in $(t, t+dt)$. The second term corresponds to $n-1$ photodetection events in (t_i, t) and one in $(t, t+dt)$. Provided that dt is chosen small enough, (4.28) is valid for any input electromagnetic field. We can recast (4.28) in the form of a differential equation as

$$\frac{dP_n(t_i, t)}{dt} = p(t)[P_{n-1}(t_i, t) - P_n(t_i, t)]. \tag{4.29}$$

The ladder of coupled differential equations implied by (4.28) may be solved starting from the $n=0$ equation: This is the probability of detecting 0 photons in the time interval (t_i, t) and satisfies a simple differential equation [$P_{n-1}(t_i, t)$ term does not exist] with an inital condition $P_0(t_i, t_i) = 1$. The solution is

$$P_0(t_i, t_i + T) = \exp\left[-\eta \int_{t_i}^{t_i+T} I(t_i, t)\,dt\right] = \exp[-\eta T \bar{I}(t_i, t_i + T)], \tag{4.30}$$

where $\bar{I}(t_i, t_i + T)$ denotes the average intensity in the time interval $(t_i, t_i + T)$. We can then substitute this solution to solve the differential equation for $P_1(t_i, t)$ using the initial condition $P_1(t_i, t_i) = 0$. The procedure can be extended to higher values of n using the initial condition $P_n(t_i, t_i) = 0, \forall n > 0$ to find

$$P_n(t_i, T) = \frac{(\eta T \bar{I}(t_i, T))^n}{n!} \exp[-\eta T \bar{I}(t_i, T)], \tag{4.31}$$

where we have let $P_n(t_i, t_i + T) \to P_n(t_i, T)$ [2], [3]. The simplest case is that of an input light with constant intensity (this is the case for a classical coherent light source). Since $\bar{I}(t_i, T) = I$ and $P_n(t_i, T) = P_n(T)$, we obtain

$$P_n(T) = \frac{\bar{n}^n}{n!} \exp[-\bar{n}], \tag{4.32}$$

where $\bar{n} = \eta I T$ is the mean number of photodetection events (photons) in the time interval of length T. This is the Poisson distribution, which can alternatively be derived by assuming that the n photon detection events are statistically independent and have identical distribution functions [9]. The variance $\overline{\Delta n^2}$ is equal to the mean $\bar{n}$. The photocount distribution of (4.32) corresponds to the diagonal elements of the density operator of the coherent state.

We note that $P_n(t_i, T)$ corresponds to the probability distribution obtained when an ensemble of identical photon-counting experiments are carried out starting at time t_i. For ergodic processes, this ensemble average may be replaced by a time average over the starting times. The latter procedure is more relevant to the actual experiments, where one obtains $P_n(T)$ by measuring $P_n(t_i, T)$ for the same particular light source at different starting times t_i. One can therefore write $P_n(T) = \langle P_n(t_i, T) \rangle$, where the brackets may denote either ensemble or time averages.

Next, we consider a thermal light source. If the source is broadband and the measurement time T is long compared to the coherence time γ^{-1}, we find $\bar{I}(t_i, T) = \bar{I}$. Therefore, for $T \gg \gamma^{-1}$, we find the Poisson distribution given in (4.32) [2]. If, however, $T \ll \gamma^{-1}$, we need to take an ensemble average over the distribution of intensities, the relevant distribution for broadband thermal light is $p(\bar{I}) = (1/\bar{I}) \exp\left[-\bar{I}(t_i, T)/\bar{I}\right]$ [2]. Setting $\bar{n} = \eta \bar{I} T$, we obtain

$$P_n(T) = \frac{\bar{n}^n}{(1 + \bar{n})^{n+1}}. \quad (4.33)$$

$P_n(T)$ of (4.33) corresponds to the diagonal matrix elements of the density operator for the single-mode thermal state. Figure 4-4 shows the classical photocount distributions obtained for coherent and thermal light for $T \ll \gamma^{-1}$. The distributions for the corresponding states in the quantum theory are identical.

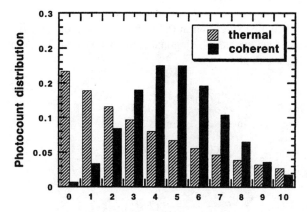

FIGURE 4-4: The photon count distributions for thermal and coherent light sources.

4.4.2. Quantum Theory of Photon Count Distribution

Despite its simplicity, the classical theory of photon counting cannot describe the experimental results if the field excitation under consideration does not have a corresponding classical state [i.e., any quantum state that gives $g^{(2)}(0) < 1$]. The quantum photo-count distribution may be derived from the first principles using a similar method [10] and is given by a trace over the field modes

$$P_n(T) = Tr_{\text{field}}\left(\hat{\rho}_{\text{field}} \mathcal{N} \frac{(\eta T \hat{I}(T))^n}{n!} \exp[-\eta T \hat{I}(T)]\right), \qquad (4.34)$$

where $\hat{\rho}_{\text{field}}$ and $\hat{I}$ correspond to the field density operator and the time-averaged intensity operator of the electromagnetic field, respectively. $\mathcal{N}$ denotes normal ordering of the operators to the right of it. Here, we shall only be interested in single-mode fields where the expression simplifies considerably since

$$\hat{I}(T) = \frac{2\epsilon_0 c}{T} \int_0^T \hat{\mathbf{E}}^{(-)}(\mathbf{r},t)\,\hat{\mathbf{E}}^{(+)}(\mathbf{r},t)\,dt = \frac{\hbar\omega c}{L^3}\hat{a}^\dagger\hat{a}. \qquad (4.35)$$

If we denote the quantum efficiency of the detector by $\zeta(T) = \hbar\omega c T \eta/L^3$, we obtain

$$P_n(T) = Tr_{\text{field}}\left(\hat{\rho}_{\text{field}} \mathcal{N} \frac{(\zeta(T)\hat{a}^\dagger\hat{a})^n}{n!} \exp[-\zeta(T)\hat{a}^\dagger\hat{a}]\right). \qquad (4.36)$$

The particular form of $\zeta(T)$ just given is valid for *traveling photons*, where a continuous-wave light source is incident on a detector [3]. If we are interested in a cavity mode, an alternative form for the quantum efficiency would be $\zeta(T) = 1 - \exp[-\hbar\omega c T \eta/L^3]$, which reduces to the original definition in the limit $\zeta(T) \ll 1$ [2].

We can evaluate the expression in (4.36) by using the Taylor series expansion of the exponential. If we denote the diagonal elements of the field density operator by $P_i = \langle i|\hat{\rho}_{\text{field}}|i\rangle$, we obtain

$$P_n(T) = \sum_{i=0}^\infty P_i \frac{\zeta(T)^n}{n!} \langle i| \sum_{m=0}^\infty (-1)^m \frac{\zeta(T)^m}{m!} \hat{a}^{\dagger n+m}\hat{a}^{n+m}|i\rangle \qquad (4.37)$$

$$= \sum_{i=n}^\infty P_i \frac{\zeta(T)^n}{n!} \sum_{m=0}^{i-n} \frac{\zeta(T)^m}{m!} \frac{i!}{(i-n-m)!} \qquad (4.38)$$

$$= \sum_{i=n}^\infty P_i \frac{i!}{n!(i-n)!} \zeta(T)^n (1-\zeta(T))^{i-n}. \qquad (4.39)$$

To obtain the last equality, we have used the fact that summation over m in the second equality corresponds to the binomial distribution. The final expression in (4.39) is known as the Bernouilli distribution [3].

We see from (4.39) that the quantum photon count distribution corresponds to the diagonal matrix elements of the field density operator only in the limit of unity quantum efficiency. For a coherent state and a thermal state in the limit of long measurement time interval, however, the resulting distribution is identical to that of the diagonal matrix elements, provided that we replace $\bar{n}$ with $\zeta(T)\bar{n}$. For a number state $|k\rangle$ ($P_i = \delta_{k,i}$), we see that the photon-count distribution $P_n(T)$ is nonzero for all $n \leq k$, unless $\zeta(T) = 1$. In fact, the resulting distribution is identical to that of an initial number state that is passed through a beam splitter with transmission $\zeta(T)$.

4.5. PHASE OPERATOR OF THE QUANTIZED ELECTROMAGNETIC FIELD

The concepts of coherence and interference in optics are closely related to the phase of an electromagnetic field. We have seen that it is possible to discuss the fundamental coherence properties of optical fields without introducing the notion of a *phase operator*. As we shall see in the forthcoming chapters, however, the phase of a quantum field is a very important concept with some inherent analytical and conceptual difficulties.

The most straightforward way to define a phase operator is to decompose the annihilation operator into an amplitude and a phase part as first suggested by Dirac [5]:

$$\hat{a} = (\hat{a}\hat{a}^\dagger)^{1/2} \exp(i\hat{\phi}), \qquad (4.40)$$

with a corresponding relation for the creation operator. As is well known, all observables in quantum mechanics are represented by Hermitian operators because all measurement results are real numbers: Therefore we would expect the *correct* phase operator to be Hermitian and this should apply to that of (4.40). However, as pointed out in the seminal work by Susskind and Glogower [11], the operator $\exp(i\hat{\phi})$ [and its Hemitian conjugate $\exp(-i\hat{\phi})$] is not unitary since it satisfies

$$\left[\exp(i\hat{\phi}), \exp(-i\hat{\phi})\right] = |0\rangle\langle 0|. \qquad (4.41)$$

This relation implies that the phase operator $\hat{\phi}$ is not Hermitian. As discussed in Chapter 3, one possible approach is to define cosine ($\hat{C}$) and sine ($\hat{S}$) operators that are Hermitian.

Despite the success of the operators $\hat{C}$ and $\hat{S}$ in mathematically describing the phase properties, the absence of a quantum analog of the classical oscillator phase has long been considered an open problem in quantum mechanics. Recently, Pegg and Barnett [12] have proposed an alternative approach that elimininates most of the difficulties associated with the Dirac and Susskind-Glogower formalism. Their approach is based on defining a complete set of orthonormal phase eigenstates and a corresponding Hermitian phase operator in a finite state-space of $s+1$ dimensions, where s denotes the maximum extent of the photon number in the number-state basis. s could be arbitrarily large, provided that it is finite; the observable physical quantities such as phase operator expectation values should be evaluated as $s \to \infty$.

The starting point for the Pegg-Barnett formalism is the definition of a normalized phase eigenstate:

$$|\theta_0\rangle = (s+1)^{-1/2} \sum_{n=0}^{s} e^{in\theta_0} |n\rangle. \quad (4.42)$$

We can obtain other phase eigenstates by using the unitary phase-shift operator $\exp(i\hat{a}^\dagger \hat{a} \varphi)$, which produces $|\theta_0 + \varphi\rangle$ starting from $|\theta_0\rangle$. We can show using (4.42) that phase eigenstates differing by $2\pi/(s+1)$ are orthogonal. Therefore, given a reference phase eigenstate $|\theta_0\rangle$, we can obtain a complete set of orthonormal phase eigenstates that span the $s+1$-dimensional Hilbert space:

$$|\theta_m\rangle = \exp\left[i\hat{n}\, m\, 2\pi/(s+1)\right] |\theta_0\rangle \quad (m = 0, 1, \ldots s), \quad (4.43)$$

where $\theta_m = \theta_0 + 2\pi m/(s+1)$. The fact that the distinct phase states can only occur for values of θ in a 2π range is built into the formalism. As $s \to \infty$, the phase eigenvalues θ_m correspond to rational numbers that differ from θ_0 by a fraction; the corresponding $|\theta_m\rangle$ then form a countably infinite set of eigenstates that have a one-to-one correspondance to the number eigenstates.

The phase operator of the electromagnetic field can now be defined using the phase eigenstates in the $s+1$-dimensional Hilbert space:

$$\hat{\phi}_s = \sum_{m=0}^{s} \theta_m |\theta_m\rangle \langle \theta_m|. \quad (4.44)$$

By construction, the eigenvalues of $\hat{\phi}_s$ are θ_m ($m = 0, 1, \ldots, s$) with corresponding eigenstates $|\theta_m\rangle$. The phase operator in (4.44) is given as a sum of projection operators into the phase eigenstates. Using (4.43), one can obtain an expression for $\hat{\phi}_s$ in terms of off-diagonal projections between photon number eigenstates.

Experimentally, the most important issue is the measurement statistics predicted by the formalism. Given any quantum state in the truncated Hilbert space $|\psi\rangle_s$, the probability of obtaining θ_m as the phase of the field is given by

$$\Pi(\theta_m) = |\langle \theta_m | \psi \rangle_s|^2. \quad (4.45)$$

In terms of the amplitudes $\langle n | \psi \rangle_s$, this probability can be rewritten as

$$\Pi(\theta_m) = \frac{1}{s+1} \left| \sum_{n=0}^{s} \exp\left(-i \frac{n m\, 2\pi}{s+1}\right) \langle n | \psi \rangle_s \right|^2. \quad (4.46)$$

The phase distribution given in (4.46) provides useful insight into the conjugate nature of the number and phase in quantum optics. For example, if we assume that the system is in a photon number eigenstate, then we find that the phase-probability distribution $\Pi(\theta)$ is flat; that is, all phase eigenstates have equal amplitudes. Conversely, for a system initally in a phase eigenstate, the photon number is completely uncertain and there is equal amplitude for all of the $s+1$ number eigenstates.

Next, we consider the commutator of the number and phase operators in the Pegg-Barnett formalism [12]. Given any state $|\psi\rangle_s$, it could be shown that

$$_s\langle\psi|\left[\hat{\phi}_s,\hat{n}\right]|\psi\rangle_s = -i\left[1 - 2\pi\,\Pi(\theta_0)\right], \tag{4.47}$$

which leads to the uncertainty relation

$$\Delta n\Delta\phi_s \geq \frac{1}{2}\left|1 - 2\pi\,\Pi(\theta_0)\right|. \tag{4.48}$$

We therefore see that the uncertainty relation depends on the choice of the reference phase. Provided that θ_0 is chosen sufficiently different from the peak of $\Pi(\theta)$, the physical properties will be insensitive to the precise choice of an otherwise arbitrary reference phase [12].

As indicated earlier, the physically relevant results could only be obtained as the limit $s \to \infty$ is taken. If we let $\theta = \lim_{s\to\infty} 2\pi m/(s+1)$, we find

$$\Pi(\theta) = \lim_{s\to\infty}\left[\frac{s+1}{2\pi}\cdot\Pi(\theta_m)\right] = \frac{1}{2\pi}\left|\sum_{n=0}^{\infty} e^{in\theta}\cdot\langle n|\psi\rangle\right|^2. \tag{4.49}$$

This phase distribution gives us useful information about the fluctuation properties of actual quantum states of light. Extending the previous discussion to the actual Hilbert space, we find that the mean value and variance of phase for a system in a number state is given by [3]

$$\begin{aligned}\langle\hat{\phi}\rangle &= \theta_0 + \pi \\ \langle\Delta\hat{\phi}^2\rangle &= \frac{2}{3}\pi,\end{aligned} \tag{4.50}$$

indicating a state with a completely random phase. For a coherent state with $\alpha = |\alpha|\exp(i\theta_i)$ where $|\alpha| \gg 1$, we obtain

$$\begin{aligned}\langle\hat{\phi}\rangle &= \theta_i \\ \langle\Delta\hat{\phi}^2\rangle &= \frac{1}{4|\alpha|^2}.\end{aligned} \tag{4.51}$$

As discussed earlier, the coherent state has a well-defined phase, with a variance that diminishes with increasing amplitude of the coherent excitation.

4.6. QUANTUM LIMITS OF OPTICAL INTERFEROMETRY

4.6.1. Standard Quantum Limits of Optical Interferometry

An interferometer is a fundamental apparatus in optical physics and technologies where the output signal is sensitive to the relative phase shift between two fields traveling in separated paths. Optical gyroscopes and gravitational wave detectors rely on the ability to resolve extremely small relative phase shifts in the two paths. The performance is in principle determined by the quantum noise of the illuminating field.

Figure 4-5 shows an optical Mach-Zehnder interferometer. The two input field operators of the interferometer are denoted by $\hat{a}$ and $\hat{b}$, where we assume the input states $\hat{a}$ and $\hat{b}$ are in a coherent state $|\alpha\rangle_a$ and vacuum state $|0\rangle_b$, respectively. The output field operators from the first 50%–50% beam splitter can be written as follows:

$$\hat{u} = \frac{1}{\sqrt{2}}(\hat{a} + \hat{b}), \tag{4.52}$$

$$\hat{\ell} = \frac{1}{\sqrt{2}}(\hat{a} - \hat{b}), \tag{4.53}$$

where $\hat{a}$ and $\hat{b}$ are the annihilation operators for the two input modes. If the lower arm of the interferometer has a differential phase shift θ with respect to the upper arm, the output field operators from the second 50%–50% beam splitter are expressed by

$$\hat{d} = \frac{1}{\sqrt{2}}(e^{i\theta}\hat{\ell} + \hat{u}), \tag{4.54}$$

$$\hat{e} = \frac{1}{\sqrt{2}}(e^{i\theta}\hat{\ell} - \hat{u}). \tag{4.55}$$

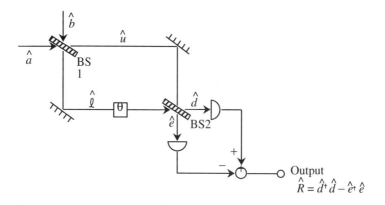

FIGURE 4-5: An optical Mach–Zehnder interferometer.

The interferometer is usually operated at one of the two bias points, either $\theta = \frac{\pi}{2}$ (bright fringe) or $\theta = 0$ (dark fringe). Here we choose the arm-length difference θ to be $\frac{\pi}{2}$ and detect the difference of the two photodetector counts $\hat{R} \equiv \hat{d}^\dagger \hat{d} - \hat{e}^\dagger \hat{e}$. If we linearize the two input operators as $\hat{a} = a_0 + \Delta \hat{a}_1 + i \Delta \hat{a}_2$ and $\hat{b} = \Delta \hat{b}_1 + i \Delta \hat{b}_2$, we obtain

$$\hat{d}^\dagger \hat{d} \simeq \frac{a_0^2}{2} + a_0(\Delta \hat{a}_1 + \Delta \hat{b}_2) + \frac{a_0^2}{2} \Delta \theta, \tag{4.56}$$

$$\hat{e}^\dagger \hat{e} \simeq \frac{a_0^2}{2} + a_0(\Delta \hat{a}_1 - \Delta \hat{b}_2) - \frac{a_0^2}{2} \Delta \theta, \tag{4.57}$$

where $\Delta \theta$ is a small phase deviation from the bias point $\theta = \frac{\pi}{2}$ that we want to measure. From the preceding expressions, we find that the signal power is proportional to $(a_0^2 \Delta \theta)^2$ while the noise power is proportional to $4 a_0^2 \langle \Delta \hat{b}_2^2 \rangle$. If we define the minimum detectable phase $\Delta \theta_{\min}$ by the unity signal-to-noise ratio, it is given by

$$\Delta \theta_{\min} \simeq \left(\frac{4 \langle \Delta \hat{b}_2^2 \rangle}{a_0^2} \right)^{1/2} = \frac{1}{\sqrt{N_a}}, \tag{4.58}$$

where $N_a = a_0^2 = |\alpha|^2$ is the average photon number of the input coherent state. Equation (4.58) is called the standard quantum limit (SQL) of an optical interferometer. The SQL originates from the quadrature amplitude noise $\langle \Delta \hat{b}_2^2 \rangle$ of the input vacuum state $|0\rangle_b$ rather than the quantum noise of the input coherent state $|\alpha\rangle_a$.

As indicated in (4.56) and (4.57), the SQL of such an optical interferometer is quite universal (i.e., independent of the choice of the quantum state for the input field $\hat{a}$). Whenever there is no input at the other port, each individual photon in the input field $\hat{a}$ is equally split into the two arms and the two probability amplitudes of a single photon interfere with each other. A photodetector count is always either one or zero: This discrete nature of the photon counting process places the minimum detectable phase shift $\Delta \theta_{\min}$.

If we compare (4.58) with the separation of the Pegg-Barnett phase eigenvalues $2\pi/(s+1)$, where s is the photon number, we find that the SQL given by (4.58) is far from the intrinsic quantum limit imposed by the number-phase uncertainty relationship. In fact, the photon number difference between the two arms, which is conjugate to the phase difference between the two arms in the preceding interferometer, is not maximized:

$$\Delta N \equiv \langle \hat{u}^\dagger \hat{u} - \hat{\ell}^\dagger \hat{\ell} \rangle = 2 a_0 \langle \Delta \hat{b}_1^2 \rangle^{1/2} = \sqrt{N_a} \ll N_a. \tag{4.59}$$

4.6.2. Squeezed State Interferometer

To improve the phase measurement sensitivity beyond the SQL (4.58), we can increase one quadrature amplitude noise $\langle \Delta \hat{b}_1^2 \rangle$ to enhance the photon number difference noise [see (4.59)] and decrease the other quadrature amplitude noise $\langle \Delta \hat{b}_2^2 \rangle$ to suppress the phase difference noise [see (4.58)]. For that purpose we can inject a squeezed vacuum state $|0, r\rangle_b$ as the input field $\hat{b}$, which is phase coherent with the main input $|\alpha\rangle_a$ [13]. The minimum detectable phase is thus decreased to

$$\Delta\theta_{min} \simeq \frac{e^{-\frac{r}{2}}}{\sqrt{N_a}}, \qquad (4.60)$$

as long as the squeezing parameter r is not too big.

When the squeezing parameter r becomes much larger than one, the squeezed vacuum state has a considerable number of photons, $N_b = \sinh^2(r)$, so the minimum detectable phase must be evaluated for the total number of photons $N_T = N_a + N_b$ used in the measurement. In this case, (4.60) takes the minimum value for a given N_T,

$$\Delta\theta_{min} \simeq \frac{1}{N_T}, \qquad (4.61)$$

at an optimum squeezing parameter $r_{opt} \simeq \ln\sqrt{2N_T + 1}$. This limit is compatible with the eigenvalue separation of the Pegg-Barnett phase operator $\frac{2\pi}{N_T+1}$. The result is also interpreted in the following way: If the interferometer is optimized, one photon count difference between the two photodetectors among the set of the possible results $\{N_T, N_T - 1, \ldots, -N_T + 1, -N_T\}$ in Fig. 4-5 should tell us one of the discrete phase values equally dividing 2π [rad]. This fundamental limit is often referred to as the Heisenberg limit.

If the input field $\hat{b}$ is a squeezed state with a finite coherent excitation $|\beta, r\rangle_b$ rather than a squeezed vacuum state $|0, r\rangle_b$, we can achieve the same sensitivity with slightly different configuration (Fig. 4-6a) [14]. Here we choose the phase difference ϕ between the two input fields to be $\frac{\pi}{2}$ and the bias point (arm length difference) θ to be 0, and detect only one output field $\hat{e}$ from the second beam splitter. The photodetector count under linearization, $\hat{a} = a_0 + \Delta\hat{a}_1 + i\Delta\hat{a}_2$ and $\hat{b} = b_0 + \Delta\hat{b}_1 + i\Delta\hat{b}_2$, is given by

$$\hat{e}^\dagger\hat{e} = \hat{b}^\dagger\hat{b} + \frac{i}{2}(e^{i\phi}\hat{a}^\dagger\hat{b} - e^{-i\phi}\hat{b}^\dagger\hat{a})\sin\theta$$
$$\simeq b_0^2 + 2b_0\Delta\hat{b}_1 - a_0 b_0 \Delta\theta. \qquad (4.62)$$

The signal power is proportional to $(a_0 b_0 \Delta\theta)^2$ and the noise power is proportional to $4b_0^2\langle\Delta\hat{b}_1^2\rangle$, so the minimum detectable phase defined by the unity signal-to-noise (S/N) ratio is given by

$$\Delta\theta_{min} \simeq \left(\frac{4\langle\Delta\hat{b}_1^2\rangle}{a_0^2}\right)^{1/2} = \frac{e^{-\frac{r}{2}}}{\sqrt{N_a}}, \qquad (4.63)$$

which is identical to (4.60) that we obtained for a squeezed vacuum state interferometer.

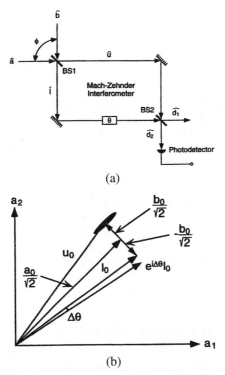

FIGURE 4-6: (a) A dual-input Mach-Zehnder interferometer. (b) The relation between input fields $\hat{a}$ and $\hat{b}$ and output fields from the first beam splitter $\hat{u}$ and $\hat{l}$.

In order to understand this interferometer intuitively, the phasor diagram shown in Fig. 4-6b is useful. The photodetector count $\hat{e}^\dagger \hat{e}$ measures the amplitude difference between $\hat{u}$ and $e^{i\Delta\theta}\hat{l}$ in the figure. Since the fluctuations in the two fields $\hat{u}$ and $e^{i\Delta\theta}\hat{l}$ due to the input field $\hat{a}$ are positively correlated, they cancel out when the two fields are combined by the second 50%–50% beam splitter. However, the fluctuations in the two fields due to the input field $\hat{b}$ are negatively correlated, and they add when the two fields are combined by the second beam splitter. Thus the phase measurement sensitivity is entirely determined by the fluctuation of the input field $\hat{b}$. Since the coherent excitation of the field $\hat{b}$ is perpendicular to that of the field $\hat{a}$, the amplitude fluctuation $\Delta \hat{b}_1$ of the field $\hat{b}$ determines the noise of this measurement.

The advantages of this interferometer, compared to the squeezed vacuum state interferometer operating at $\theta = \frac{\pi}{2}$, are as follows: (1) The photodetector needs to detect only weak squeezed state $\hat{e}$, which avoids the detector saturation problem; and (2) the strong output $\hat{d}$ can be recirculated by coherently combining with the input field $\hat{a}$ to enhance the effective number of photons N_a circulating in the interferometer.

96 COHERENCE OF THE ELECTROMAGNETIC FIELDS

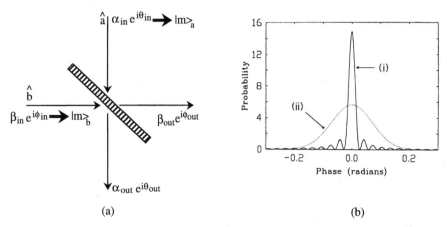

FIGURE 4-7: (a) A 50%-50% beam splitter illuminated by classical fields $\alpha_{in} e^{i\theta_{in}}$ and $\beta_{in} e^{i\phi_{in}}$ or dual-photon number eigenstates $|m\rangle_a |m\rangle_b$. (b) The phase difference distribution $P(\Delta\theta)$ for two photon number eigenstates $|m\rangle_a |m\rangle_b$ (i) and two coherent states $|\alpha\rangle_a |\alpha\rangle_b$ (ii) with the same photon number $m = |\alpha|^2 = 50$.

4.6.3. Photon Number Eigenstate Interferometer

The Heisenberg limit (4.61) can also be achieved by illuminating the two input ports of the Mach-Zehnder interferometer by photon number eigenstates [15]. In order to understand the principle of this interferometer, consider two classical fields with definite amplitudes α_{in} and β_{in} and with definite phase θ_{in} and ϕ_{in} incident on a 50%–50% beam splitter as illustrated in Fig. 4-7(a). The phase difference and intensity difference between the two output fields are given by

$$\tan(\phi_{out} - \theta_{out}) = \frac{\alpha_{in} - \beta_{in}}{2\alpha_{in}\beta_{in}\cos(\phi_{in} - \theta_{in})}, \quad (4.64)$$

$$(\alpha_{out})^2 - (\beta_{out})^2 = 2\alpha_{in}\beta_{in}\sin(\phi_{in} - \theta_{in}). \quad (4.65)$$

If the input fields have equal amplitudes, $\alpha_{in} = \beta_{in}$, the 50%–50% beam splitter produces the output fields with zero phase difference, independent of the input phase θ_{in} and ϕ_{in}. This remarkable result suggests that if the two input classical fields are replaced by two photon number eigenstates, the phase difference noise can be minimized even though the phases of the input photon number eigenstates are completely random. Alternatively, if the two input classical fields are replaced by two phase eigenstates with equal eigenvalues, the photon number difference noise can be minimized.

The phase difference distribution is evaluated by applying the unitary evolution operator of the 50%–50% beam splitter on the dual photon number eigenstates, projecting onto the phase states and tracing over the possible absolute phase values [15]:

$$P(\Delta\theta) = \sum_{\ell=0}^{s} a \langle \theta_\ell | b \langle \theta_{\ell-\Delta\theta/\varepsilon} | e^{i\frac{\pi}{4}(\hat{a}^\dagger \hat{b} + \hat{a}\hat{b}^\dagger)} |m\rangle_a |m\rangle_b |^2$$

$$= \frac{1}{2^{2m}(s+1)} \left| \sum_{r=0}^{m} \sqrt{\frac{2(m-r)!}{(m-r)!^2}} \sqrt{\frac{2r!}{r!^2}} e^{2ir\Delta\theta} \right|^2. \quad (4.66)$$

Here $|\theta_\ell\rangle = \frac{1}{\sqrt{s+1}} \sum_{n=0}^{s} e^{in\ell\varepsilon} |n\rangle$ ($\ell = 0, \ldots, s$) is the Pegg-Barnett phase state, $\varepsilon = \frac{2\pi}{s+1}$, and $s = 2m$ is the total number of photons in the system. Figure 4-7(b) illustrates $P(\Delta\theta)$ for $m = 50$ photons. The distribution is well localized around a phase difference of zero at the Heisenberg limit of $\Delta\theta \simeq \frac{1}{2m}$. Localization of the relative phase indicates that the beam splitter establishes strong correlation for the phases of the two output fields and that the phase measurement sensitivity reaches the Heisenberg limit. Figure 4-7(b) also shows $P(\Delta\theta)$ for two input coherent states $|\alpha\rangle_a |\alpha\rangle_b$ with average photon number of $|\alpha|^2 = 50$: The width in this case is much broader.

4.6.4. Photonic de Broglie Wave Interferometer

The Heisenberg limit (4.61) can also be achieved by using a quantum logic gate that splits the incident photons into the two paths as a whole rather than a simple 50%–50% beam splitter, which *divides* each individual photon one by one [16]. Consider the schematic setup of Fig. 4-8. Here an atom incident on two cavities controls the transmission/reflection properties of these cavities depending on whether or not the atom is in the ground or excited state due to an atomic state dependent index of refraction [17]. Consider the case when an excited state atom $|e\rangle$ is driven by a $\frac{\pi}{2}$ optical pulse that transforms the atom state as $|e\rangle \longrightarrow \frac{1}{\sqrt{2}}(|e\rangle + |g\rangle)$. Simultaneously, a coherent state $|\alpha\rangle$ and a vacuum state $|0\rangle$ are incident on the two input ports of the

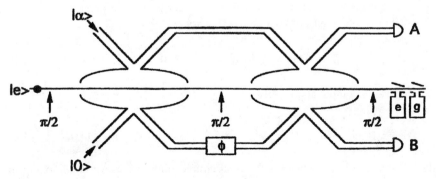

FIGURE 4-8: Schematic of a Mach-Zehnder interferometer for measuring the de Broglie wavelength of an optical pulse.

cavity. The input $|\psi\rangle_1$ and the output $|\psi\rangle_2$ states of the first cavity are written as

$$|\psi\rangle_1 = \frac{1}{\sqrt{2}}(|\alpha, 0, e\rangle + |\alpha, 0, g\rangle), \tag{4.67}$$

$$|\psi\rangle_2 = \frac{1}{\sqrt{2}}(|0, \alpha, e\rangle + |\alpha, 0, g\rangle), \tag{4.68}$$

where the state $|a, b, x\rangle$ stands for the state vector of the two photon field modes a and b, and the atom state x. Equation (4.68) results from the fact that the presence of an excited state atom in the cavity causes the incident photon states to be exchanged between mode a and mode b (i.e., reflection), whereas the presence of a ground state atom causes no mode exchange (i.e., transmission). The state after the phase shifter ϕ in the lower arm but before the second $\frac{\pi}{2}$ optical pulse is given by

$$|\psi\rangle_3 = \frac{1}{\sqrt{2}}(|0, \alpha, e\rangle + |\alpha e^{-i\phi}, 0, g\rangle), \tag{4.69}$$

where the free evolution of a coherent state

$$\hat{U}|\alpha\rangle = e^{-i\phi \hat{a}^\dagger \hat{a}}|\alpha\rangle = |\alpha e^{-i\phi}\rangle \tag{4.70}$$

is used. The second $\frac{\pi}{2}$ optical pulse transforms the excited state as $|e\rangle \longrightarrow \frac{1}{\sqrt{2}}(|e\rangle + |g\rangle)$ and the ground state as $|g\rangle \longrightarrow \frac{1}{\sqrt{2}}(|g\rangle - |e\rangle)$.

Using the preceding rules we can calculate the final state after the second $\frac{\pi}{2}$ pulse, the second cavity, and the third $\frac{\pi}{2}$ pulse:

$$|\psi\rangle_4 = \frac{1}{2\sqrt{2}}(|\alpha, 0, e\rangle + |\alpha, 0g\rangle + |0, \alpha, g\rangle - |0, \alpha, e\rangle$$
$$+ |\alpha e^{-i\phi}, 0, g\rangle - |\alpha e^{-i\phi}, 0, e\rangle - |0, \alpha e^{-i\phi}, e\rangle - |0, \alpha e^{-i\phi}, g\rangle). \tag{4.71}$$

As shown in Fig. 4-8, we measure the final state of the atom. If we are interested in the case that the measurement result is $|e\rangle$, the corresponding photon field state is obtained by projecting (4.64) onto the bra vector $\langle e|$,

$$|\psi_f\rangle_e = \frac{1}{2}(|\alpha, 0\rangle - |0, \alpha\rangle - |\alpha e^{-i\phi}, 0\rangle - |0, \alpha e^{-i\phi}\rangle), \tag{4.72}$$

and the average output at port A is

$$N_A \equiv {}_e\langle\psi_f|\hat{a}^\dagger \hat{a}|\psi_f\rangle_e = \frac{|\alpha|^2}{2}\left[1 - e^{|\alpha|^2(\cos\phi - 1)}\cos(\phi + |\alpha|^2 \sin\phi)\right]. \tag{4.73}$$

If we were simply to trace over the final atomic state, instead of conditioning the photon output at port A on a specific atom state ($|e\rangle$ or $|g\rangle$), the output of the interferometer would be ${}_e\langle\psi_f|\hat{a}^\dagger \hat{a}|\psi_f\rangle_e + {}_g\langle\psi_f|\hat{a}^\dagger \hat{a}|\psi_f\rangle_g = |\alpha|^2$, yielding no interference.

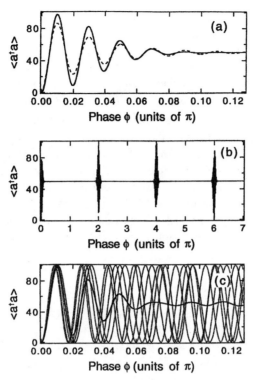

FIGURE 4-9: (a, solid line), (b) The expectation number $\langle \hat{a}^\dagger \hat{a} \rangle$ output of the de Broglie wavelength Mach-Zehnder interferometer for an incident coherent state with average photon number $N = 100$. (a, dashed line) An approximate model of the output indicating the coherence length of the coherent state; see text. (c, dashed) The expectation number output of the de Broglie wavelength Mach-Zehnder interferometer for several different incident Fock states. (c, solid) The ensemble-averaged output for 500 incident Fock states with photon numbers distributed according to a Poisson distribution centered at $N = 100$.

The conditional interferometer output of (4.73) is plotted as a function of ϕ in Fig. 4-9, where the average photon number of the input coherent state is $|\alpha|^2 = 100$. We see that the output has an oscillation period of $2\pi/|\alpha|^2$, indicating that the effective wavelength in this interferometer is not λ but $\lambda/|\alpha|^2 = \lambda/100$. The origin of this rapid oscillation is the fact that the free evolution operator $\hat{U} = e^{-i\omega_0 \hat{a}^\dagger \hat{a} t}$ introduces the phase shift $e^{-i\omega_0 n t}$ (instead of $e^{-i\omega_0 t}$) for a photon number eigenstate $|n\rangle$. A one photon eigenstate $|1\rangle$ has an ordinary phase shift determined by the optical wavelength $\lambda_0 = \frac{2\pi c}{\omega_0}$ but n photon number eigenstate $|n\rangle$ acquires a larger phase shift determined by the photonic de Broglie wavelength $\lambda_{dB} = \frac{h}{P} = \frac{2\pi c}{n\omega_0}$ of the photon wavepacket as a whole.

We may understand the decay of the interference for long phase delays in Fig. 4-9(a) by recalling that the coherent state consists of linear superpositions of different photon number eigenstates and each photon number eigenstate has a different photonic de Broglie wavelength as shown in Fig. 4-9(c). The ensemble average output for photon numbers distributed according to a Poisson distribution centered at $N = 100$ shown in Fig. 4-9(c) is identical to the interference pattern for the coherent state $|\alpha = 10\rangle$ shown in Fig. 4-9(a). Figure 4-9(b) shows that all of the wavelength components of the coherent state are rephased at periods of 2π.

Defining the interferometer signal as $\langle \hat{a}^\dagger \hat{a} \rangle_{\phi+\delta\phi} - \langle \hat{a}^\dagger \hat{a} \rangle_\phi$ and the interferometer noise as $\sqrt{\langle \hat{a}^\dagger \hat{a} \hat{a}^\dagger \hat{a} \rangle_\phi - \langle \hat{a}^\dagger \hat{a} \rangle_\phi^2}$, we may solve for the minimum detectable phase shift $\Delta\phi_{\min}$ by requiring the S/N ratio to be unity. For a phase setting of ϕ close to zero (dark fringe), the interferometer is operating in the Heisenberg limit (i.e., $\Delta\phi_{\min} \simeq \frac{1}{N}$). This result may be understood by recalling that a normal interferometer when fed a single photon in one input port and the vacuum in the other port is both Heisenberg and standard quantum limited as $\Delta\phi_{\min} = \frac{1}{N} = \frac{1}{\sqrt{N}}$ for $N = 1$. The incident coherent state photon packet is in essence a single particle when incident on the quantum logic gate of the photonic de Broglie wave interferometer.

REFERENCES

[1] C. Cohen-Tannoudji, J. Dupont-Roc, and G. Grynberg, *Atom-Photon Interactions* (Wiley-Interscience, New York, 1992).
[2] R. Loudon, *The Quantum Theory of Light* (Oxford University Press, New York, 1985).
[3] D. F. Walls and G. J. Milburn, *Quantum Optics* (Springer-Verlag, Berlin, 1994).
[4] G. S. Agarwal, Phys. Rev. A **18**, 1490 (1978).
[5] P. A. M. Dirac, *The Principles of Quantum Mechanics* (Oxford University Press, Oxford, UK, 1958).
[6] R. F. Pfleegor and L. Mandel, Phys. Rev. **159**, 1084 (1967).
[7] R. J. Glauber, Phys. Rev. **130**, 2529 (1963).
[8] C. W. Gardiner, *Quantum Noise* (Springer-Verlag, Berlin, 1991).
[9] J. W. Goodman, *Statistical Optics* (Wiley-Interscience, New York, 1985).
[10] P. L. Kelley and W. H. Kleiner, Phys. Rev. **136**, 316 (1964).
[11] L. Susskind and J. Glogower, Physics **1**, 49 (1964).
[12] D. T. Pegg and S. M. Burnett, Phys. Rev. A **39**, 1665 (1989).
[13] C. M. Caves, Phys. Rev. D**23**, 1693 (1981).
[14] S. Inoue, G. Björk, and Y. Yamamoto, Proc. SPIE **2378**, 99 (1995).
[15] M. J. Holland and K. Burnett, Phys. Rev. Lett. **71**, 1355 (1993).
[16] J. M. Jacobson, G. Björk, I. Chuang, and Y. Yamamoto, Phys. Rev. Lett. **74**, 4835 (1995).
[17] L. Davidovich, A. Maali, M. Brune, J. M. Raimond, and S. Haroche, Phys. Rev. Lett. **71**, 2360 (1993).

5 Quantum States of Atoms

In Chapter 3 we studied two particular quantum states of the electromagnetic field. The photon number eigenstates (Fock states), which are eigenstates of the free-field Hamiltonian, form a complete orthonormal set for the corresponding Hilbert space. These states are, however, poorly suited for the description of high-intensity laser fields, which contain a large number of photons with a finite variance. The coherent states (Glauber states), which are eigenstates of the (non-Hermitian) annihilation operator, form an overcomplete set for the corresponding Hilbert space. The coherent states are the stationary minimum uncertainty wavepackets of a harmonic oscillator and are well suited for the description of high-intensity laser fields. The coherent state wavepackets "cohere" (i.e., the uncertainties of the generalized coordinate and generalized momentum operators are preserved at all times). Another important feature of coherent states is that they are obtained from the vacuum state by a simple unitary translation operator, so they preserve the same quantum fluctuations as the vacuum state. These properties of the coherent states make them unique and useful in the quantum description of the electromagnetic fields.

Many problems in quantum optics are based on the interaction of an assembly of two-level atoms or free carriers in semiconductors with an electromagnetic field. If the interatomic distance $N^{-\frac{1}{3}}$ is much smaller than the optical wavelength λ_0 (where N is the density of atoms), the atoms collectively interact with the electromagnetic field. In such a case, the internal state of the assembly of two-level atoms can be described by (collective) quasi-bosonic operators irrespective of whether the atoms are fermions or bosons. The spectra of the quasi-bosonic operators are harmonic oscillator-like but consist of a finite number of eigenstates. A particular set of quantum states can be used for the description of the internal states of the atoms in analogy with the quantum states of the electromagnetic field. This chapter demonstrates that collective atomic internal states with completely analogous properties to the photon number eigenstates and coherent states of light can be defined. Throughout this chapter, it is assumed that the interatomic distance $N^{-\frac{1}{3}}$ is larger than the thermal de Broglie wavelength $\lambda_T = \sqrt{\frac{2\pi\hbar^2}{mk_BT}}$ of the atoms, where m is the mass of the atom. Thus, the external state of each individual atom can still be considered independent. Because of this assumption, the quantum state of the assembly of two-level atoms can be described by (collective) quasi-bosonic operators even if the particular atom is a fermion; this is not the case once the thermal de Broglie wavelength λ_T becomes larger than the interatomic distance $N^{-\frac{1}{3}}$.

101

5.1. ANGULAR MOMENTUM ALGEBRA

5.1.1. Quantization of Angular Momentum

Classically, the angular momentum of a particle about an origin 0 is defined by

$$\vec{l} = \vec{q} \times \vec{p}, \tag{5.1}$$

where $\vec{q}$ is the radius vector from the origin 0 to the position of the particle and $\vec{p}$ is its linear momentum. We know that $\hat{\vec{l}}$ is a Hermitian operator since it is an observable and is defined by (5.1), where $\hat{\vec{q}}$ and $\hat{\vec{p}}$ are the coordinate and the momentum operators. We let $\hat{\vec{q}} = [\hat{q}_x, \hat{q}_y, \hat{q}_z]$ and $\hat{\vec{p}} = [\hat{p}_x, \hat{p}_y, \hat{p}_z]$ be the coordinate and momentum operators that obey the commutation relations

$$[\hat{q}_i, \hat{p}_j] = i\hbar\delta_{ij},$$
$$[\hat{q}_i, \hat{q}_j] = [\hat{p}_i, \hat{p}_j] = 0, \tag{5.2}$$

where $i, j = x, y, z$.

Using the expansions of $\hat{\vec{l}}, \hat{\vec{q}}$ and $\hat{\vec{p}}$ in (5.2), we obtain

$$\hat{\ell}_x = \hat{q}_y \hat{p}_z - \hat{q}_z \hat{p}_y$$
$$\hat{\ell}_y = \hat{q}_z \hat{p}_x - \hat{q}_x \hat{p}_z$$
$$\hat{\ell}_z = \hat{q}_x \hat{p}_y - \hat{q}_y \hat{p}_x. \tag{5.3}$$

Using (5.2), we can easily show that

$$[\hat{\ell}_x, \hat{\ell}_y] = i\hbar\hat{\ell}_z$$
$$[\hat{\ell}_y, \hat{\ell}_z] = i\hbar\hat{\ell}_x$$
$$[\hat{\ell}_z, \hat{\ell}_x] = i\hbar\hat{\ell}_y. \tag{5.4}$$

No additional postulate is necessary in order to quantize the angular momentum. The total angular momentum $\hat{l}^2$ is defined by

$$\hat{l}^2 = \hat{\ell}_x^2 + \hat{\ell}_y^2 + \hat{\ell}_z^2, \tag{5.5}$$

which satisfies

$$[\hat{l}^2, \hat{l}_i] = 0 \quad (i = x, y, z). \tag{5.6}$$

Let us define the non-Hermitian operators $\hat{l}_+$ and $\hat{l}_-$ by

$$\hat{l}_+ = \hat{\ell}_x + i\hat{\ell}_y \quad (\longleftrightarrow \hat{a}^\dagger), \tag{5.7}$$
$$\hat{l}_- = \hat{\ell}_x - i\hat{\ell}_y \quad (\longleftrightarrow \hat{a}). \tag{5.8}$$

where we put the corresponding operators in the harmonic oscillator algebra to the right of the angular momentum operator for the purpose of comparison. We also note that $\hat{\ell}_z$ corresponds to $\hat{n} = a^\dagger a$. The following relations are easily derived:

$$(\hat{l}_+)^\dagger = \hat{l}_-, \tag{5.9}$$

$$\hat{l}^2 = \hat{l}_z^2 + \frac{1}{2}(\hat{l}_+\hat{l}_- + \hat{l}_-\hat{l}_+), \tag{5.10}$$

$$[\hat{l}_+, \hat{l}_-] = 2\hbar\hat{l}_z, \tag{5.11}$$

$$\hat{l}_+\hat{l}_- = \hat{l}^2 - \hat{l}_z^2 + \hbar\hat{l}_z, \tag{5.12}$$

$$\hat{l}_-\hat{l}_+ = \hat{l}^2 - \hat{l}_z^2 - \hbar\hat{l}_z, \tag{5.13}$$

$$[\hat{\ell}_z, \hat{l}_\pm] = \pm\hbar\hat{l}_\pm. \tag{5.14}$$

5.1.2. Angular Momentum Operators and Eigenstates

Since $\hat{l}^2$ and $\hat{\ell}_z$ commute [(5.6)], a set of simultaneous eigenstates for the two Hermitian operators $\hat{l}^2$ and $\hat{\ell}_z$ should exist. The eigenvalue problem is written as

$$\hat{\ell}_z|\mu, \nu\rangle = \hbar\mu|\mu, \nu\rangle, \tag{5.15}$$

$$\hat{l}^2|\mu, \nu\rangle = \hbar^2\nu|\mu, \nu\rangle, \tag{5.16}$$

where $|\mu, \nu\rangle$ is a simultaneous eigenstate of $\hat{l}^2$ and $\hat{\ell}_z$ with the eigenvalues of $\hbar^2\nu$ and $\hbar\mu$. Since $[\hat{\ell}^2, \ell_\pm] = 0$, we have

$$\hat{l}^2\hat{l}_\pm|\mu, \nu\rangle = \hat{l}_\pm\hat{l}^2|\mu, \nu\rangle = \hbar^2\nu\hat{l}_\pm|\mu, \nu\rangle. \tag{5.17}$$

This relation indicates that $|\mu, \nu\rangle$, $\hat{l}_+|\mu, \nu\rangle$, and $\hat{l}_-|\mu, \nu\rangle$ are all eigenstates of $\hat{l}^2$ with the same eigenvalue $\hbar^2\nu$. Using (5.14), we obtain

$$\hat{\ell}_z\hat{l}_\pm|\mu, \nu\rangle = l_\pm\hbar\mu|\mu, \nu\rangle \pm \hbar l_\pm|\mu, \nu\rangle = \hbar(\mu \pm 1)l_\pm|\mu, \nu\rangle. \tag{5.18}$$

This relation indicates that $|\mu, \nu\rangle$, $\hat{l}_+|\mu, \nu\rangle$, and $\hat{l}_-|\mu, \nu\rangle$ are all eigenstates of $\hat{\ell}_z$ with different eigenvalues $\hbar\mu$, $\hbar(\mu + 1)$ and $\hbar(\mu - 1)$.

The inner product (norm) of eigenstate $\hat{l}_+|\mu, \nu\rangle$ is calculated by

$$\langle\mu, \nu|\hat{l}_-\hat{l}_+|\mu, \nu\rangle = \langle\mu, \nu|\hat{l}^2 - \hat{\ell}_z^2 - \hbar\hat{\ell}_z|\mu, \nu\rangle$$
$$= \hbar^2(\nu - \mu^2 - \mu)\langle\mu, \nu|\mu, \nu\rangle. \tag{5.19}$$

The norm of $|\mu, \nu\rangle$ and $\hat{l}_+|\mu, \nu\rangle$ must be nonnegative, so the coefficient of the right-hand side of (5.19) should be nonnegative:

$$\nu \geq \mu(\mu + 1). \tag{5.20}$$

An upper and lower bound on the value of μ exists for a given value of ν. If the maximum value of μ is denoted by l, $\nu = l(l+1)$ then we have an upper constraint,

$$\hat{l}_+|l, \nu\rangle = 0. \tag{5.21}$$

Next consider the application of the lowering operation of $\hat{l}_-$ k times on the maximum eigenstate $|l, \nu\rangle$, which results in $|l-k, \nu\rangle$. The norm of $\hat{l}_-|l-k, \nu\rangle$ is calculated by

$$\langle l-k, \nu|\hat{l}_+\hat{l}_-|l-k, \nu\rangle = \langle l-k, \nu|\hat{l}^2 - \hat{\ell}_z^2 + \hbar\hat{\ell}_z|l-k, \nu\rangle$$
$$= \hbar^2[l(l+1) - (l-k)^2 + (l-k)]\langle l-k|l-k\rangle. \tag{5.22}$$

The norm of $\hat{l}_-|l-k, \nu\rangle$ and the norm of $|l-k, \nu\rangle$ must be nonnegative, which results in the constraint on the maximum number of lowering operations,

$$k_{\max} = 2l. \tag{5.23}$$

Since k is a positive integer, l takes only $0, \frac{1}{2}, 1, \frac{3}{2}, \ldots$. The following examples [using ℓ instead of $\nu = \ell(\ell+1)$] illustrate the structure of the angular momentum eigenstates:

$$l = \frac{1}{2}, \nu = \frac{3}{4} \implies |\mu, l\rangle = |-\frac{1}{2}, \frac{1}{2}\rangle, |\frac{1}{2}, \frac{1}{2}\rangle$$
$$l = 1, \nu = 2 \implies |\mu, l\rangle = |-1, 1\rangle, |0, 1\rangle, |1, 1\rangle$$
$$l = \frac{3}{2}, \nu = \frac{15}{4} \implies |\mu, l\rangle = |-\frac{3}{2}, \frac{3}{2}\rangle, |-\frac{1}{2}, \frac{3}{2}\rangle, |\frac{1}{2}, \frac{3}{2}\rangle, |\frac{3}{2}, \frac{3}{2}\rangle.$$

From now on we use m instead of μ and j instead of l, so the angular momentum eigenstate is denoted by $|m, j\rangle$, which satisfies

$$\hat{\ell}_z|m, j\rangle = \hbar m|m, j\rangle, \tag{5.24}$$

$$\hat{l}^2|m, j\rangle = \hbar^2 j(j+1)|m, j\rangle, \tag{5.25}$$

$$\langle m', j'|m, j\rangle = \delta_{mm'}\delta_{jj'}. \tag{5.26}$$

The non-Hermitian operator $\hat{l}_+$ raises the eigenvalue m to $m+1$ and we can thus write

$$\hat{l}_+|m, j\rangle = \hbar\lambda_{mj}|m+1, j\rangle, \tag{5.27}$$

where λ_{mj} is a constant to be determined. From (5.27), we have

$$\langle m+1, j|\hat{l}_+|m, j\rangle = \hbar\lambda_{mj}. \tag{5.28}$$

The adjoint of (5.28) is

$$\langle m, j|\hat{l}_-|m+1, j\rangle = \hbar \lambda_{mj}^*. \quad (5.29)$$

Therefore, we have the following relation for the lowering operator:

$$\hat{l}_-|m+1, j\rangle = \hbar \lambda_{mj}^*|m, j\rangle. \quad (5.30)$$

Using (5.27) and (5.30), we have

$$\begin{aligned}\hat{l}_-\hat{l}_+|m, j\rangle &= \hbar^2|\lambda_{mj}|^2|m, j\rangle \\ &= (\hat{l}^2 - \hat{l}_z^2 - \hbar\hat{l}_z)|m, j\rangle \\ &= \hbar^2[j(j+1) - m^2 - m]|m, j\rangle. \end{aligned} \quad (5.31)$$

From this relation, we can determine the constant λ_{mj} as

$$\lambda_{mj} = \sqrt{j(j+1) - m(m+1)}. \quad (5.32)$$

Using (5.32) in (5.27) and (5.30), we obtain

$$\hat{l}_+|m, j\rangle = \hbar\sqrt{j(j+1) - m(m+1)}\,|m+1, j\rangle$$
$$(\longleftrightarrow \hat{a}^\dagger|n\rangle = \sqrt{n+1}\,|n+1\rangle), \quad (5.33)$$
$$\hat{l}_-|m, j\rangle = \hbar\sqrt{j(j+1) - m(m-1)}\,|m-1, j\rangle$$
$$(\longleftrightarrow \hat{a}|n\rangle = \sqrt{n}\,|n-1\rangle). \quad (5.34)$$

5.2. ASSEMBLY OF TWO-LEVEL ATOMS

5.2.1. Pauli Spin Operators

If l is to represent a classical orbital angular momentum, the eigenvalue of $\hat{\ell}_z$ should be an integer times $\hbar$. The half-integral eigenvalue of $\hat{\ell}_z$ does not have a classical analog; that is, there is no coordinate and momentum representation for the half-integral angular momentum eigenvalue. The reason we did not discard the half-integral values for the angular momentum eigenvalue m is based on the experimental fact that the measurement results of the electron energy spectrum in a *dc* magnetic field demand the half-integral eigenvalues. An electron has inherent angular momentum of $j = \frac{1}{2}$ and projection $m = \pm\frac{1}{2}$. This intrinsically quantum mechanical angular momentum, which does not have a classical analog, is called a *spin angular momentum*,

$$\hat{s} = \frac{1}{2}\hbar\hat{\sigma}, \quad (5.35)$$

where $\hat{\sigma}$ is the Pauli spin operator. A detailed discussion concerning spin angular momentum can be found in the Refs. [1] and [2]. If we use $\hat{\sigma}_i = \frac{2}{\hbar}\hat{\ell}_i$ ($i = x, y, z$) in (5.4) and (5.6), we obtain the following commutation relations for the Pauli spin operator:

$$[\hat{\sigma}_i, \hat{\sigma}_j] = 2i\sigma_k \quad [i, j, k = \text{permutation of } (x, y, z)], \tag{5.36}$$

$$[\hat{\sigma}^2, \hat{\sigma}_i] = 0, \tag{5.37}$$

where $\hat{\sigma}^2 = \hat{\sigma}_x^2 + \hat{\sigma}_y^2 + \hat{\sigma}_z^2$. In order to restrict the eigenvalues of $\hat{\sigma}_z$ to ± 1 (the eigenvalues of the spin angular momentum $\hat{s}_z$ to $\pm\frac{1}{2}\hbar$), we require

$$[\hat{\sigma}_i, \hat{\sigma}_j]_+ \equiv \hat{\sigma}_i\hat{\sigma}_j + \hat{\sigma}_j\hat{\sigma}_i = 2\delta_{ij}, \tag{5.38}$$

or, equivalently,

$$\hat{\sigma}_i^2 = 1. \tag{5.39}$$

From the commutation relation (5.36) and the anticommutation relation (5.38), we obtain

$$\hat{\sigma}_i\hat{\sigma}_j = i\hat{\sigma}_k \quad (i \neq j). \tag{5.40}$$

The raising and lowering operators are defined by

$$\hat{\sigma}_+ = \frac{1}{2}(\hat{\sigma}_x + i\hat{\sigma}_y), \tag{5.41}$$

$$\hat{\sigma}_- = \frac{1}{2}(\hat{\sigma}_x - i\hat{\sigma}_y), \tag{5.42}$$

which satisfy the following fermionic operator characteristics:

$$[\hat{\sigma}_+, \hat{\sigma}_-]_+ = 1, \tag{5.43}$$

$$\hat{\sigma}_+^2 = 0, \tag{5.44}$$

$$\hat{\sigma}_-^2 = 0. \tag{5.45}$$

5.2.2. Collective Angular Momentum Operators

A two-level atom is mathematically equivalent to a spin-$\frac{1}{2}$ system. Suppose the upper and lower states are denoted by $|2\rangle$ and $|1\rangle$ and $\hat{a}_i$ and $\hat{a}_i^\dagger$ are the annihilation and creation operators for the states $i = 1$ or 2. The Hamiltonian is given by

$$\hat{\mathcal{H}} = \frac{\hbar\omega}{2}|2\rangle\langle 2| - \frac{\hbar\omega}{2}|1\rangle\langle 1| = \frac{\hbar\omega}{2}(\hat{a}_2^+\hat{a}_2 - \hat{a}_1^+\hat{a}_1). \tag{5.46}$$

The corresponding spin operators are defined by

$$\hat{\sigma}_x = |2\rangle\langle 1| + |1\rangle\langle 2| = \hat{a}_2^\dagger \hat{a}_1 + \hat{a}_1^\dagger \hat{a}_2, \tag{5.47}$$

$$\hat{\sigma}_y = \frac{1}{i}[|2\rangle\langle 1| - |1\rangle\langle 2|] = \frac{1}{i}(\hat{a}_2^\dagger \hat{a}_1 - \hat{a}_1^\dagger \hat{a}_2), \tag{5.48}$$

$$\hat{\sigma}_z = |2\rangle\langle 2| - |1\rangle\langle 1| = \hat{a}_2^\dagger \hat{a}_2 - \hat{a}_1^\dagger \hat{a}_1, \tag{5.49}$$

$$\hat{\sigma}_+ = \frac{1}{2}(\sigma_x + i\sigma_y) = |2\rangle\langle 1| = \hat{a}_2^\dagger \hat{a}_1, \tag{5.50}$$

$$\hat{\sigma}_- = \frac{1}{2}(\sigma_x - i\sigma_y) = |1\rangle\langle 2| = \hat{a}_1^\dagger \hat{a}_2. \tag{5.51}$$

For an assembly of N two-level atoms, the corresponding Hilbert space is spanned by the set of 2^N product states:

$$|\phi\rangle = \prod_{n=1}^{N} |\psi_n\rangle. \tag{5.52}$$

Collective angular momentum operators are introduced for this case:

$$\hat{J}_\mu = \frac{1}{2} \sum_{n=1}^{N} \hat{\sigma}_{n\mu} \quad (\mu = x, y, z), \tag{5.53}$$

$$\hat{J}^2 = \hat{J}_x^2 + \hat{J}_y^2 + \hat{J}_z^2. \tag{5.54}$$

The effect of different spatial positions of atoms $1, 2, \ldots, N$ is ignored because of the assumption $N^{-\frac{1}{3}} \ll \lambda_0$. As mentioned earlier, quantum statistical effects as well as direct dipole-dipole interaction between the atoms are neglected.

In order to clarify the analogies between the free-field quantization and the free-atom quantization, we combine the corresponding equations for the angular momentum and the harmonic oscillator algebra. The angular momentum operators $\hat{J}_x$ and $\hat{J}_y$ satisfy the commutation relation

$$[\hat{J}_x, \hat{J}_y] = i\hat{J}_z \quad \longleftrightarrow \quad [\hat{q}, \hat{p}] = i\hbar. \tag{5.55}$$

The lowering and raising operators, defined by

$$\hat{J}_- = \hat{J}_x - i\hat{J}_y \quad \longleftrightarrow \quad \hat{a} = \frac{1}{\sqrt{2\hbar\omega}}(\omega\hat{q} + i\hat{p}), \tag{5.56}$$

$$\hat{J}_+ = \hat{J}_x + i\hat{J}_y \quad \longleftrightarrow \quad \hat{a}^\dagger = \frac{1}{\sqrt{2\hbar\omega}}(\omega\hat{q} - i\hat{p}), \tag{5.57}$$

$$\hat{J}_z = \frac{1}{2}(\hat{J}_+\hat{J}_- - \hat{J}_-\hat{J}_+) \quad \longleftrightarrow \quad \hat{n} = \hat{a}^\dagger \hat{a}, \tag{5.58}$$

obey

$$[\hat{J}_-, \hat{J}_+] = -2\hat{J}_z \quad \longrightarrow \quad [\hat{a}, \hat{a}^\dagger] = 1, \quad (5.59)$$

$$[\hat{J}_-, \hat{J}_z] = \hat{J}_- \quad \longrightarrow \quad [\hat{a}, \hat{n}] = \hat{a}, \quad (5.60)$$

$$[\hat{J}_+, \hat{J}_z] = -\hat{J}_+ \quad \longrightarrow \quad [\hat{a}^\dagger, \hat{n}] = -\hat{a}^\dagger. \quad (5.61)$$

When all the atoms are in the ground state, the eigenvalue of $\hat{J}_z$ is $-J = -\frac{N}{2}$ so that the commutation relation (5.59) is reduced to a bosonlike one, $[\hat{J}_-, \hat{J}_+] \simeq N$.

5.2.3. Angular Momentum Eigenstates (Dicke States)

We consider a subspace of degenerate eigenstates of $\hat{J}^2$ with a constant eigenvalue $J(J+1)$. The Dicke states are defined as the simultaneous eigenstates of the Hermitian operators $\hat{J}_z$ and $\hat{J}^2$; that is, $\hat{J}_z|M, J\rangle = M|M, J\rangle$ and $\hat{J}^2|M, J\rangle = J(J+1)|M, J\rangle$. These states are mathematically constructed by operating $\hat{J}_+$ on the ground state, $|-J, J\rangle$, $(M+J)$ times,

$$\hat{J}_+|-J, J\rangle = \sqrt{2J}|-J+1, J\rangle,$$

$$\hat{J}_+|-J+1, J\rangle = \sqrt{2(2J-1)}|-J+2, J\rangle,$$

$$\hat{J}_+|-J+2, J\rangle = \sqrt{3(2J-2)}|-J+3, J\rangle,$$

$$\vdots$$

$$\hat{J}_+|-J+N, J\rangle = \sqrt{(N+1)(2J-N)}|-J+N+1, J\rangle,$$

where (5.33) is used. The product of the preceding equations results in the following relation:

$$\hat{J}_+^N|-J, J\rangle = \sqrt{2J}\sqrt{2(2J-1)}\sqrt{3(2J-2)}\cdots\sqrt{N(2J-N-1)}|-J+N, J\rangle. \quad (5.62)$$

If we replace N in (5.62) by $J + M$, we obtain

$$|M, J\rangle = \frac{1}{(M+J)!}\binom{2J}{M+J}^{-\frac{1}{2}} \hat{J}_+^{(M+J)}|-J, J\rangle \quad \longleftrightarrow \quad |n\rangle = \frac{1}{\sqrt{n!}}(\hat{a}^\dagger)^n|0\rangle.$$

$$(M = -J, -J+1, \cdots J-1, J) \qquad (n = 0, 1, 2, \cdots)$$

$$(5.63)$$

The set of Dicke states $|M, J\rangle$ span the space of the angular momentum quantum number J. The ground state is defined by

$$\hat{J}_-|-J, J\rangle = 0 \quad \longleftrightarrow \quad \hat{a}|0\rangle = 0. \quad (5.64)$$

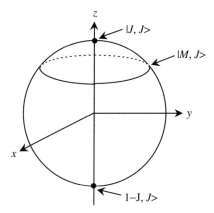

FIGURE 5-1: Angular momentum eigenstates (Dicke states) of the assembly of non-interacting two-level atoms.

To avoid any misunderstanding, we note that the quantum number J has nothing to do with the total angular momentum of an atom. The Dicke state is the counterpart of the Fock state: for example, the state $|M, J\rangle$ denotes an atomic ensemble where exactly $J + M$ atoms are in the excited state out of $N = 2J$ atoms. A schematic representation of the Dicke states is shown in Fig. 5-1.

5.2.4. Coherent Atomic States (Bloch States)

Let us consider the rotation operator, which produces a rotation by an angle θ about an axis $\vec{n} = (\sin\varphi, -\cos\varphi, 0)$ in the x-y plane, as shown in Fig. 5-2.

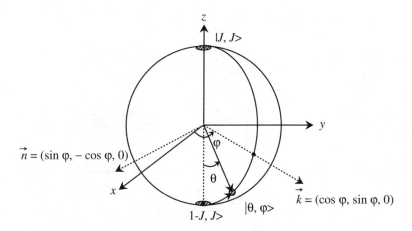

FIGURE 5-2: A schematic representation of the Bloch state.

110   QUANTUM STATES OF ATOMS

The rotation of the x and y axes by an angle φ results in the transformation of the angular momentum operators:

$$\hat{J}_z = \hat{J}_z$$
$$\hat{J}_n = \hat{J}_x \sin\varphi - \hat{J}_y \cos\varphi = \frac{i}{2}(\hat{J}_+ e^{-i\varphi} - \hat{J}_- e^{i\varphi}) \quad (5.65)$$
$$\hat{J}_k = \hat{J}_x \cos\varphi + \hat{J}_y \sin\varphi = \frac{1}{2}(\hat{J}_+ e^{-i\varphi} + \hat{J}_- e^{i\varphi}).$$

The rotation by an angle θ with respect to $\vec{n}$ is desribed by

$$\begin{aligned}\hat{R}(\theta,\varphi) &= e^{-i\theta \hat{J}_n} \\ &= e^{-i\theta(\hat{J}_x \sin\varphi - \hat{J}_y \cos\varphi)} \\ &= e^{-\frac{i\theta}{2}(\sin\varphi + i\cos\varphi)\hat{J}_+ - \frac{\theta}{2}(\sin\varphi - i\cos\varphi)\hat{J}_-},\end{aligned} \quad (5.66)$$

where (5.56) and (5.57) are used to derive the third line. Equation (5.66) is rewritten as

$$\hat{R}(\theta,\varphi) = \exp(\zeta \hat{J}_+ - \zeta^* \hat{J}_-) \quad \longleftrightarrow \quad \hat{D}(\alpha) = \exp(\alpha \hat{a}^\dagger - \alpha^* \hat{a})$$
$$\zeta = \frac{1}{2}\theta e^{-i\varphi} \qquad\qquad \alpha = \frac{1}{\sqrt{2\hbar\omega}}(\omega\langle\hat{q}\rangle + i\langle\hat{p}\rangle). \quad (5.67)$$

A coherent atomic state (or Bloch state) $|\theta,\varphi\rangle$ is obtained by the rotation of the ground state $|-J,J\rangle$ by an angle θ around $\vec{n}$:

$$|\theta,\varphi\rangle = \hat{R}(\theta,\varphi)|-J,J\rangle \quad \longleftrightarrow \quad |\alpha\rangle = \hat{D}(\alpha)|0\rangle, \quad (5.68)$$

where θ determines the population of the excited state atoms and φ represents the phase of the atomic dipole. Referring to Fig. 5-2, we obtain

$$\begin{aligned}\hat{R}(\theta,\varphi)\hat{J}_- \hat{R}(\theta,\varphi)^{-1} &= \hat{R}(\theta,\varphi)\left[\left(\hat{J}_k + i\hat{J}_n\right)\right]\hat{R}(\theta,\varphi)^{-1} \\ &= \left[\hat{J}_k \cos\theta + \hat{J}_z \sin\theta + i\hat{J}_n\right]e^{-i\varphi} \quad (5.69) \\ &= e^{-i\varphi}\left[\hat{J}_- e^{i\varphi}\cos^2\frac{\theta}{2} - \hat{J}_+ e^{-i\varphi}\sin^2\frac{\theta}{2} + \hat{J}_z \sin\theta\right].\end{aligned}$$

Multiplying both sides of (5.68) by $\hat{R}_{\theta,\varphi}\hat{J}_-\hat{R}_{\theta,\varphi}^{-1}$, we obtain the eigenvalue equation:

$$\left[\hat{J}_- e^{i\varphi}\cos^2\frac{\theta}{2} - \hat{J}_+ e^{-i\varphi}\sin^2\frac{\theta}{2} + \hat{J}_z \sin\theta\right]|\theta,\varphi\rangle = 0 \quad \longleftrightarrow \quad (\hat{a} - \alpha)|\alpha\rangle = 0,$$
$$(5.70)$$

together with

$$\hat{J}^2|\theta, \varphi\rangle = J(J+1)|\theta, \varphi\rangle. \tag{5.71}$$

Equations (5.70) and (5.71) uniquely determine the Bloch states.

Using the disentangling theorem for angular momentum operators [3], the rotation operator $\hat{R}(\theta, \varphi)$ given by (5.67) becomes

$$\hat{R}(\theta, \varphi) = e^{\tau \hat{J}_+} e^{\ln(1+|\tau|^2)\hat{J}_z} e^{-\tau^*\hat{J}_-} \quad \longleftrightarrow \quad D(\alpha) = e^{-\frac{|\alpha|^2}{2}} e^{\alpha \hat{a}^\dagger} e^{-\alpha^* \hat{a}}, \tag{5.72}$$

where

$$\tau = e^{-i\varphi} \tan \frac{\theta}{2} \simeq e^{-i\varphi} \frac{\theta}{2} = \zeta. \tag{5.73}$$

Using (5.72) in (5.68), we obtain an expression for the Bloch state $|\theta, \varphi\rangle$ in terms of the (Dicke) ground state

$$|\theta, \varphi\rangle = \frac{e^{\tau \hat{J}_+}}{(1+|\tau|^2)^J}|-J, J\rangle \quad \longleftrightarrow \quad |\alpha\rangle = e^{-\frac{|\alpha|^2}{2}} e^{\alpha \hat{a}^\dagger}|0\rangle. \tag{5.74}$$

Expanding the exponential and using (5.63), we obtain

$$|\theta, \varphi\rangle = \sum_{M=-J}^{J} \frac{1}{(1+|\tau|^2)^J} \frac{\tau^{M+J}}{(M+J)!} \hat{J}_+^{M+J}|-J, J\rangle \tag{5.75}$$

$$= \sum_{M=-J}^{J} \binom{2J}{M+J}^{\frac{1}{2}} \frac{\tau^{M+J}}{(1+|\tau|^2)^J}|M, J\rangle \quad \longleftrightarrow \quad |\alpha\rangle$$

$$= \sum_{n} \left(e^{-\frac{|\alpha|^2}{2}} \alpha^n / \sqrt{n!}\right)|n\rangle. \tag{5.76}$$

The inner product of two Bloch states is obtained from (5.75) using the completeness property of Dicke states $\sum_M |M, J\rangle\langle M, J| = \hat{I}$,

$$|\langle\theta, \varphi|\theta', \varphi'\rangle|^2 = \cos^{4J} \frac{\Phi}{2} \quad \longleftrightarrow \quad |\langle\alpha|\beta\rangle|^2 = e^{-|\alpha-\beta|^2}, \tag{5.77}$$

where the angle between $|\theta, \varphi\rangle$ and $|\theta', \varphi'\rangle$ is given by $\cos\Phi = \cos\theta\cos\theta' + \sin\theta\sin\theta'\cos(\varphi-\varphi')$. The Bloch states are not orthogonal with each other, but become orthogonal in the limit $J \to \infty$ or $\Phi = \pi$.

If we define new angular momentum operators using the rotation operator

$$(\hat{J}_\xi, \hat{J}_\eta, \hat{J}_\zeta) = \hat{R}(\theta, \varphi)(\hat{J}_x, \hat{J}_y, \hat{J}_z)\hat{R}(\theta, \varphi)^{-1}, \tag{5.78}$$

these angular momentum operators satisfy the commutation relation and the Heisenberg uncertainty relation:

$$[\hat{J}_\xi, \hat{J}_\eta] = i\hat{J}_\zeta, \tag{5.79}$$

$$\langle \Delta \hat{J}_\xi^2 \rangle \langle \Delta \hat{J}_\eta^2 \rangle \geq \frac{1}{4} |\langle \hat{J}_\zeta \rangle|^2. \tag{5.80}$$

The Bloch state $|\theta, \varphi\rangle$ satisfies the equality of (5.80), so it is a minimum uncertainty wavepacket. [Note that the eigenvalue equation, $(\hat{J}_\xi + i J_\eta)|\psi\rangle = (\langle \hat{J}_\xi \rangle + i\langle \hat{J}_\eta \rangle)|\psi\rangle$, for the minimum uncertainty wavepacket is identical to the eigenvalue equation for the Bloch state (5.70).]

The completeness properties of the Bloch states can be obtained by using (5.75) and $\sum_M |M, J\rangle \langle M, J| = \hat{I}$:

$$(2J+1) \int \frac{d\Omega}{4\pi} |\theta, \varphi\rangle \langle \theta, \varphi| = \sum_M |M, J\rangle \langle M, J| = \hat{I} \longleftrightarrow \int \frac{d^2\alpha}{\pi} |\alpha\rangle \langle \alpha| = \hat{I}. \tag{5.81}$$

The Bloch states form an overcomplete set. A schematic representation of the Bloch states is shown in Fig. 5-2.

When $\theta = 0$ or $\theta = \pi$, the Bloch states are identical to the ground state $|-J, J\rangle$ or the highest excited state $|J, J\rangle$, which are simultaneously the Dicke states. The population uncertainty is $\langle \Delta \hat{N}^2 \rangle = 4\langle \Delta \hat{J}_z^2 \rangle = 0$ but the atomic dipole phase is totally random because $\langle \Delta \hat{J}_x^2 \rangle = \langle \Delta \hat{J}_y^2 \rangle = \frac{J}{2}$ and $\langle \hat{J}_x \rangle = \langle \hat{J}_y \rangle = 0$. When $\varphi = 0$ and $\theta = \frac{\pi}{2}$, the Bloch state $|\frac{\pi}{2}, 0\rangle$ has a finite population noise $\langle \Delta \hat{N}^2 \rangle = 4\langle \Delta \hat{J}_z^2 \rangle = 2J = N$ and a finite dipole phase noise $\langle \Delta \hat{\Phi}^2 \rangle = \frac{\langle \Delta \hat{J}_x^2 \rangle}{\langle \hat{J}_y \rangle^2} = \frac{1}{2J} = \frac{1}{N}$, which satisfy the minimum uncertainty product $\langle \Delta \hat{N}^2 \rangle \langle \Delta \hat{\Phi}^2 \rangle = 1$. The Bloch state $|\frac{\pi}{2}, \varphi\rangle$, created by the rotation $(\theta = \frac{\pi}{2}, \varphi)$, has the same property as the Bloch state $|\frac{\pi}{2}, 0\rangle$.

5.3. SQUEEZED ATOMIC STATES

Consider a coherent atomic state $|\theta, \varphi\rangle$ where $\theta = \frac{\pi}{2}$ and $\varphi = 0$. The population difference noise is $\langle \Delta \hat{N}^2 \rangle = 4\langle \Delta \hat{J}_z^2 \rangle = N$ and the dipole phase noise is $\langle \Delta \hat{\Phi}^2 \rangle = \langle \Delta \hat{J}_y^2 \rangle / \langle \hat{J}_x \rangle^2 = \frac{1}{N}$ at this point (Fig. 5-3(a)).

The Heisenberg uncertainty principle for collective angular momentum, $\langle \Delta \hat{J}_y^2 \rangle \langle \Delta \hat{J}_z^2 \rangle \geq \frac{1}{4}|\langle \hat{J}_x \rangle|^2$, does not prohibit the value of $\langle \Delta \hat{J}_y^2 \rangle$ being smaller than that for the coherent atomic state $\langle \Delta \hat{J}_y^2 \rangle = \frac{J}{2} = \frac{N}{4}$. This should be allowed as long as $\langle \Delta \hat{J}_z^2 \rangle$ is increased at the same time. The reduction of $\langle \Delta \hat{J}_y^2 \rangle$ requires the formation of negative correlation among $\Delta \hat{\sigma}_y^{(k)}$ of constituent atoms (k), which leads to the

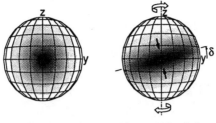

(a) Coherent Atomic State   (b) Squeezed Atomic State

FIGURE 5-3: Twisting schemes for atomic squeezing.

mutual cancellation of $\Delta\hat{\sigma}_y^{(k)}$ of each atom. At the same time, positive correlation is established among $\Delta\hat{\sigma}_z^{(k)}$ of constituent atoms, which results in the enhanced $\langle\Delta\hat{J}_z^2\rangle$. The isotropic noise distribution of the original coherent atomic state can be squeezed into the y direction and anti-squeezed into the z-direction. Such a state is called a squeezed atomic state [4] and can be generated by the interaction Hamiltonian

$$\hat{\mathcal{H}}_J = \hbar\chi\hat{J}_z^2, \tag{5.82}$$

which modulates the atomic dipole phase $\hat{J}_+$ (or $\hat{J}_-$) in proportion to the population difference $\hat{J}_z$. The dipole-dipole interaction among the constituent atoms can be utilized for this purpose. The isotropic noise distribution of the initial coherent atomic state is squeezed due to the quantum correlation between the population differenece and the dipole phase, as shown in Fig. 5-3(b).

REFERENCES

[1] C. Cohen-Tannoudji et al., *Quantum Mechanics*, Vols. 1 and 2 (Wiley, New York, 1977).

[2] W. H. Louisell, *Quantum Statistical Properties of Radiation* (Wiley, New York, 1973).

[3] F. T. Arrechi et al., Phys. Rev. **A6**, 2211 (1972).

[4] M. Kitagawa and M. Ueda, Phys. Rev. Lett. **67**, 1852 (1991).

6 Interaction between Atoms and Fields

In Chapters 3 and 5, we have seen the properties of quantized free electromagnetic fields and assembly of two-level atoms. The goal of this chapter is to start from where we left in Chapter 2: by transforming the atom-field Lagrangian into a more convenient form, we obtain in Sec. 6-1 the new conjugate momenta and the desired multipolar form of the Hamiltonian. Once again, the text by C. Cohen-Tannoudji, J. Dupont-Roc, and G. Grynberg [1] gives an excellent detailed treatment of the transformation and the quantization procedure for the interacting fields.

The second section of this chapter is devoted to the simple *second quantized* form for the atom-field Hamiltonian: here, we introduce the Jaynes–Cummings model for a single two-level atom coupled to a single-mode field in a number-state and then extend this model to N two-level atoms coupled to a single-mode field. The collective vacuum Rabi oscillation (normal mode splitting) and the concepts of dressed fermions and dressed bosons are also introduced in this section, along with a discussion of the related subjects, such as the Mollow's triplet of resonance fluorescence and the radiation trapped state. Section 6-3 treats a single two-level atom coupled to a single-mode coherent state field. The Cummings collapse and revival due to the discreteness of photon number eigenvalues are explained. Section 6-4 studies a single two-level atom interacting with a multimode vacuum state. In the limit of a continuum of radiation field modes, this model gives the Wigner-Weisskopf theory of spontaneous emission. Finally, in Section 6.5, we discuss the collective interaction of N two-level atoms and continuum vacuum states, which is the so-called superradiance problem.

6.1. ATOM-FIELD INTERACTION IN THE LENGTH GAUGE: POWER–ZIENAU–WOOLLEY TRANSFORMATION

The principal goal of this section is to obtain the electric-dipole interaction Hamiltonian that forms the basis of the Jaynes–Cummings model that we shall study subsequently. Starting from the so-called minimal coupling formalism of Chapter 2, the desired form of the atom-field interaction Hamiltonian can be rigourously obtained by employing the Power–Zienau–Woolley (PZW) transformation [2]. We start by recalling the Lagrangian of the total atom-field system given in (2.22) of Chapter 2:

$$L = L_{\text{particles}} + L_{\text{field}} + \int d^3r (\mathbf{j}(\mathbf{r}) \cdot \mathbf{A}(\mathbf{r}) - \rho(\mathbf{r}) U(\mathbf{r})), \quad (6.1)$$

where $L_{\text{particles}}$ and L_{field} represent the previously given particle and field Lagrangian, respectively; and $\mathbf{j}(\mathbf{r})$ and $\rho(\mathbf{r})$ denote the current and the charge density of the system of nonrelativistic particles.

We now assume that the system of charges form a globally neutral system, which makes the electric displacement $\mathbf{D}(\mathbf{r})$ a transverse field. In this limit, the contribution to the Lagrangian of (6.1) from the longitudinal part of the electric field gives the Coulomb energy of the particle system. Additionally, the total current density $\mathbf{j}(\mathbf{r})$ for this globally neutral system can be written as a sum of polarization and magnetization currents [1]:

$$\mathbf{j}(\mathbf{r}) = \mathbf{j}_p(\mathbf{r}) + \mathbf{j}_m(\mathbf{r}) = \dot{\mathbf{P}}(\mathbf{r}) + \nabla \times \mathbf{M}(\mathbf{r}) \tag{6.2}$$

where

$$\mathbf{P}(\mathbf{r}) = \sum_i \int_0^1 du \, q_i \mathbf{r}_i \delta(\mathbf{r} - u\mathbf{r}_i) \tag{6.3}$$

$$\mathbf{M}(\mathbf{r}) = \sum_i \int_0^1 du \, u q_i \mathbf{r}_i \times \dot{\mathbf{r}}_i \delta(\mathbf{r} - u\mathbf{r}_i). \tag{6.4}$$

Here, $\mathbf{r}_i$ denotes the coordinate of the ith particle that constitutes the particle system. The details of the derivation of the particular expansion of the total particle current and the expressions for the polarization and magnetization currents may be found elsewhere [1].

We pointed out in Chapter 2 that the Lagrangian or the Lagrangian density that results in the correct equations of motion is not unique. More specifically, if we add the total time derivative of an analytical function of *coordinates* to the Lagrangian, the Lagrange equations remain unchanged. The most well-known example is that of a gauge transformation. The PZW transformation that we want to employ changes the minimal-coupling Lagrangian by

$$L^{\text{PZW}} = L + \frac{dF}{dt} = L - \frac{d}{dt} \int d^3r \mathbf{P}(\mathbf{r}) \cdot \mathbf{A}(\mathbf{r}). \tag{6.5}$$

We note that the analytic function F depends only on the coordinates $\mathbf{r}_i$ and $\mathbf{A}(\mathbf{r})$. The contribution from the total time derivative modifies the interaction Lagrangian to give

$$L_I^{\text{PZW}} = \int d^3r \mathbf{j}(\mathbf{r}) \cdot \mathbf{A}(\mathbf{r}) - \int d^3r \left(\dot{\mathbf{P}}(\mathbf{r}) \cdot \mathbf{A}(\mathbf{r}) + \mathbf{P}(\mathbf{r}) \cdot \dot{\mathbf{A}}(\mathbf{r}) \right)$$

$$= \int d^3r \left(\mathbf{P}(\mathbf{r}) \cdot \mathbf{E}_\perp(\mathbf{r}) + \mathbf{M}(\mathbf{r}) \cdot \mathbf{B}(\mathbf{r}) \right), \tag{6.6}$$

where we have used (6.2) and (2.23) of Chapter 2 and the fact that the contribution from the last term of (6.1) vanishes for the assumed globally neutral system. Finally, the total Lagrangian of the interacting system after the transformation is

$$L^{\text{PZW}} = \sum_i \frac{1}{2} m_i \dot{\mathbf{r}}_i^2 - \sum_{i<j} \frac{q_i q_j}{4\pi\epsilon_0 |\mathbf{r}_i - \mathbf{r}_j|} - V_{\text{self}} \qquad (6.7)$$

$$+ \frac{\epsilon_0}{2} \int d^3r \left(\mathbf{E}_\perp^2(r) - c^2 \mathbf{B}^2(r) \right) + \int d^3r \left(\mathbf{P}(r) \cdot \mathbf{E}_\perp(r) + \mathbf{M}(r) \cdot \mathbf{B}(r) \right),$$

where the second term corresponds to the Coulomb energy of the interacting particles and V_{self} denotes the self Coulomb energy [1].

Even though the equations of motion that follow the new Lagrangian of (6.7) remain unchanged, the same assessment is not true in general for the conjugate momenta and the Hamiltonian. For example, the momentum conjugate with respect to the position $\mathbf{r}_i$ of the ith particle in the PZW formulation is given by

$$\mathbf{p}_i^{\text{PZW}} = \mathbf{p}_i + \nabla F^{\text{PZW}}. \qquad (6.8)$$

The wavefunction and the observables in the quantum theory that follows the new formulation will therefore be different from those of the standard formulation. More specifically, the new physical variables may be obtained from the original ones by employing the unitary transformation [1]:

$$\hat{T}^{\text{PZW}} = \exp\left[\frac{i}{\hbar} \hat{F}^{\text{PZW}} \right]. \qquad (6.9)$$

The quantum physics predicted by the standard and the PZW formulations are naturally identical: A straightforward derivation shows that the transition amplitudes remain unchanged under the action of the unitary transformation of (6.9).

To derive the new Hamiltonian, we first need to evaluate the new conjugate momenta associated with the particle and field coordinates. Using (2.5) and (6.7), we find that the momenta conjugate to $\mathbf{r}_i$ is

$$\mathbf{p}_i^{\text{PZW}} = \mathbf{p}_i = m_i \dot{\mathbf{r}}_i + \int_0^1 du\, u\, q_i \mathbf{B}(\mathbf{r}_i \cdot u) \times \mathbf{r}_i, \qquad (6.10)$$

which is in contrast to the conjugate momenta in the minimal-coupling form given by $\mathbf{p}_i = m_i \dot{\mathbf{r}}_i + q_i \mathbf{A}(\mathbf{r}_i)$. We note that the second term in the expression for $\mathbf{p}_i$ in the new Lagrangian is proportional to $|\mathbf{r}_i|$: Therefore, for a system that is localized in the length scale of an optical wavelength, the conjugate particle momenta is given by the kinetic momentum of the particle. This is a desired result that simplifies the calculations in many cases and is directly related to the long wavelength approximation that we shall introduce shortly.

Next, we evaluate the momenta $\Pi(\mathbf{r})$ conjugate to the vector potential: A straightforward calculation in the reciprocal space gives [1]

$$\Pi^{\text{PZW}}(\mathbf{r}) = \Pi(\mathbf{r}) = -\mathbf{D}(\mathbf{r}), \qquad (6.11)$$

which simply states that for the globally neutral system, the conjugate field momenta is the electric displacement. In the presence of free charges, the right-hand side is replaced by the transverse component of the displacement.

Instead of giving an expression for the new Hamiltonian, we next turn to the so-called long-wavelength approximation (LWA). To this end, we first evaluate the integral over d^3r using the delta functions coming from the expressions for the polarization and magnetization [(6.3) and (6.4), respectively]. This gives [1]

$$L_I^{PZW} = \sum_i q_i \int_0^1 du[\mathbf{r}_i \cdot \mathbf{E}_\perp(u\mathbf{r}_i) + (\mathbf{r}_i \times \dot{\mathbf{r}}_i) \cdot \mathbf{B}(u\mathbf{r}_i)u]. \qquad (6.12)$$

We then expand the electric and magnetic field terms appearing in the interaction Lagrangian of (6.12) in power series around the origin. Keeping only the first three terms, we obtain

$$L_I^{PZW} \approx \sum_i q_i \mathbf{r}_i \cdot \mathbf{E}_\perp(0) + \sum_i \frac{q_i}{2}(\mathbf{r}_i \times \dot{\mathbf{r}}_i) \cdot \mathbf{B}(0)$$
$$+ \sum_{\substack{k,\ell=x,y,z \\ i}} \frac{q_i}{2}\left(r_{i,k}r_{i,\ell} - \frac{1}{3}\delta_{k\ell}r_i^2\right)\frac{\partial}{\partial r_{i,k}}E_\ell(0). \qquad (6.13)$$

The LWA as stated previously is justified in practically all the cases that we are interested in: For atomic systems, the relevant length scale for the particles is approximately determined by the atomic Bohr radius ($a_B = 0.52$ Å), which is typically four orders of magnitude smaller than the optical wavelengths that determine the characteristic length scale of the optical fields. A similar consideration applies to the strongly allowed transitions of the Rydberg atoms as well. In the case of semiconductor-field interactions, the relevant length scale for electric-dipole transitions is approximately given by the lattice constant $a_{\text{lattice}} \simeq 4$ Å. Even though the optical wavelengths are shorter by a factor $\simeq 3.5$ due to large dielectric constants, LWA is still easily satisfied. These considerations apply for both exciton and free electron-hole transitions.

Finally, keeping only the first term in the expansion of (6.13) and applying the canonical quantization procedure, we obtain the matter-field Hamiltonian in the electric-dipole approximation

$$\hat{H}^{PZW} = H = \sum_i \frac{\hat{\mathbf{p}}_i^2}{2m_i} + \hat{V}_{\text{self}} + \sum_{i<j} \frac{q_i q_j}{4\pi\epsilon_0|\hat{\mathbf{r}}_i - \hat{\mathbf{r}}_j|} + \int d^3r \frac{|\hat{\mathbf{P}}_\perp(r)|^2}{\epsilon_0}$$
$$+ \sum_{\mathbf{k},\epsilon} \hbar\omega_\mathbf{k}\left[\hat{a}_\epsilon^\dagger(\mathbf{k})\hat{a}_\epsilon(\mathbf{k}) + \frac{1}{2}\right] - \sum_i q_i\hat{\mathbf{r}}_i \cdot \hat{\mathbf{D}}_\perp(0). \qquad (6.14)$$

The first line of (6.14) gives the Hamiltonian of the particles with the last term denoting the dipole self-energy. Equation (6.14) will form the basis of our discussion of the second-quantized atom-field Hamiltonian and the Jaynes–Cummings model that we consider in the next section. Naturally, this Hamiltonian can be directly extended to treat the interactions of second quantized matter fields with the radiation field. We shall use this many-body form of (6.14) in deriving the Hamiltonian of the interacting semiconductor-field system in Chapter 10.

6.2. JAYNES–CUMMINGS HAMILTONIAN

Consider the special case where the system of particles described in the previous section consists of a single atom with a predetermined spectrum. We assume that its eigenstates, obtained by diagonalizing the particle Hamiltonian $\hat{H}_a$ of the first line of (6.14), are denoted by $|i\rangle$, where $i = g, e, \ldots$. The calculation of this spectrum is a problem of atomic physics and is not of direct interest to us here. Without losing generality, we can write

$$\hat{H}_a |i\rangle = \hbar \omega_i |i\rangle, \tag{6.15}$$

where $\hbar \omega_i$ denotes the eigenenergy corresponding to $|i\rangle$. Using the completeness property of the eigenstates, we obtain

$$\hat{H}_a = \sum_i \hbar \omega_i |i\rangle \langle i|. \tag{6.16}$$

Next, we consider the interaction term in (6.14): If we let $\mu_{ij} = \langle i|q\hat{\mathbf{r}}|j\rangle$, we obtain [3]

$$\hat{H} = \sum_i \hbar \omega_i |i\rangle \langle i| + \sum_{\mathbf{k},\epsilon} \hbar \omega_\mathbf{k} \left[\hat{a}^\dagger_\epsilon(\mathbf{k}) \hat{a}_\epsilon(\mathbf{k}) + \frac{1}{2} \right]$$
$$+ i\hbar \sum_{\mathbf{k},\epsilon} \sum_{i,j} g_{\mathbf{k},i,j} \left[\hat{a}_\epsilon(\mathbf{k}) - \hat{a}^\dagger_\epsilon(\mathbf{k}) \right] |i\rangle \langle j|, \tag{6.17}$$

where

$$g_{\mathbf{k},i,j} = \left(\frac{\omega_\mathbf{k}}{2\epsilon_0 \hbar V} \right)^{1/2} \epsilon \cdot \mu_{ij} \tag{6.18}$$

with $V = L^3$ denoting the quantization volume as introduced earlier in Chapter 2. The interaction part of the Hamiltonian of (6.17) describes processes where the atomic transition from state $|j\rangle$ to $|i\rangle$ is accompanied by both absorption and emission of photons. The interaction coefficient $g_{\mathbf{k},i,j}$, on the other hand, is orders of magnitude smaller than the atomic transition frequency $\omega_{ij} = \omega_i - \omega_j$. Therefore, we can safely assume that only the transitions that (approximately) conserve energy are important: This is the rotating wave approximation (RWA) discussed earlier. If, in addition to RWA, we assume that only interactions of a single optical cavity mode with two atomic states $|g\rangle$ and $|e\rangle$ are important, we obtain

$$\hat{H} = \frac{1}{2}\hbar\omega|e\rangle\langle e| - \frac{1}{2}\hbar\omega|g\rangle\langle g| + \hbar v \left(a^\dagger a + \frac{1}{2}\right)$$
$$+ \hbar g \left(|e\rangle\langle g|\hat{a} + \hat{a}^\dagger|g\rangle\langle e|\right), \quad (6.19)$$

where we have taken $\omega = \omega_{eg}$ and introduced the cavity-mode energy $\hbar v$. In addition, we have introduced a trivial phase factor in the interaction term. The coupling coefficient g directly follows (6.18). This is the celebrated Jaynes–Cummings model. In the following sections, we will discuss the physics and extensions of this model.

6.3. INTERACTION BETWEEN TWO-LEVEL ATOM AND SINGLE-MODE PHOTON NUMBER STATE

6.3.1. Vacuum Rabi Oscillation

The eigenstates of the unperturbed Hamiltonian, the first and second terms of (6.19), are given by

$$\hat{H}_0|e, n\rangle = \hbar \left(\frac{1}{2}\omega + \left(n + \frac{1}{2}\right)v\right)|e, n\rangle, \quad (6.20)$$

$$\hat{H}_0|g, n+1\rangle = \hbar \left(-\frac{1}{2}\omega + \left(n + \frac{3}{2}\right)v\right)|g, n+1\rangle, \quad (6.21)$$

where $|e, n\rangle$ denotes the state with the excited state atom and n photons, and $|g, n+1\rangle$ represents the state with the ground state atom and $n + 1$ photons. The interaction Hamiltonian, the third term of (6.19), couples the atom-field bare states $|e, n\rangle$ and $|g, n + 1\rangle$ in the same ($n + 1$ excitation) manifold, but does not couple states in different manifolds due to the rotating wave approximation. Therefore, the combined state in the ($n + 1$) excitation manifold is represented by

$$|\psi(t)\rangle = \left[C_{e,n}(t)|e, n\rangle + C_{g,n+1}(t)|g, n+1\rangle\right]e^{-iv(n+1)t}, \quad (6.22)$$

where $C_{e,n}(t)$ and $C_{g,n+1}(t)$ are the slowly varying excitation amplitudes. The Schrödinger equation for the combined state vector is

$$i\hbar\frac{d}{dt}|\psi(t)\rangle = \hat{H}|\psi(t)\rangle. \quad (6.23)$$

Using the relations, $\hat{a}^\dagger\hat{\sigma}_-|e, n\rangle = \sqrt{n+1}|g, n+1\rangle$ and $\hat{a}\hat{\sigma}_+|g, n+1\rangle = \sqrt{n+1}|e, n\rangle$ in (6.23), where $\hat{\sigma}_- = |g\rangle\langle e|$ and $\hat{\sigma}_+ = |e\rangle\langle g|$, we obtain

120 INTERACTION BETWEEN ATOMS AND FIELDS

$$i\hbar \left[\dot{C}_{e,n}|e,n\rangle + \dot{C}_{g,n+1}|g,n+1\rangle \right]$$
$$= \hbar \left[\frac{\omega - \nu}{2} C_{e,n}|e,n\rangle + \frac{\nu - \omega}{2} C_{g,n+1}|g,n+1\rangle \right]$$
$$- \hbar g \sqrt{n+1} \left[C_{e,n}|g,n+1\rangle + C_{g,n+1}|e,n\rangle \right]. \quad (6.24)$$

Projecting the states $|e,n\rangle$ and $|g,n+1\rangle$ onto (6.24), we have

$$\dot{C}_{e,n} = -i\frac{\delta}{2} C_{e,n} - ig\sqrt{n+1}\, C_{g,n+1}, \quad (6.25)$$

$$\dot{C}_{g,n+1} = i\frac{\delta}{2} C_{g,n+1} - ig\sqrt{n+1}\, C_{e,n}. \quad (6.26)$$

We can write (6.25) and (6.26) as a single-vector equation

$$\frac{d}{dt}[C] = \frac{i}{2}[M][C], \quad (6.27)$$

where

$$[C] = \begin{pmatrix} C_{e,n} \\ C_{g,n+1} \end{pmatrix}, \quad (6.28)$$

$$[M] = \begin{pmatrix} -\delta & -R_0 \\ -R_0 & \delta \end{pmatrix}. \quad (6.29)$$

Here, $\delta = \omega - \nu$ is the atomic transition frequency detuning from the cavity resonant frequency and $\frac{R_0}{2} = g\sqrt{n+1}$ is the (quantum) Rabi oscillation frequency. When $n = 0$ (vacuum field), $\frac{R_0}{2}$ is reduced to the vacuum Rabi frequency Ω_R. Equation (6.27) has a solution of the form $\exp\left(\frac{1}{2}i\lambda t\right)$, which diagonalizes the matrix. Substituting $[C(t)] = e^{\frac{1}{2}i\lambda t}[C(0)]$ into (6.27), we obtain

$$\det[M - \lambda I] = 0, \quad (6.30)$$

using the generalized Rabi oscillation frequency $\frac{R}{2} = \frac{1}{2}\sqrt{R_0^2 + \delta^2}$. The general solution for coupled differential equations (6.25) and (6.26) can be written as

$$\begin{bmatrix} C_{e,n}(t) \\ C_{g,n+1}(t) \end{bmatrix} = \begin{bmatrix} \cos\frac{R}{2}t - i\frac{\delta}{R}\sin\frac{R}{2}t & i\frac{R_0}{R}\sin\frac{R}{2}t \\ i\frac{R_0}{R}\sin\frac{R}{2}t & \cos\frac{R}{2}t + i\frac{\delta}{R}\sin\frac{R}{2}t \end{bmatrix} \begin{bmatrix} C_{e,n}(0) \\ C_{g,n+1}(0) \end{bmatrix}.$$
$$(6.31)$$

The excitation of the two bare states $|e, n\rangle$ and $|g, n+1\rangle$ changes periodically. In the special case of zero detuning, $\delta = 0$, the probability $|C_{e,n}|^2$ of finding the atom in the excited state $|e\rangle$ and the field in the $|n\rangle$ state and the probability $|C_{g,n+1}|^2$ of finding the atom in the ground state $|g\rangle$ and the field in the $|n+1\rangle$ state oscillate according to

$$|C_{e,n}(t)|^2 = \cos^2(g\sqrt{n+1}\,t), \tag{6.32}$$

$$|C_{g,n+1}(t)|^2 = \sin^2(g\sqrt{n+1}\,t). \tag{6.33}$$

Here, $C_{e,n}(0) = 1$ and $C_{g,n+1}(0) = 0$ are assumed. When the initial cavity field is a vacuum state $|0\rangle$, the vacuum field fluctuation effectively stimulates the excited atom to emit a photon. This process is spontaneous emission, because there is no existing photon stimulating the photon emission. However, this spontaneous emission process is different from the usual irreversible spontaneous emission process with exponential decay. In the usual process an excited atom in free space couples with a continuous spectrum of vacuum field fluctuations that attempt to create Rabi oscillations with slightly different frequencies. The resulting upper-level probabilities interfere destructively due to the different Rabi oscillation frequencies $\frac{R}{2} = \frac{1}{2}\sqrt{R_0^2 + (\omega - \nu)^2}$, providing the exponential decay. On the other hand, if an atom interacts with a single field mode in a high-Q cavity, the phase coherence is preserved between the atomic dipole and the field, so the reversible spontaneous emission occurs.

6.3.2. Normal Mode Splitting

The (total) Hamiltonian of this system is given in the bare state $\{|e, n\rangle |g, n+1\rangle\}$ basis by the matrix form

$$\hat{H} = \hbar(n+1)\nu \begin{bmatrix} 1 & 0 \\ 0 & 1 \end{bmatrix} + \frac{\hbar}{2} \begin{bmatrix} \delta & 2g\sqrt{n+1} \\ 2g\sqrt{n+1} & -\delta \end{bmatrix}. \tag{6.34}$$

We can diagonalize this matrix in the same manner as the vector equation (6.30). The new energy eigenvalues for such a diagonalized Hamiltonian are

$$E_{2n} = \hbar(n+1)\nu - \frac{1}{2}\hbar R, \tag{6.35}$$

$$E_{1n} = \hbar(n+1)\nu + \frac{1}{2}\hbar R. \tag{6.36}$$

The eigenvectors corresponding to (6.35) and (6.36) are

$$|2n\rangle = \cos\theta_n |e, n\rangle - \sin\theta_n |g, n+1\rangle, \tag{6.37}$$

$$|1n\rangle = \sin\theta_n |e, n\rangle + \cos\theta_n |g, n+1\rangle, \tag{6.38}$$

where

$$\cos 2\theta_n = -\frac{\delta}{R}, \qquad (6.39)$$

$$\sin 2\theta_n = \frac{R_0}{R}. \qquad (6.40)$$

At resonance ($\omega = \nu$), the $|e, n\rangle$ and $|g, n+1\rangle$ states have equal contributions for states $|1n\rangle$ and $|2n\rangle$:

$$|1n\rangle = \frac{1}{\sqrt{2}}(|e,n\rangle + |g, n+1\rangle) \quad \text{(even mode)}, \qquad (6.41)$$

$$|2n\rangle = \frac{1}{\sqrt{2}}(|e,n\rangle - |g, n+1\rangle) \quad \text{(odd mode)}. \qquad (6.42)$$

States $|1n\rangle$ and $|2n\rangle$ are termed dressed states (or normal modes), while states $|e, n\rangle$ and $|g, n+1\rangle$ are termed bare states (or coupled modes).

Although bare states feature (quantum) Rabi oscillation (as shown in the previous section), dressed states are eigenstates of the total Hamiltonian and the state vectors simply have phase rotation at their eigenfrequencies $(n+1)\nu + \frac{R}{2}$ and $(n+1)\nu - \frac{R}{2}$. The energy exchange behavior of bare states is simply the quantum interference effect of the two phase rotating dressed states; that is, if the atom-field system is prepared in the bare state $|e, n\rangle$ at $t = 0$ and the interaction Hamiltonian is switched on, the two dressed states $|1n\rangle$ and $|2n\rangle$ are excited in phase with equal amplitudes. After a time $t = \frac{\pi}{2g\sqrt{n+1}}$, the probability amplitudes of states $|1n\rangle$ and $|2n\rangle$ become out of phase due to their different eigenfrequencies. If the combined state is projected onto the bare states, the full excitation of $|g, n+1\rangle$ and zero excitation of $|e, n\rangle$ are obtained.

6.4. INTERACTION BETWEEN N TWO-LEVEL ATOMS AND SINGLE-MODE PHOTON NUMBER STATE

6.4.1. Collective Rabi Oscillation and Normal Mode Splitting

Consider an assembly of N two-level atoms interacting with a single-mode field in a photon number eigenstate. The (collective) angular momentum eigenstates (Dicke states) that describe the N two-level atoms are denoted as

$$|-J\rangle = |g\rangle_1|g\rangle_2\cdots|g\rangle_N, \qquad (6.43)$$

$$|-J+1\rangle = \frac{1}{\sqrt{N}}\{|e\rangle_1|g\rangle_2\cdots|g\rangle_N + |g\rangle_1|e\rangle_2\cdots|g\rangle_N + \cdots + |g\rangle_1|g\rangle_2\cdots|e\rangle_N\}, \qquad (6.44)$$

where $J = \frac{N}{2}$. The state $|-J\rangle$ corresponds to the case in which all the atoms are in the ground state $|g\rangle$ and the state $|-J+1\rangle$ corresponds to the case in which only one atom is in the excited state $|e\rangle$. The total Hamiltonian is written as

$$\hat{\mathcal{H}} = \frac{1}{2}\hbar\omega \hat{J}_z + \hbar\nu\left(\hat{a}^\dagger \hat{a} + \frac{1}{2}\right) + \hbar g(\hat{a}\hat{J}_+ + \hat{a}^\dagger \hat{J}_-), \qquad (6.45)$$

where the single-atom vacuum Rabi frequency g is as given in (6.18). The Hamiltonian of (6.45) is a simple N-atom extension of (6.19), obtained in the high-temperature limit where quantum statistical effects are negligible. The collective angular momentum operators $(\hat{J}_z, \hat{J}_+, \hat{J}_-)$ are as defined in Chapter 5.

The combined atom-field state for the one excitation manifold is expanded by

$$|\psi(t)\rangle = \left[C_{-J,1}(t)|-J, 1\rangle + C_{-J+1,0}(t)|-J+1, 0\rangle\right] e^{-i\frac{3}{2}\nu t}. \qquad (6.46)$$

Here the state $|-J, 0\rangle$ stands for the combined Dicke state $|-J\rangle$ and Fock (vacuum) state $|0\rangle$. It is assumed that the atomic transition frequency is identical to the field frequency, $\omega = \nu$. Using the recursion relations for the Dicke and Fock states, the following projection rules are obtained:

$$\hat{a}\hat{J}_+|-J, 1\rangle = \sqrt{N}|-J+1, 0\rangle, \qquad (6.47)$$

$$\hat{a}^\dagger \hat{J}_-|-J+1, 0\rangle = \sqrt{N}|-J, 1\rangle. \qquad (6.48)$$

Using (6.46), (6.47), and (6.48) in the Schrödinger equation, we obtain

$$i\hbar \left[\dot{C}_{-J,1}|-J, 1\rangle + \dot{C}_{-J+1,0}|-J+1, 0\rangle\right]$$
$$= -\hbar g \sqrt{N} \left[C_{-J+1,0}|-J, 1\rangle + C_{-J,1}|-J+1, 0\rangle\right]. \qquad (6.49)$$

By projecting $\langle -J, 1|$ and $\langle -J+1, 0|$ onto both sides of (6.49), the coupled mode equations for the two probability amplitudes are obtained:

$$\dot{C}_{-J,1} = ig\sqrt{N}\, C_{-J+1,0}, \qquad (6.50)$$

$$\dot{C}_{-J+1,0} = ig\sqrt{N}\, C_{J,1}. \qquad (6.51)$$

The vacuum Rabi frequency is enhanced by a factor of $\sqrt{N}$. When N unexcited atoms are probed by a weak external field, the normal mode splitting is equal to

$$R_{\text{collective}} = 2g\sqrt{N}. \qquad (6.52)$$

This is a collective vacuum Rabi oscillation or normal mode splitting. A single-atom vacuum Rabi splitting $R_0 = 2g$ is often too small to observe. Due to the $\sqrt{N}$ enhancement, on the other hand, the collective vacuum Rabi splitting $R_{\text{collective}} = 2g\sqrt{N}$ is large enough to observe using moderately good optical cavities.

6.4.2. Dressed Fermions and Dressed Bosons

When the excitation manifold is larger than one, the combined system cannot be expanded by only two atom-field state vectors as is done in (6.46). However, if the excitation manifold n is much smaller than the total number of atoms N, the system still behaves linearly and the simple dressed boson picture is applicable. Figure 6-1(a) shows the bare states and the dressed states for the $n = 0$ (system ground state), $n = 1$ (one excitation), and $n = 2$ (two excitation) manifolds of N two-level atoms and the single-mode field. The two degenerate bare states $|-J, 1\rangle$ and $|-J+1, 0\rangle$ for the $n = 1$ excitation manifold form the two nondegenerate dressed states at energies $\hbar(\nu + \sqrt{N}g)$ and $\hbar(\nu - \sqrt{N}g)$:

$$|+, 1\rangle = \frac{1}{\sqrt{2}}[|-J, 1\rangle + |-J+1, 0\rangle], \tag{6.53}$$

$$|-, 1\rangle = \frac{1}{\sqrt{2}}[|-J, 1\rangle - |-J+1, 0\rangle]. \tag{6.54}$$

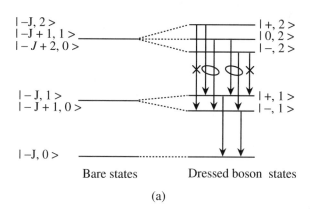

(a)

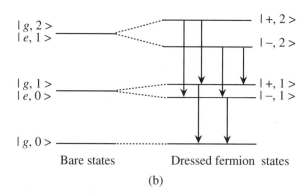

(b)

FIGURE 6-1: Dressed boson states (a) and dressed fermion states (b).

The three degenerate bare states $|-J, 2\rangle$, $|-J+1, 1\rangle$, and $|-J+2, 0\rangle$ for the $n = 2$ excitation manifold form the three nondegenerate dressed states at three energies $\hbar(2\nu + 2\sqrt{N} g)$, $\hbar(2\nu)$, and $\hbar(2\nu - 2\sqrt{N} g)$ [4]:

$$|+, 2\rangle = \frac{1}{2}|-J+2, 0\rangle + \frac{1}{\sqrt{2}}|-J+1, 1\rangle + \frac{1}{2}|-J, 2\rangle, \qquad (6.55)$$

$$|0, 2\rangle = -\frac{1}{\sqrt{2}}|-J+2, 0\rangle + \frac{1}{\sqrt{2}}|-J, 2\rangle, \qquad (6.56)$$

$$|-, 2\rangle = \frac{1}{2}|-J+2, 0\rangle - \frac{1}{\sqrt{2}}|-J+1, 1\rangle + \frac{1}{2}|-J, 2\rangle. \qquad (6.57)$$

The vertical transition due to spontaneous decay from the $n = 2$ excitation manifold to the $n = 1$ excitation manifold has the six transition lines indicated by the arrows in Fig. 6-1(a).

However, the matrix elements for the two outlying transitions (highest and lowest transition energies) are identically equal to zero [4]:

$$\langle -, 1|\hat{J}_-|+, 2\rangle = 0$$
$$\langle +, 1|\hat{J}_-|-, 2\rangle = 0. \qquad (6.58)$$

The remaing four transition lines are pairwise degenerate and the two allowed transition frequencies, $\nu + \sqrt{N} g$ and $\nu - \sqrt{N} g$, are identical to those from the $n = 1$ excitation manifold to the $n = 0$ ground state. It is straightforward to show that the same conclusion applies to the transition from the arbitrary n excitation manifold to the $n - 1$ excitation manifold as long as $n \ll N$. When the excitation n is much smaller than the total number of atoms N, the system behaves linearly and the emission spectrum is identical to the doublet obtained from the $n = 1$ to $n = 0$ manifold transition. In fact, the angular momentum operators approximately satisfy the bosonic commutation relation in this limit (i.e., $[\hat{J}_-, \hat{J}_+] = 2\hat{J}_z \simeq N$). We will see in Chapter 11 that this dressed boson picture is a useful concept for describing the emission properties of the exciton polaritons in a microcavity.

When a single two-level atom interacts with a single-mode field in a photon number eigenstate, the system is inherently nonlinear because of the fermionic feature of the Pauli spin operator. For instance, the two bare states $|e, n\rangle$ and $|g, n+1\rangle$ for the $(n+1)$ excitation manifold form the two nondegenerate dressed states $|+, n+1\rangle$ and $|-, n+1\rangle$ at energies $\hbar[(n+1)\nu + \sqrt{n+1} g]$ and $\hbar[(n+1)\nu - \sqrt{n+1} g]$. This is a dressed fermion. There are four allowed transition lines from the $(n+1)$ excitation manifold to n excitation manifold for $n \geq 1$, as shown in Fig. 6-1(b).

6.4.3. Atomic Cavity Quantum Electrodynamics (QED)

There are two types of experimental schemes to observe vacuum Rabi oscillation and normal mode splitting. One scheme is atom-type spectroscopy, in which a probe wave excites an atom by side illumination, as shown in Fig. 6-2(a). In this case, the

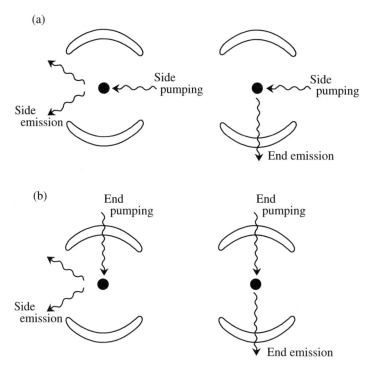

FIGURE 6-2: (a) The configurations in an atom-type spectroscopy. (b) The configurations in a cavity-type spectroscopy.

bare state $|-J+1, 0\rangle$ is initially excited at $t=0$, which corresponds to the simultaneous excitation of the two dressed states $|+, 1\rangle$ and $|-, 1\rangle$ with the probability amplitudes $\sin\theta_n$ and $\cos\theta_n$, respectively. In the case of zero detuning, the two normal modes are equally excited because $\theta_n = \frac{\pi}{4}$. In the case of nonzero detuning, state $|+, 1\rangle$ is dominantly excited for $\omega > \nu$ and state $|-, 1\rangle$ is dominantly excited for $\omega < \nu$ (that is, the atomlike state is primarily excited) and the scattered light features the asymmetric split spectrum. The side emission in the configuration shown in Fig. 6-2(a) corresponds to the decay of the bare atom state, so the transition amplitude is calculated by $\langle -J, 0|\hat{J}_-|\pm, 1\rangle$. On the other hand, if the emission is measured along the cavity axis (end emission), the transition amplitude corresponding to the decay of the bare photon state, $\langle -J, 0|\hat{a}|\pm, 1\rangle$, should be used. The emission spectra for the two cases are generally different.

The other experimental scheme is cavity-type spectroscopy, in which a probe wave excites a cavity internal field by end illumination, as shown in Fig. 6-2(b). In this case, the bare state $|-J, 1\rangle$ is initially excited at $t=0$. The two dressed states $|+, 1\rangle$ and $|-, 1\rangle$ are simultaneously excited with the probability amplitudes $\cos\theta_n$ and $-\sin\theta_n$, respectively. In this case, the cavity-like state is primarily excited. In the case of nonzero detuning, the scattered light features the opposite asymmetry.

The split spectrum has complementary physical interpretation. From the atom viewpoint, the atom is dressed by the electromagnetic vacuum field and periodically emits and absorbs a virtual photon that introduces a frequency split [1]. From the cavity viewpoint, the atom in the cavity modulates the resonant frequency of the cavity through its refractive index (the atomic dispersion characteristics), which introduces a split resonant frequency of the cavity.

The N-dependent normal mode splitting was experimentally observed in various experimental configurations. The recent text by P. R. Berman [5] provides an excellent overview of the current state of the atomic cavity QED experiments.

6.4.4. Mollow's Triplet

When the excitation manifold n becomes much greater than one for a dressed fermion system (single two-level atom), the normal mode splitting for the $(n + 1)$ excitation manifold and that for the n excitation manifold are approximately identical (i.e., $2g\sqrt{n+1} \simeq 2g\sqrt{n}$). In such a case the two central transition lines from $|+, n+1\rangle$ to $|+, n\rangle$ and from $|-, n+1\rangle$ to $|-, n\rangle$ are almost degenerate. Therefore, the emission spectrum features the three peaks (instead of four peaks in the weak excitation regime) with frequency separation of $2g\sqrt{n}$ (Fig. 6-3). The intensity ratios of the three peaks are $1 : 2 : 1$.

An atom in free space, which is illuminated by a strong coherent state field, also features the three peak emission spectrum [6]. This is because the highly excited coherent state $|\alpha\rangle$ has a relatively narrow (Poisson) photon number distribution around its average value $|\alpha|^2$, and thus the emission peak separation is approximately equal to $2g|\alpha|$. This is called the *Mollow's triplet*. As is evident from the preceding argument, the Mollow's triplet is the emission characteristic of a single two-level atom in the classical limit. In Chapter 7, we will derive the resonance fluorescence spectrum in this limit using optical Bloch equations and the quantum regression theorem.

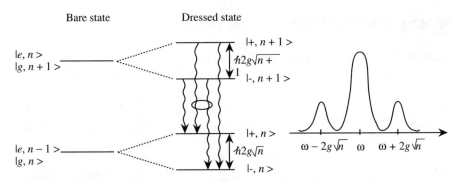

FIGURE 6-3: A dressed fermion in high-excitation manifold and the Mollow's triplet.

6.4.5. Radiation Trapped State

When two two-level atoms (a, b) interact with a single-mode field, there are three relevant angular momentum eigenstates $|1, 1\rangle = |e\rangle_a |e\rangle_b$, $|0, 1\rangle = \frac{1}{\sqrt{2}}(|e\rangle_a |g\rangle_b + |g\rangle_a |e\rangle_b)$, and $|-1, 1\rangle = |g\rangle_a |g\rangle_b$. These Dicke states are coupled with each other by the electric dipole interaction $\hat{\mathcal{H}}_I = \hbar g(\hat{a}\hat{J}_+ + \hat{a}^\dagger \hat{J}_-)$. We can also construct the antisymmetric linear superposition state according to

$$|0, 0\rangle = \frac{1}{\sqrt{2}}(|e\rangle_a |g\rangle_b - |g\rangle_a |e\rangle_b). \tag{6.59}$$

Even though the symmetric linear superposition state $|0, 1\rangle$ couples to the $|1, 1\rangle$ and $|-1, 1\rangle$ states, this antisymmetric linear superposition state $|0, 0\rangle$ does not couple to the other states because the electric dipole matrix elements are identically equal to zero:

$$\langle -1, 1|\hat{J}_-|0, 0\rangle = 0,$$
$$\langle 1, 1|\hat{J}_+|0, 0\rangle = 0. \tag{6.60}$$

When one prepares the atom a in an excited state $|e\rangle_a$ and the atom b in a ground state $|g\rangle_b$ and put them together into the same cavity at $t = 0$, the two normal modes $|0, 1\rangle$ and $|0, 0\rangle$ are excited with equal probabilities. The symmetric state $|0, 1\rangle$ decays to the ground $|-1, 1\rangle$ state by emitting a photon, but the antisymmetric state $|0, 0\rangle$ does not. The two atoms emit and absorb the photon in between them and the photon never escapes from the two-atom system. This is called a radiation trapped state or dark state [7].

The mutually coupled states $|1, 1\rangle$, $|0, 1\rangle$ and $|-1, 1\rangle$ are called triplet states, and the isolated state $|0, 0\rangle$ is called a singlet state. A detailed discussion on the subject can be found in Ref. [1].

6.5. CUMMINGS COLLAPSE AND REVIVAL

Suppose that a two-level atom initially in an excited state interacts with a single-mode field initially in a coherent state. The combined state is represented by

$$|\psi(t)\rangle = \sum_{n=0}^{\infty} C_{e,n}(t) \frac{e^{-\frac{|\alpha|^2}{2}} \alpha^n}{\sqrt{n!}} |e, n\rangle + \sum_{n=0}^{\infty} C_{g,n+1}(t) \frac{e^{-\frac{|\alpha|^2}{2}} \alpha^n}{\sqrt{n!}} |g, n+1\rangle. \tag{6.61}$$

At $t = 0$, $C_{e,n} = 1$ and $C_{g,n+1} = 0$ for all n and at $t \neq 0$, the probability amplitude $C_{e,n}$ decreases according to $C_{e,n} = \cos(g\sqrt{n+1}\,t)$ if there is no detuning (i.e.,

$\omega = \nu$). The probability of finding the atom in the excited state at $t \neq 0$ is

$$P_e = e^{-|\alpha|^2} \sum_{n=0}^{\infty} \frac{|\alpha|^{2n}}{n!} \cos^2(g\sqrt{n+1}\,t). \qquad (6.62)$$

It may be shown that the Rabi oscillation for $\omega = \nu$ (zero detuning) collapses in a short time range $t \ll |\alpha|/g$ according to [8]

$$P_e \simeq \frac{1}{2} + \frac{1}{2} \cos(2|\alpha|gt) e^{-\frac{1}{2}g^2 t^2}. \qquad (6.63)$$

Intuitively, this result can be understood as follows: The initial field in a coherent state simultaneously occupies different photon number eigenstates, and thus the Rabi oscillation frequency ranges from approximately $g\sqrt{\langle n \rangle + \Delta n} = g\sqrt{|\alpha|^2 + |\alpha|}$ to $g\sqrt{\langle n \rangle - \Delta n} = g\sqrt{|\alpha|^2 - |\alpha|}$ as shown in Fig. 6-4(a), because the photon number variance of the coherent state is given by $\Delta n = \langle n \rangle^{1/2} = |\alpha|$. The probability (6.62) dephases in a time scale

$$t_{\text{collapse}}^{-1} \simeq g\sqrt{|\alpha|^2 + |\alpha|} - g\sqrt{|\alpha|^2 - |\alpha|} \simeq g, \qquad (6.64)$$

since the phase difference between the photon number components $\langle n \rangle - \Delta n$ and $\langle n \rangle + \Delta n$ becomes one radian. The Rabi oscillation is damped with a Gaussian envelope independent of the average photon number, as shown in Fig. 6-4(b). This damped oscillation is termed the Cummings collapse and is due to the destructive interference among the probability amplitudes at different Rabi oscillation frequencies.

The collapse function for the off-resonant case ($\omega \neq \nu$) has an $\langle n \rangle$ dependence and is considerably slower than the on-resonant case ($\omega = \nu$) [9]:

$$\left\{\sqrt{g^2(|\alpha|^2 + |\alpha|) + \frac{\delta^2}{4}} - \sqrt{g^2(|\alpha|^2 - |\alpha|) + \frac{\delta^2}{4}}\right\} t = 1 \longrightarrow t_{\text{collapse}}^{-1}$$

$$\simeq \frac{g|\alpha|}{\sqrt{g^2|\alpha|^2 + \frac{\delta^2}{4}}} \cdot g, \qquad (6.65)$$

so the Gaussian damping term becomes

$$\exp\left(-\frac{1}{2}g^2 t^2\right) \longrightarrow \exp\left[\left(-\frac{2\langle n \rangle g^2}{\delta^2 + 4\langle n \rangle g^2}\right) g^2 t^2\right]. \qquad (6.66)$$

The slowing down of the Cummings collapse indicated by (6.66) stems from the fact that the Rabi oscillation frequency becomes insensitive to the photon number when the detuning δ becomes large.

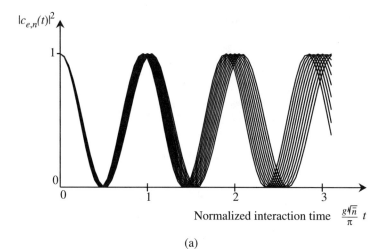

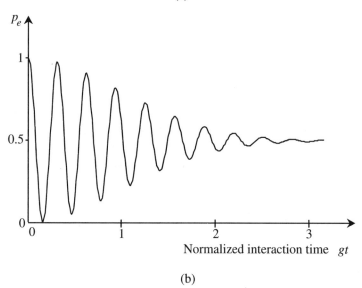

FIGURE 6-4: (a) The Rabi oscillations $|C_{e,n}(t)|^2$ for different photon number eigenstate components. (b) The probability P_e for the excited atom interacting with a coherent state field (Cummings collapse).

An even more striking feature of this coupled two-level atom and coherent-state system is found in the longer time regime $t \geq \frac{|\alpha|}{g}$ [9]. The Rabi oscillation revives in a time

$$2\sqrt{g^2(n+1) + \frac{\delta^2}{4}}\, t - 2\sqrt{g^2 n + \frac{\delta^2}{4}}\, t = 2\pi \longrightarrow t_{\text{revival}} \simeq \frac{\pi}{g^2}\sqrt{\delta^2 + 4g^2 \langle n \rangle}.$$

(6.67)

At the first and the successive revival times given by $k t_{\text{revival}} (k = 1, 2, 3, \ldots,)$ the probability amplitudes for different photon number eigenstates approximately rephase and constructive interference occurs among them. In fact, (6.67) gives the condition that the phase difference between the probability amplitudes for the two adjacent photon number eigenstates $|n\rangle$ and $|n-1\rangle$ is equal to 2π.

This revival property is an unambiguous quantum effect. The discrete Rabi frequencies in (6.62), arising from the discrete eigenvalues of the photon number operator, allow neighboring terms to recover the original phase relation at the revival time (6.67). If the field intensity had a continuous spread (like a classical intensity fluctuation), the Rabi oscillation would collapse but never revive because the Rabi oscillation frequency would be continuously spread in such a case. The revival effect was experimentally observed in a single two-level atom coupled to a single-mode thermal field [5]. The second-order coherence of the field does not have any effect on the revival property. The statistical mixture of photon number states and the linear superposition state induces identical collapse and revival properties.

6.6. WEISSKOPF–WIGNER THEORY OF SPONTANEOUS EMISSION

In the previous sections, we discussed an excited atom interacting with a single-mode field. In free space an atom interacts with a continuum of radiation field modes, which leads to an irreversible exponential decay of the excitation. The physical origin for such irreversible spontaneous emission rather than reversible vacuum Rabi oscillation is somewhat similar to the Cummings collapse. Each field mode intends to create reversible spontaneous emission at a different Rabi oscillation frequency, which leads to destructive interference between the probability amplitudes for different modes. The field mode spectrum is continuous for spontaneous emission in free space, so that the Rabi oscillation never rephases to revive, and thus irreversible spontaneous decay occurs. In this section, the Weisskopf–Wigner theory of spontaneous emission for a single two-level atom and the superradiance from N atoms in free space are reviewed.

The total Hamiltonian for this system in the rotating wave approximation (RWA) is given by

INTERACTION BETWEEN ATOMS AND FIELDS

$$\hat{H} = \frac{1}{2}\hbar\omega\hat{\sigma}_z + \sum_s \hbar v_s \left(\hat{a}_s^\dagger \hat{a}_s + \frac{1}{2}\right) - \sum_s \hbar g_s \left(\hat{a}_s\hat{\sigma}_+ + \hat{a}_s^\dagger \hat{\sigma}_-\right), \tag{6.68}$$

where $\hat{\sigma}_z = |e\rangle\langle e|$, $\hat{\sigma}_+ = |e\rangle\langle g|$, and $\hat{\sigma}_- = |g\rangle\langle e|$. The combined atom-field system is represented by

$$|\psi(t)\rangle = C_{e,0}(t)e^{-i\omega_a t}|e,0\rangle + \sum_s C_{g,1_s}(t)e^{-i(\omega_b + v_s)t}|g,1_s\rangle, \tag{6.69}$$

where $|g,1_s\rangle$ represents the state in which the atom is in the lower state and the field mode s has one photon while all the other modes are in the vacuum state. Substituting (6.69) into the Schrödinger equation and projecting into each state, we obtain

$$\dot{C}_{e,0} = -i \sum_s g_s e^{-i(v_s - \omega)t} C_{g,1_s}, \tag{6.70}$$

$$\dot{C}_{g,1_s} = -i g_s e^{i(v_s - \omega)t} C_{e,0}. \tag{6.71}$$

Carrying out the formal integration of (6.71) and inserting into (6.70), we obtain

$$\dot{C}_{e,0} = -\sum_s |g_s|^2 \int_0^t dt' \, e^{-i(v_s - \omega)(t - t')} C_{e,0}(t'). \tag{6.72}$$

Consider a cubic cavity with a volume $V = L^3$, where the resonant modes have discrete wavenumbers:

$$k_x = \frac{2\pi}{L} n_x, \tag{6.73}$$

$$k_y = \frac{2\pi}{L} n_y, \tag{6.74}$$

$$k_z = \frac{2\pi}{L} n_z. \tag{6.75}$$

$(n_x, n_y, n_z:$ integers$)$

The number of resonant modes in the wavenumber range between k and $k + dk$ is

$$dn = \left(\frac{L}{2\pi}\right)^3 d^3k. \tag{6.76}$$

In the large volume limit, $V = L^3 \to \infty$, the summation $\sum_s$ in (6.72) is replaced by the integral

$$\sum_s g_s^2 \longrightarrow \frac{V}{(2\pi)^3} \int d^3k \, g^2(k). \tag{6.77}$$

In the polar coordinates (k, θ, φ),

$$\begin{aligned} d^3k &= dk \cdot k\, d\theta \cdot k \sin\theta \cdot d\varphi \\ &= k^2 \sin\theta\, dk\, d\theta\, d\varphi \\ &= \frac{v_s^2}{c^3} \sin\theta\, dv_s\, d\theta\, d\varphi, \end{aligned} \qquad (6.78)$$

where $k = \frac{v_s}{c}$ is used. Equation (6.77) is then rewritten as

$$\sum_s g_s^2 = \frac{V}{(2\pi)^3} \int dv_s \frac{v_s^2}{c^3} \int_0^\pi d\theta \sin\theta \int_0^{2\pi} d\varphi\, g^2(v,\theta), \qquad (6.79)$$

where θ is the angle between the wavevector of the field and the atomic dipole moment oriented along z-axis. Inserting (6.79) into (6.72), we obtain

$$\begin{aligned} \dot{C}_{e,0} &= -\frac{V}{(2\pi c)^3} \int dv_s\, v_s^2 \int_0^\pi d\theta \sin\theta \int_0^{2\pi} d\varphi\, |g_s|^2 \int_0^t dt'\, e^{-i(v_s-\omega)(t-t')} C_{e,0}(t') \\ &= -\frac{1}{6\varepsilon_0 \pi^2 \hbar c^3} \int dv_s\, v_s^3 \mu^2 \int_0^t dt'\, e^{-i(v_s-\omega)(t-t')} C_{e,0}(t'). \end{aligned} \qquad (6.80)$$

In order to solve (6.80), we use the Laplace transform defined by

$$\overline{C(s)} = \int_0^\infty e^{-st} C(t)\, dt. \qquad (6.81)$$

The Laplace transform on the left-hand side (LHS) of (6.80) is given by

$$\int_0^\infty e^{-st} \left[\frac{d}{dt} C_{e,0}(t) \right] = s\overline{C_{e,0}(s)} - 1. \qquad (6.82)$$

The Laplace transform of the integral on the right-hand side (RHS) of (6.80) is given by

$$\int_0^\infty e^{-st} \left[\int_0^t dt'\, e^{i\Omega(t-t')} C_{e,0}(t)' \right] dt = \frac{\overline{C_{e,0}(s)}}{s - i\Omega}. \qquad (6.83)$$

Using these two relations, we have

$$\overline{C_{e,0}(s)} = \left[s + i\frac{M^2}{\omega - v_s + is} \right]^{-1} \equiv \Delta(s)^{-1}, \qquad (6.84)$$

where $\frac{M^2}{w - v_s + is} = \frac{1}{6\pi^2 \varepsilon_0 \hbar c^3} \int dv_s \frac{\mu^2 v_s^3}{w - v_s + is}$. The inverse Laplace transform is

$$C_{e,0}(t) = \frac{1}{2\pi i} \int_{\varepsilon-i\infty}^{\varepsilon+i\infty} e^{st} \overline{C_{e,0}(s)} \, ds$$

$$= \exp\left[-\left(\frac{1}{2}\Gamma + i\Delta\omega\right)t\right], \tag{6.85}$$

which is the desired result. In order to solve the pole of $\overline{C_{e,0}(s)}$ [the zero of $\Delta(s)$], we first assumed that, if the atom-field interaction is weak ($M^2 \to 0$), $\Delta(s) \simeq s$ and thus $\Delta(s=0) = 0$, which is the zeroth-order solution. The first-order approximation is [10]

$$\Delta(s) - s \simeq \lim_{s\to 0+} i \frac{M^2}{\omega - v_s + is} = \frac{1}{2}\Gamma + i\Delta\omega. \tag{6.86}$$

Using $\lim_{s\to +0} \frac{1}{x+is} = \mathcal{P}\left[\frac{1}{x}\right] - i\pi\delta(x)$ with $\mathcal{P}$ denoting the principal value, we obtain

$$\Gamma = \frac{1}{3\pi\varepsilon_0\hbar c^3} \int dv_s \, v_s^3 \mu^2 \delta(\omega - v_s)$$

$$= \frac{\omega^3 \mu^2}{3\pi\varepsilon_0\hbar c^3}, \tag{6.87}$$

$$\Delta\omega = \frac{1}{6\pi^2}\varepsilon_0\hbar c^3 \mathcal{P} \int dv_s \frac{v_s^3 \mu^2}{\omega - v_s}. \tag{6.88}$$

The spontaneous decay rate (6.87) gives the Einstein's A coefficient, and the frequency shift (6.88) represents the Lamb shift (i.e., alternating current (AC) Stark shift of atomic transition frequency due to vacuum field fluctuations). An excellent description of QED calculation of the Lamb shift (6.88) is found in the text by J. J. Sakurai [11].

6.7. SUPERRADIANCE

Next, we consider the spontaneous emission rate for an assembly of N atoms that is initially either in a Dicke or a Bloch state. The total Hamiltonian is

$$\hat{\mathcal{H}} = \frac{1}{2}\hbar\omega \hat{J}_z + \sum_s \hbar v_s \left(\hat{a}_s^\dagger \hat{a}_s + \frac{1}{2}\right) - \sum_s \hbar g_s(\hat{a}_s \hat{J}_+ + \hat{a}_s^\dagger \hat{J}_-). \tag{6.89}$$

Note the difference between (6.68) and (6.89). In (6.68), $\hat{\sigma}_+$ and $\hat{\sigma}_-$ are the Pauli spin operators for a single atom that have a fermionic character; in (6.89), $\hat{J}_+$ and $\hat{J}_-$ are the collective angular momentum operators for N atoms. The transition probability from an initial atomic state to a final atomic state with an emission of one photon is calculated by the Fermi's golden rule:

$$W = \sum_s \frac{2\pi}{\hbar} \delta(\omega - \nu_s) |\langle f, n_s + 1 | \hat{\mathcal{H}}_I | i, n_s \rangle|^2, \quad (6.90)$$

where the first and second symbols in the state labeling indicate the atomic state and the photon number eigenstate of the field. For a single atom initially in the upper state and a field in the vacuum state, $|i, n_s\rangle = |e, 0\rangle$ and $|f, n_s + 1\rangle = |g, 1_s\rangle$, the emission probability W is reduced to the normal spontaneous emission rate $W^{(0)} = \Gamma$ discussed in the previous section.

When N atoms are initially prepared in the Dicke state, $|i\rangle = |M, J\rangle$, and all the field modes are in vacuum states $|0\rangle_s$, the spontaneous emission probability is

$$W^{(D)} = |\langle M - 1, J | \hat{J}_- | M, J \rangle|^2 W^{(0)} = (J + M)(J - M + 1) W^{(0)}, \quad (6.91)$$

where the matrix element $\langle M - 1, J | \hat{J}_- | M, J \rangle = [(J + M)(J - M + 1)]^{1/2}$ is used. If $M \simeq 0$ (approximately half of the atoms are excited), the spontaneous emission probability is proportional to the square of the atomic number, $J(J + 1) \simeq \frac{N^2}{4}$. This is Dicke's superradiance [12]. The decay rate of the N atoms is enhanced by a factor of N. If $M \simeq J$ (i.e., most of the atoms are excited), the spontaneous emission probability is proportional to the atomic number, $2J = N$, which is the simple sum of individual atom's decay rate (normal spontaneous emission). If $M = -J + 1$ (only one atom is excited), the spontaneous emission probability is proportional to the atomic number $2J + 1 \simeq N$. This is also a signature of Dicke's superradiance [12]. The decay rate of the single atom is enhanced by a factor of N; this enhancement can be traced back to the collective enhancement of atom-field coupling strength discussed in the previous section.

On the other hand, the (net) stimulated emission probability into mode s is proportional to

$$|_s\langle n + 1 | \langle M - 1, J | \hat{a}_s^+ \hat{J}_- | M, J \rangle | n \rangle_s|^2$$
$$- |_s\langle n - 1 | \langle M + 1, J | \hat{a}_s \hat{J}_+ | M, J \rangle | n \rangle_s|^2 = 2Mn, \quad (6.92)$$

which is the normal stimulated emission rate proportional to population inversion $2M$ and the photon number n but is independent of N. If $M \simeq 0$, the (net) stimulated emission rate vanishes. Here, we see the breakdown of the relation between Einstein's A (spontaneous emission) and B (stimulated emission) coefficients.

When N atoms are initially prepared in the Bloch state, $|i\rangle = |\theta, \varphi\rangle$, the emission probability is

$$W^{(B)} = \sum_M |\langle M, J | \hat{J}_- | \theta, \varphi \rangle|^2 W^{(0)} = |\langle \theta, \varphi | \hat{J}_+ \hat{J}_- | \theta, \varphi \rangle|^2 W^{(0)}$$

$$= \left(J^2 \sin^2 \theta + 2J \sin^4 \frac{\theta}{2} \right) W^{(0)}, \quad (6.93)$$

where $\sum_M |M, J\rangle\langle M, J| = \hat{I}$ is used. Equation (6.93) indicates that the Dicke superradiance occurs at $\theta \simeq \pi/2$ since $W^{(B)} \simeq \frac{N^2}{4} W^{(0)}$. Equation (6.91), with the replacement of $M = -J \cos \theta$, may be rewritten as

$$W^{(D)} = \left(J^2 \sin^2 \theta + 2J \sin^2 \frac{\theta}{2} \right) W^{(0)}, \quad (6.94)$$

which is almost identical to (6.93) for $J = \frac{N}{2} \gg 1$. The stimulated emission probability is proportional to

$$\sum_M \left(|\langle M, J|\hat{J}_-|\theta, \varphi\rangle|^2 - |\langle M, J|\hat{J}_+|\theta, \varphi\rangle|^2 \right) = -2J \cos \theta = 2M, \quad (6.95)$$

which is identical to the stimulated emission rate obtained for the Dicke state. For the totally inverted state, both states have the identical spontaneous emission rate $2JW^{(0)} = NW^{(0)}$, which is the normal spontaneous emission from N independent atoms.

Although the Dicke state has a zero average dipole moment and classical emission is not expected, the preceding quantum mechanical treatment shows that it has the same emission rate as the Bloch state, which has a nonzero average dipole moment.

REFERENCES

[1] C. Cohen-Tannoudji, J. Dupont-Roc, and G. Grynberg, *Photons and Atoms, Introduction to Quantum Electrodynamics* (Wiley-Interscience, New York, 1989).
[2] E. A. Power, *Introductory Quantum Electrodynamics* (Longman, London, 1964).
[3] R. Loudon, *The Quantum Theory of Light* (Oxford University Press, New York, 1985).
[4] Y. Yamamoto, G. Björk, J. Jacobson, S. Pau, and H. Cao, in *Quantum Coherence and Decoherence*, eds. K. Fujikawa and Y. A. Ono (North-Holland, Amsterdam, 1996), p. 5.
[5] P. R. Berman, *Cavity Quantum Electrodynamics* (Academic Press, Boston, 1994).
[6] B. R. Mollow, Phys. Rev. **188**, 1969 (1969).
[7] J. H. Eberly, American J. Phys. **40**, 1374 (1972).
[8] F. W. Cummings, Phys. Rev. **140**, A105 (1965).
[9] J. H. Eberly et al., Phys. Rev. Lett. **44**, 1323 (1980).
[10] V. Weisskopf and E. Wigner, Z. Phys. **63**, 54 (1930).
[11] J. J. Sakurai, *Advanced Quantum Mechanics* (Addison-Wesley, Reading, Mass., 1967).
[12] R. H. Dicke, Phys. Rev. **93**, 99 (1954).

7 Mathematical Methods for System-Reservoir Interaction

Our treatment of the quantum theory of fields and atoms has predominantly involved isolated systems such as the electromagnetic field in a perfect cavity. On the other hand, in most of the experimentally relevant cases, the quantum system of interest couples to an environment. This environment, which might simply be the electromagnetic vacuum itself, normally has many more degrees of freedom compared to the system of interest. Even though the system dynamics are profoundly changed by this coupling, in general we can assume that the environment remains unaffected. In other words, the energy dissipated or gained by the system due to this interaction does not change the quantum statistical state (i.e., density operator) of the environment. This assumption will allow us in this chapter to introduce irreversibility and dissipation into the quantum dynamics in a relatively simple fashion. The topic of such open quantum systems is of significant interest as it is considered to be key in understanding how the classical world appears from an underlying quantum description. Almost equally important is the fact that any information gathering process (i.e., measurement) from a quantum system necessarily takes place via a system-environment coupling.

We shall start by showing the inconsistency of the naive method of incorporating the dissipation into system dynamics. Suppose damping of the field is described by introducing a phenomenological decay term in the Heisenberg equations of motion; the operators would then have the solutions

$$\frac{d}{dt}\hat{a}(t) = \left(-i\omega - \frac{\gamma}{2}\right)\hat{a}(t) \quad \longrightarrow \quad \hat{a}(t) = \hat{a}(0)e^{-(i\omega+\frac{\gamma}{2})t}$$
$$\frac{d}{dt}\hat{a}^\dagger(t) = \left(i\omega - \frac{\gamma}{2}\right)\hat{a}^\dagger(t) \quad \longrightarrow \quad \hat{a}^\dagger(t) = \hat{a}^\dagger(0)e^{(i\omega-\frac{\gamma}{2})t}. \quad (7.1)$$

The field would then decay in a time on the order of $1/\gamma$ due to, for instance, leakage loss from imperfect mirrors or absorption loss by two-level atoms. This result violates the fundamental postulate of quantum mechanics since the commutator bracket

$$[\hat{a}(t), \hat{a}^\dagger(t)] = [\hat{a}(0), \hat{a}^\dagger(0)]e^{-\gamma t} \quad (7.2)$$

approaches zero and thus the uncertainty principle is violated. A similar problem is encountered if amplification of the field is described by introducing a phenomenological gain term γ' in the Heisenberg equations of motion; in such a case, the commutator bracket

$$[\hat{a}(t), \hat{a}^\dagger(t)] = [\hat{a}(0), \hat{a}^\dagger(0)]e^{\gamma' t} \tag{7.3}$$

goes to infinity.

The difficulty with this model lies in the fact that there are quantum fluctuations in reservoirs responsible for either damping or amplification of the field, and they have not yet been taken into account in the naive formulation of (7.1). Yet another contradiction is obtained when one describes a lossless beam splitter without keeping track of the appropriate environment coupling: If the field mode of interest is divided in half by a 50%-50% beam splitter without additional noise, the two quadrature amplitudes $\hat{a}_{s1}$ and $\hat{a}_{s2}$ could be measured independently by two homodyne detectors. The uncertainty product for the two measurement results satisfies $\langle \Delta a_{s1}^2 \rangle \langle \Delta a_{s2}^2 \rangle = \frac{1}{16}$ if the field is in a minimum uncertainty state. This result violates the generalized uncertainty relation for the simultaneous measurement of two conjugate observables. The resolution in this case lies in the fact that in the 50%-50% beam splitter configuration, the vacuum field fluctuation $\hat{a}_i$ is incident upon the beam splitter from an open port, which guarantees that the commutator bracket for the output modes is preserved:

$$\hat{b} = \frac{1}{\sqrt{2}}(\hat{a}_s + \hat{a}_i) \quad \longrightarrow \quad [\hat{b}, \hat{b}^\dagger] = 1$$

$$\hat{d} = \frac{1}{\sqrt{2}}(\hat{a}_s - \hat{a}_i) \quad \longrightarrow \quad [\hat{d}, \hat{d}^\dagger] = 1, \tag{7.4}$$

where $\hat{a}_s = \hat{a}_I \otimes \hat{I}_{II}$, $\hat{a}_i = \hat{I}_I \otimes \hat{a}_{II}$, $\hat{a}_I$, and $\hat{a}_{II}$ are the input field operators from the two open ports. The generalized uncertainty relation for the readout operators $\hat{a}_{s1,\text{eff}} \equiv \sqrt{2}\,\hat{b}_1$ and $\hat{a}_{s2,\text{eff}} \equiv \sqrt{2}\,\hat{d}_2$ is also satisfied

$$\begin{aligned}\langle \Delta \hat{a}_{s1,\text{eff}}^2 \rangle &\equiv 2\langle \Delta \hat{b}_1^2 \rangle = \langle \Delta \hat{a}_{s1}^2 \rangle + \langle \Delta \hat{a}_{i1}^2 \rangle \\ \langle \Delta \hat{a}_{s2,\text{eff}}^2 \rangle &\equiv 2\langle \Delta \hat{d}_2^2 \rangle = \langle \Delta \hat{a}_{s2}^2 \rangle + \langle \Delta \hat{a}_{i2}^2 \rangle\end{aligned} \quad \searrow \atop \nearrow \quad \langle \Delta \hat{a}_{s1,\text{eff}}^2 \rangle \langle \Delta \hat{a}_{s2,\text{eff}}^2 \rangle \geq \frac{1}{4}. \tag{7.5}$$

Finally, if it had been possible to amplify the field without additional noise, then the two quadrature amplitudes $\hat{a}_1$ and $\hat{a}_2$ could have been measured by a beam splitter after the amplifier. This is due to the fact that once the field is amplified to a macroscopic level, the vacuum field fluctuation introduced by a beam splitter is not important since it is much smaller than the amplified signal noise. In reality, additional noise is added to the signal inside the amplifier to preserve the commutator bracket:

$$\hat{a}_\text{out} = \sqrt{G}\,\hat{a}_\text{in} + \sqrt{G-1}\,\hat{f}^\dagger \quad \longrightarrow \quad [\hat{a}_\text{out}, \hat{a}_\text{out}^\dagger] = 1. \tag{7.6}$$

The noise operator $\hat{f}$ is the dipole fluctuation operator for a laser amplifier and the vacuum field fluctuation operator at the idler frequency for a nondegenerate parametric amplifier. The readout operators $\hat{a}_{1,\text{eff}} \equiv \hat{a}_{\text{out},1}/\sqrt{G}$ and $\hat{a}_{2,\text{eff}} \equiv \hat{a}_{\text{out},2}/\sqrt{G}$ satisfy the generalized uncertainty relation

$$\langle \Delta \hat{a}_{1,\text{eff}}^2 \rangle \equiv \frac{\langle \Delta \hat{a}_{\text{out},1}^2 \rangle}{G} = \langle \Delta \hat{a}_{\text{in},1}^2 \rangle + \left(1 - \frac{1}{G}\right) \langle \Delta \hat{f}_1^2 \rangle \searrow$$
$$\langle \Delta \hat{a}_{2,\text{eff}}^2 \rangle \equiv \frac{\langle \Delta \hat{a}_{\text{out},2}^2 \rangle}{G} = \langle \Delta \hat{a}_{\text{in},2}^2 \rangle + \left(1 - \frac{1}{G}\right) \langle \Delta \hat{f}_2^2 \rangle \nearrow \quad \langle \Delta \hat{a}_{1,\text{eff}}^2 \rangle \langle \Delta \hat{a}_{2,\text{eff}}^2 \rangle \geq \frac{1}{4}.$$

(7.7)

The preceding examples demonstrate that a damped or amplified signal is subject to additional noise; that is, any dissipation or amplification process must be accompanied by a fluctuation. This relation is expressed mathematically in the dissipation-fluctuation theorem, which plays an important role in many physical systems. As described earlier, the problem of dissipation/amplification in quantum mechanics is usually attacked by separating the complete system into a system of primary interest (termed *system*) and a system of secondary interest with a large number of degrees of freedom (termed *reservoir*). To obtain the stochastic quantum dynamics of the system alone, we need to eliminate the reservoir degrees of freedom. In the Schrödinger or interaction pictures, this is achieved via the master equation techniques for the reduced system density operator. In the Heisenberg picture, the system dynamics are governed by Heisenberg-Langevin equations; this approach is alternatively referred to as the noise operator method. We will start our discussion with the latter since it provides additional insight into the intertwined roles of fluctuation and dissipation.

7.1. NOISE OPERATOR METHOD

7.1.1. Field Damping by Field Reservoirs

Since our principal goal in this section is to introduce the fluctuation-dissipation theorem, we will restrict ourselves to the simplest of all dissipative system models; that is, a simple harmonic oscillator with frequency ω_s and annihilation operator $\hat{a}_s$ interacting with a bath of harmonic oscillators (reservoir). We will let $\omega_\mathbf{k}$ and $\hat{b}_\mathbf{k}$ denote the frequency and annihilation operator for each mode $\mathbf{k}$ that constitutes the bath. Despite its simplicity, this model accurately describes the field damping by field reservoirs (i.e., decay of an optical cavity mode due to its coupling to the radiation field reservoir via imperfect mirrors).

The Hamiltonian of the complete system is

$$\hat{H}_{s-r} = \hbar \omega_s \hat{a}_s^\dagger \hat{a}_s + \sum \hbar \omega_\mathbf{k} \hat{b}_\mathbf{k}^\dagger \hat{b}_\mathbf{k} + \hbar \sum (g_\mathbf{k} \hat{b}_\mathbf{k} + g_\mathbf{k}^* \hat{b}_\mathbf{k}^\dagger)(\hat{a}_s + \hat{a}_s^\dagger), \quad (7.8)$$

where $g_\mathbf{k}$ denotes the coupling strength. For simplicity, we will let $\mathbf{k} \to k$ from now on, keeping in mind that, in general, k is a vector quantity denoting not only the three components of the wavevector of the field mode but also its polarization $\sigma_\mathbf{k}$. Before writing down the coupled Heisenberg equations for $\hat{a}_s$ and $\hat{b}_k$, we introduce the first important approximation that greatly simplifies the problem of dissipation in quantum optics: We will assume that the frequency of the oscillator ω_s is much larger than the coupling strength g_k and the dissipation rate that we obtain as a result of this interaction. This assumption essentially says that the quantum system undergoes a large number of free oscillations (with period $2\pi/\omega_s$) before the effects of reservoir coupling becomes significant (i.e., the system is in the underdamped motion regime). In such a weak coupling situation, the change in the energy of either the system or the reservoir quanta due to interactions is extremely small and only the interaction terms that approximately conserve the total bare-state energies are significant. This is the rotating wave approximation (RWA) that we discussed earlier. The validity of RWA greatly simplifies the problem of dissipation in quantum optics [1]–[5] as compared to that of quantum Brownian motion [6]–[9].

We start out by writing the Heisenberg equation for the field mode in the RWA:

$$\frac{d}{dt}\hat{a}_s(t) = -i\omega_s \hat{a}_s(t) - \sum_k ig_k \hat{b}_k(t), \tag{7.9}$$

which forms a set of coupled operator equations with the corresponding equation for each reservoir mode k:

$$\frac{d}{dt}\hat{b}_k(t) = -i\omega_k \hat{b}_k(t) - ig_k \hat{a}_s(t). \tag{7.10}$$

The next step in the derivation is to integrate (7.10) and substitute it in (7.9) to obtain the integro-differential equation for the system mode:

$$\frac{d}{dt}\hat{a}_s(t) = -i\omega_s \hat{a}_s(t) - \int_0^t dt' \sum_k |g_k|^2 \exp[-i\omega_k(t-t')]\hat{a}_s(t')$$
$$- \sum_k ig_k \exp[-i\omega_k t]\hat{b}_k(0), \tag{7.11}$$

where we assume that the system-reservoir interactions are turned on at time $t = 0$. In all cases where one encounters dissipative phenomena, the reservoir modes should be considered as having a continuous energy spectrum; we then need to replace the summation over the reservoir modes with an integral

$$\sum_k \to \frac{V}{(2\pi)^3} \sum_{\sigma_k} \int d^3k, \tag{7.12}$$

where we explicitly used the fact that we are concentrating on electromagnetic field reservoir in a three-dimensional box with volume V (Fig. 7-1). In what follows, we will discard the sum over the independent polarizations for simplicity. By switching

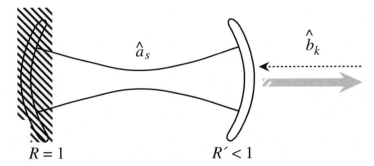

FIGURE 7-1: A single-mode cavity field with a finite damping rate due to output coupling.

to a rotating frame defined by $\hat{a}_s(t) = \hat{a}'_s(t)e^{-i\omega_s t}$, we obtain

$$\frac{d}{dt}\hat{a}'_s(t) = -\frac{V}{(2\pi)^3}\int d^3k |g_k|^2 \int_0^t dt' \, \hat{a}'_s(t')e^{-i(\omega_k-\omega_s)(t-t')}$$
$$-\frac{V}{(2\pi)^3}\int d^3k \, g_k \hat{b}_k(0) e^{-i(\omega_k-\omega_s)t}$$
$$= -\int_0^t d\tau \, \Gamma(\tau)\hat{a}'_s(t-\tau) - \hat{f}(t), \tag{7.13}$$

where $\tau = t - t'$ and

$$\Gamma(\tau) = \frac{V}{(2\pi)^3}\int d^3k |g_k|^2 e^{-i(\omega_k-\omega_s)\tau} = \int_0^\infty d\omega \, R(\omega) e^{-i(\omega_k-\omega_s)\tau}$$
$$\hat{f}(\tau) = \frac{V}{(2\pi)^3}\int d^3k \, g_k \hat{b}_k(0) e^{-i(\omega_k-\omega_s)\tau}. \tag{7.14}$$

The two new terms introduced in (7.14), $R(\omega)$ and $\hat{f}(t)$, correspond to the reservoir spectral function (imaginary part of the system self-energy) and the reservoir fluctuation operator, respectively.

We remark that at all temperatures, the reservoir average of the fluctuation operators vanish: $\langle \hat{f}(t) \rangle = Tr_R(\hat{f}(t)\rho_R) = \langle \hat{f}^\dagger(t) \rangle = 0$, with ρ_R denoting the reservoir density operator. At zero temperature, the dissipation kernel $\Gamma(\tau)$ and $\hat{f}(t)$ satisfy the relations

$$\langle \hat{f}(t+\tau)\hat{f}^\dagger(t) \rangle = \Gamma(\tau), \tag{7.15}$$
$$\langle \hat{f}^\dagger(t+\tau)\hat{f}(t) \rangle = 0, \tag{7.16}$$
$$\langle \hat{f}(t+\tau)\hat{f}(t) \rangle = 0, \tag{7.17}$$
$$\langle \hat{f}^\dagger(t+\tau)\hat{f}^\dagger(t) \rangle = 0. \tag{7.18}$$

The relation in (7.15) demonstrates that the the two effects arising from the reservoir interactions, dissipation and fluctuations, are related to each other; in fact, (7.15) should be regarded as a *fluctuation-dissipation theorem* where memory effects have been taken into account. There is no need to invoke Born approximation since the system-reservoir interaction that we consider here is linear. In fact, the only approximation introduced so far is the RWA. It should also be noted here that if the initial reservoir operator is Gaussian, then the first- and second-order moments given in (7.15)–(7.18) completely determine the stochastic process [9].

As mentioned earlier, in most quantum optics applications, the reservoir spectral function $R(\omega) \propto \omega^\alpha$ is essentially a flat function of ω in the frequency range $(\omega_s - \Delta\omega/2, \omega_s + \Delta\omega/2)$ of interest. This is due to the fact that the time scales over which the system observables change are much longer than the free oscillation period (i.e., $\Delta\omega \ll \omega_s$). When this is the case, we can assume that

$$\Gamma(\tau) \simeq R(\omega_s)\delta(\tau). \tag{7.19}$$

Strictly speaking, (7.19) is not correct; there is an additional imaginary term that corresponds to an energy change introduced by the reservoir (i.e., the Lamb shift that we discussed in the previous chapter). Nevertheless, we can see that the flat reservoir or Markov approximation allows us to eliminate the time and frequency integrals in (7.13) and to obtain a simple stochastic differential operator equation. Conversely, the fact that $R(\omega)$ can be considered as being independent of frequency in the frequency range of interest tells us that the reservoir operators have a very short correlation time τ_c.

We now go back to (7.13) and introduce the Markov approximation. By invoking (7.19), we can assume that $\hat{a}'_s(t - \tau)$ will remain unchanged in the time period τ during which $\Gamma(\tau)$ is appreciably different from zero. In that case we can set $\hat{a}'_s(t - \tau) = \hat{a}'_s(t)$ and take $\hat{a}'_s(t)$ out of the integrals. When we go back to the original nonrotating basis, we obtain

$$\frac{d}{dt}\hat{a}(t) = -\left(i\omega_s + i\Delta + \frac{\gamma}{2}\right)\hat{a}(t) + \hat{f}(t), \tag{7.20}$$

which is the (quantum) Langevin equation for the field mode in the Markov approximation. The dissipation rate and the reservoir induced energy shift are given by

$$\frac{\gamma}{2} + i\Delta = \frac{V}{(2\pi)^3}\int d^3k |g_k|^2 \int_0^t dt'\, e^{-i(\omega_k - \omega_s)(t-t')}$$
$$= \pi D(\omega_s)|g(\omega_s)|^2 + i\mathcal{P}\int_0^\infty d\omega\, \frac{D(\omega_s)|g(\omega_s)|^2}{\omega_s - \omega}, \tag{7.21}$$

where we have introduced the density of states $D(\omega)$ and set $g_k \to g(\omega)$. The principal value integral in (7.21) gives the energy shift Δ; from now on we will assume that this energy shift is incorporated into the renormalized system mode energy ω_s.

Next, we consider some elementary properties of the reservoir fluctuation operators, which would enable us to justify some of the assumptions we have introduced earlier. Finite temperature correlation functions of the noise operators are easily calculated in the Markovian limit:

$$\langle \hat{f}^\dagger(t)\hat{f}(t')\rangle_R = \gamma \bar{n}(\omega_s)\delta(t-t') = 2\langle D_{\hat{a}_s^\dagger \hat{a}_s}\rangle_R \delta(t-t'), \qquad (7.22)$$

where $\langle D_{\hat{a}_s^\dagger \hat{a}_s}\rangle_R$ is termed the diffusion coefficient. In (7.22),

$$\bar{n}(\omega_s) = \langle \hat{b}^\dagger_{|k|=\omega_s/c}\hat{b}_{|k|=\omega_s/c}\rangle = \frac{1}{\exp\left[\frac{\hbar\omega_s}{k_B T_R}\right] - 1}$$

gives the mean occupancy of the reservoir modes with energy ω_s and temperature T_R. The only other nonzero second-order correlation function is

$$\langle \hat{f}(t)\hat{f}^\dagger(t')\rangle_R = \gamma(1+\bar{n}(\omega_s))\delta(t-t') = 2\langle D_{\hat{a}_s \hat{a}_s^\dagger}\rangle_R \delta(t-t'). \qquad (7.23)$$

To demonstrate that the commutator bracket for the system operator $\hat{a}_s$ is properly conserved by the quantum Langevin equation, we need to evaluate the correlation function between the reservoir fluctuation and system operators. We first introduce the trivial identity

$$\hat{a}_s(t) = \hat{a}_s(t-\Delta t) + \int_{t-\Delta t}^{t} dt'\, \dot{\hat{a}}_s(t'). \qquad (7.24)$$

Now, the reservoir averages of $\hat{a}_s^\dagger(t)\hat{f}(t)$ and its adjoint can be easily calculated as

$$\langle \hat{a}_s^\dagger(t)\hat{f}(t)\rangle_R = \langle \hat{a}_s^\dagger(t-\Delta t)\hat{f}(t)\rangle_R + \int_{t-\Delta t}^{t} dt' \langle \left[-\frac{\gamma}{2}\hat{a}_s^\dagger(t') + \hat{f}^\dagger(t')\right]\hat{f}(t)\rangle_R$$

$$= \int_{t-\Delta t}^{t} dt' \langle \hat{f}^\dagger(t')\hat{f}(t)\rangle_R$$

$$= \int_{-\infty}^{0} ds\, \langle \hat{f}^\dagger(s)\hat{f}(0)\rangle_R$$

$$= \frac{1}{2}\gamma\bar{n}(\omega_s) = \langle D_{\hat{a}_s^\dagger \hat{a}_s}\rangle_R, \qquad (7.25)$$

where we set the first term in the first line equal to zero, since the system operator cannot depend on the value of the (δ-correlated) reservoir fluctuation operator at a later time. Similarly, we obtain

$$\langle \hat{f}^\dagger(t)\hat{a}_s(t)\rangle_R = \langle D_{\hat{a}_s^\dagger \hat{a}_s}\rangle_R, \tag{7.26}$$

$$\langle \hat{a}_s(t)\hat{f}^\dagger(t)\rangle_R = \langle \hat{f}(t)\hat{a}_s^\dagger(t)\rangle_R = \langle D_{\hat{a}_s \hat{a}_s^\dagger}\rangle_R. \tag{7.27}$$

Using these relations, we can check the time dependence of the commutator bracket:

$$\frac{d}{dt}[\hat{a}_s(t), \hat{a}_s^\dagger(t)] = \dot{\hat{a}}_s(t)\hat{a}_s^\dagger(t) + \hat{a}_s(t)\dot{\hat{a}}_s^\dagger(t) - \dot{\hat{a}}_s^\dagger(t)\hat{a}_s(t) - \hat{a}_s^\dagger(t)\dot{\hat{a}}_s^\dagger(t)$$

$$= -\gamma[\hat{a}_s(t), \hat{a}_s^\dagger(t)] + \langle \hat{f}(t)\hat{a}_s^\dagger(t) + \hat{a}_s(t)\hat{f}^\dagger(t)\rangle_R$$

$$- \langle \hat{f}^\dagger(t)\hat{a}_s(t) + \hat{a}_s^\dagger(t)\hat{f}(t)\rangle_R$$

$$= 0, \tag{7.28}$$

which clearly indicates that the $[\hat{a}_s(t), \hat{a}_s^\dagger(t)] = 1$ is satisfied at all times. Finally, the time evolution of the photon number operator $\hat{a}_s^\dagger \hat{a}_s$ obeys

$$\frac{d}{dt}\langle \hat{a}_s^\dagger(t)\hat{a}_s(t)\rangle_R = -\gamma \langle \hat{a}_s^\dagger(t)\hat{a}_s(t)\rangle_R + \langle \hat{f}^\dagger(t)\hat{a}_s(t)\rangle_R + \langle \hat{a}_s^\dagger(t)\hat{f}(t)\rangle_R$$

$$= -\gamma \langle \hat{a}_s^\dagger(t)\hat{a}_s(t)\rangle_R + \gamma \bar{n}(\omega_s). \tag{7.29}$$

We see that although the field amplitude $\langle \hat{a}_s(t)\rangle_R$ decays to zero [as shown in (7.20)], the photon number $\langle \hat{a}_s^\dagger(t)\hat{a}_s(t)\rangle_R$ approaches a steady-state value $\bar{n}(\omega_s)$.

7.1.2. Einstein Relation between Drift and Diffusion Coefficients

We have seen that in a simple dissipative system, the Heisenberg-Langevin equation has the general form

$$\frac{d}{dt}\hat{A}_\mu = \hat{D}_\mu(t) + \hat{F}_\mu(t), \tag{7.30}$$

where $\hat{A}_\mu$ is a system operator; for example, $\hat{A}_\mu = \{\hat{a}, \hat{a}^\dagger\}$ for a harmonic oscillator and $\hat{A}_\mu = \{\hat{\sigma}_z, \hat{\sigma}_+, \hat{\sigma}_-\}$ for a spin $\frac{1}{2}$ system. $\hat{D}_\mu(t)$ is the drift term [in the previous example, $\hat{D}_\mu(t) = -\frac{1}{2}\gamma \hat{A}$] and $\hat{F}_\mu(t)$ is the noise operator that satisfies $\langle \hat{F}_\mu(t)\rangle_R = \langle \hat{F}_\mu(t)\rangle = 0$. Therefore, the drift term is given by

$$\langle \hat{D}_\mu(t)\rangle_R = \frac{d}{dt}\langle A_\mu(t)\rangle_R. \tag{7.31}$$

The diffusion coefficient in the Markovian case is defined by

$$\langle \hat{F}_\mu(t)\hat{F}_\nu(t')\rangle_R = 2\langle D_{\mu\nu}\rangle_R \delta(t-t'). \tag{7.32}$$

In the preceding example,

$$2\langle D_{\hat{A}^\dagger \hat{A}}\rangle_R = \gamma n_{\text{th}} = \gamma \langle \hat{A}^\dagger(t)\hat{A}(t)\rangle_R + \frac{d}{dt}\langle \hat{A}^\dagger(t)\hat{A}(t)\rangle_R \quad (7.33)$$

$$= -\langle \hat{D}_{\hat{A}^\dagger}(t)\hat{A}(t)\rangle_R - \langle \hat{A}^\dagger(t)\hat{D}_{\hat{A}}(t)\rangle_R + \frac{d}{dt}\langle \hat{A}^\dagger(t)\hat{A}(t)\rangle_R.$$

In general, the diffusion coefficient $\langle D_{\mu\nu}\rangle_R$ and the drift coefficient $\langle \hat{D}_\mu(t)\rangle_R$ are related by the Einstein relation:

$$2\langle D_{\mu\nu}\rangle_R = -\langle \hat{D}_\mu(t)\hat{A}_\nu(t)\rangle_R - \langle \hat{A}_\mu(t)\hat{D}_\nu(t)\rangle_R + \frac{d}{dt}\langle \hat{A}_\mu(t)\hat{A}_\nu(t)\rangle_R, \quad (7.34)$$

which is referred to as the Einstein relation between drift and diffusion coefficients.

7.2. DENSITY OPERATOR METHOD

7.2.1. Derivation of the Master Equation

Despite its obvious appeal arising mainly from a direct correspondence to classical description of stochastic systems, in general it is extremely difficult to deal with the Heisenberg–Langevin equations. In many cases of interest, the underlying Heisenberg equations will be nonlinear, and we then have to deal with nonlinear stochastic differential equations for operators. Even though we will discuss linearization techniques that are used to handle these equations, it is clearly appealing to develop a Schrödinger or interaction picture analysis that gives us a linear deterministic differential equation for the reduced system density operator. Naturally, as the quantum system is open, there is statistical as well as quantum uncertainty and a true wavefunction description is no longer possible.

Complete rigorous treatments of the master equation derivation already exist in the literature [9], [10]. Our goal here is to present a short discussion that enables us to consider the specific quantum systems of interest.

Once again, we consider a system S interacting with a reservoir R via the interaction Hamiltonian $\hat{V}$. The combined density operator is denoted by $\hat{\rho}(t)$. We assume that at an initial time t, the two systems are uncorrelated and, hence, the initial density operator $\hat{\rho}(0)$ is given by the simple outer product

$$\hat{\rho}(0) = \hat{\rho}_S(0) \otimes \hat{\rho}_R(0). \quad (7.35)$$

The dynamics of $\hat{\rho}(t)$ in the interaction picture are governed by the Liouville–von Neumann equation

$$\frac{d}{dt}\hat{\rho}(t) = \frac{1}{i\hbar}[\hat{H}_{\text{int}}(t), \hat{\rho}(t)]. \quad (7.36)$$

The principal idea underlying master equation approach is that the number of degrees of freedom of the reservoir is very large, which makes it impossible to keep track of its quantum evolution. The goal then is to obtain an equation that only involves the system operators. The reduced density operator for the system, obtained by tracing over the reservoir degrees of freedom, satisfies

$$\frac{d}{dt}\hat{\rho}_S(t) = \frac{1}{i\hbar}\text{Tr}_R[\hat{H}_{\text{int}}(t), \hat{\rho}(t)]. \tag{7.37}$$

Formal integration of (7.36) and subsequent resubstitution gives

$$\frac{d}{dt}\hat{\rho}_S(t) = \left(\frac{1}{i\hbar}\right)^2 \int_0^t dt'\, \text{Tr}_R\Big([\hat{H}_{\text{int}}(t), [\hat{H}_{\text{int}}(t'), \hat{\rho}(t')]]\Big). \tag{7.38}$$

The first-order term $\text{Tr}_R[\hat{H}_{\text{int}}(t), \hat{\rho}(0)]$ that comes from the formal integration of the equation of motion of the density operator is identically zero, since $\text{Tr}_R[\hat{\rho}_R(0)\hat{H}_{\text{int}}(t)]$ vanishes for all the interaction Hamiltonians of interest in quantum optics.

Note that we have not made any approximations so far and therefore (7.38) is exact. The derivation of a simpler master equation, however, requires that four key approximations are introduced: (1) RWA; (2) Born approximation: $\hat{\rho}(t') = \hat{\rho}_S(t') \otimes \hat{\rho}_R(t')$; (3) the initial radiation field density operator commutes with the free Hamiltonian, and the reservoir is not affected by the interactions with the system [i.e., $\hat{\rho}_R(t) = \text{Tr}_S[\hat{\rho}(t)] = \hat{\rho}_R(0)$]; and (4) Markov approximation: $\hat{\rho}_S(t') \simeq \hat{\rho}_S(t)$. We will refer to these four approximations as *Born-Markov approximation*; in this limit, we find

$$\frac{d\hat{\rho}_S(t)}{dt} = \left(\frac{1}{i\hbar}\right)^2 \int_0^t dt'\, \text{Tr}_R\Big([\hat{H}_{\text{int}}(t), [\hat{H}_{\text{int}}(t'), \hat{\rho}_S(t) \otimes \hat{\rho}_R(0)]]\Big). \tag{7.39}$$

We now evaluate the time integral in (7.39) for specific system-reservoir interactions. Since our main goal is to analyze dissipation, we discard the principal value terms with the assumption that they could in principle be included in the (Hermitian) free system Hamiltonian.

7.2.2. Field Damping by Atomic Reservoirs

We first consider the damping of an optical cavity mode by a two-level atomic beam reservoir (Fig. 7-2). In some sense, this is the reverse problem of a laser, where one is interested in amplification of a field mode by atomic reservoirs. Since the atoms are assumed to form a reservoir, their initial atomic density matrix is a statistical mixture of the upper $|e\rangle$ and lower $|g\rangle$ states determined by the Boltzmann distribution:

$$\hat{\rho}_{\text{atom}}(t) = \begin{pmatrix} \rho_{ee} & 0 \\ 0 & \rho_{gg} \end{pmatrix} = \rho_{ee}|e\rangle\langle e| + \rho_{gg}|g\rangle\langle g|, \tag{7.40}$$

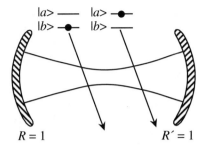

FIGURE 7-2: A single-mode cavity field damped by atomic beam reservoirs.

where

$$\rho_{ee} = \frac{1}{1 + \exp\left(\frac{\hbar\omega_{eg}}{k_B T}\right)}, \tag{7.41}$$

$$\rho_{gg} = \frac{\exp\left(\frac{\hbar\omega_{eg}}{k_B T}\right)}{1 + \exp\left(\frac{\hbar\omega_{eg}}{k_B T}\right)}. \tag{7.42}$$

Equivalently, we are assuming that there is no quantum coherence between the upper and lower states, $\rho_{eg} = \rho_{ge} = 0$. The frequency $\omega_{eg} = (E_e - E_g)/\hbar$. We will assume for simplicity that these two-level atoms couple individually to a cavity mode that is on resonance with the atomic transition: $\omega_{cav} \simeq \omega_{eg}$. The interaction Hamiltonian for a single atom is then

$$\hat{H}_{\text{int}} = \hbar g(\hat{\sigma}_- \hat{a}^\dagger + \hat{\sigma}_+ \hat{a}) = \hbar g \begin{pmatrix} 0 & \hat{a} \\ \hat{a}^\dagger & 0 \end{pmatrix}, \tag{7.43}$$

where $\sigma_+ = \sigma_{eg} = |e\rangle\langle g|$ and $\sigma_- = \sigma_{ge} = |g\rangle\langle e|$ are the atomic raising and lowering operators that we have already encountered.

Before discussing the validity of Born-Markov approximation, we consider the atom-field density operator $\hat{\rho}$ and its commutator with $\hat{H}_{\text{int}}$. At time $t = 0$, we have

$$\hat{\rho}(t) = \hat{\rho}_f(t) \otimes \begin{pmatrix} \rho_{ee} & 0 \\ 0 & \rho_{gg} \end{pmatrix} = \begin{pmatrix} \rho_{ee}\hat{\rho}_f(t) & 0 \\ 0 & \rho_{gg}\hat{\rho}_f(t) \end{pmatrix}. \tag{7.44}$$

The first-order commutator $[\hat{H}_{\text{int}}, \hat{\rho}(t)]$ only has off-diagonal elements

$$[\hat{H}_{\text{int}}, \hat{\rho}(t)] = \hbar g \begin{pmatrix} 0 & \hat{a}\rho_{ee}\hat{\rho}_f(t) - \rho_{gg}\hat{\rho}_f(t)\hat{a} \\ \hat{a}^\dagger \rho_{ee}\hat{\rho}_f(t) - \rho_{gg}\hat{\rho}_f(t)\hat{a}^\dagger & 0 \end{pmatrix}$$

$$\tag{7.45}$$

and therefore the trace of this term over the atomic parameters vanishes. On the other hand, the second-order commutator is given by

$$\left[\hat{H}_{\text{int}}, [\hat{H}_{\text{int}}, \hat{\rho}(t)]\right]$$

$$= \hbar^2 g^2 \begin{pmatrix} \hat{a}\hat{a}^\dagger \rho_{ee}\hat{\rho}_f(t) - \hat{a}\rho_{gg}\hat{\rho}_f(t)\hat{a}^\dagger & 0 \\ 0 & \hat{a}^\dagger\hat{a}\rho_{gg}\hat{\rho}_f(t) - \hat{a}^\dagger\rho_{ee}\hat{\rho}_f(t)\hat{a} \end{pmatrix} + \text{adjoint}. \quad (7.46)$$

So far, we have assumed single-frequency atoms coupled to a single-frequency field mode; since there is no continuum of reservoir states, Born-Markov approximation cannot be justified in such an idealized interaction. In practical absorbers or amplifiers, on the other hand, the atoms have a broadening; in many cases the origin of this broadening is the dephasing collisions between atoms. A less common but more illustrative example is the transit-time broadening of an atomic transition, in which atoms are assumed to be injected into the cavity, where they spend an average time of τ seconds. If r atoms are injected into the cavity per second, we can obtain the coarse-grained time rate of change of $\hat{\rho}_f(t)$ by substituting (7.46) in (7.39). This yields

$$\dot{\hat{\rho}}_f(t) = -\frac{1}{2}R_e[\hat{a}\hat{a}^\dagger\hat{\rho}_f - \hat{a}^\dagger\hat{\rho}_f\hat{a}] - \frac{1}{2}R_g[\hat{a}^\dagger\hat{a}\hat{\rho}_f - \hat{a}\hat{\rho}_f\hat{a}^\dagger] + \text{adjoint}, \quad (7.47)$$

where $R_e = r\rho_{ee}g^2\tau^2$ and $R_g = r\rho_{gg}g^2\tau^2$ are the rate coefficients for photon emission and photon absorption by atoms per second. Born-Markov approximation is valid provided $g \ll \tau^{-1}$.

The (cavity) photon decay rate $\frac{\nu}{Q_0}$ and the thermal equilibrium photon number n_{th} can be defined by

$$\frac{\nu}{Q_0} = R_g - R_e, \quad (7.48)$$

$$R_e(1 + n_{\text{th}}) = R_g n_{\text{th}} \Longrightarrow n_{\text{th}} = \frac{R_e}{R_g - R_e} = \frac{1}{\exp\left(\frac{\hbar\omega}{k_B T}\right) - 1}. \quad (7.49)$$

Here, the thermal equilibrium photon number n_{th} is given by the condition that the sum of spontaneous and stimulated emission rate per second $[R_e(1 + n_{\text{th}})]$ is equal to the (stimulated) absorption rate per second $[R_g n_{\text{th}}]$. Using (7.48) and (7.49) in (7.47), we obtain

$$\frac{d}{dt}\hat{\rho}(t) = -\frac{1}{2}\frac{\nu}{Q_0}\left\{n_{\text{th}}(\hat{a}\hat{a}^\dagger\hat{\rho}_f - \hat{a}^\dagger\hat{\rho}_f\hat{a}) + (n_{\text{th}} + 1)(\hat{a}^\dagger\hat{a}\hat{\rho}_f - \hat{a}\hat{\rho}_f\hat{a}^\dagger)\right\} + \text{adjoint}. \quad (7.50)$$

The density matrix element ρ_{nm} in the photon number representation is obtained by projecting $\langle n|$ from the left and $|m\rangle$ from the right of both sides of (7.50):

$$\dot{\rho}_{nm}(t) = -\frac{1}{2}\frac{\nu}{Q_0}[2n_{\text{th}}(n+m+1)+n+m]\rho_{nm} + \frac{\nu}{Q_0}n_{\text{th}}\sqrt{nm}\,\rho_{n-1,m-1}$$
$$+ \frac{\nu}{Q_0}(n_{\text{th}}+1)\sqrt{(n+1)(m+1)}\,\rho_{n+1,m+1}. \qquad (7.51)$$

In particular, the diagonal elements satisfy

$$\dot{\rho}_{nn}(t) = \left[-\frac{\nu}{Q_0}n_{\text{th}}(n+1) - \frac{\nu}{Q_0}(n_{\text{th}}+1)n\right]\rho_{nn}$$
$$+ \frac{\nu}{Q_0}n_{\text{th}}\,n\rho_{n-1,n-1} + \frac{\nu}{Q_0}(n_{\text{th}}+1)(n+1)\rho_{n+1,n+1}. \qquad (7.52)$$

Equation (7.52) can be understood in terms of Fig. 7-3, in which the flow of the photon number probability is depicted. The terms $\frac{\nu}{Q_0}n_{\text{th}} = R_e$ represent photon emission into state $|n\rangle$ (both spontaneous and stimulated emission) and correspond to the arrows pointing up in the diagram, while the terms $\frac{\nu}{Q_0}(n_{\text{th}}+1) = R_g$ correspond to photon absorption and the arrows pointing down.

Equilibrium is obtained when the net flow between all pairs of levels vanishes; for example,

$$\frac{\nu}{Q_0}(n_{\text{th}}+1)n\rho_{nn} = \frac{\nu}{Q_0}n_{\text{th}}n\rho_{n-1,n-1} \implies \rho_{nn} = \frac{n_{\text{th}}}{n_{\text{th}}+1}\rho_{n-1,n-1}. \qquad (7.53)$$

This condition is referred to as detailed balance. The solution of (7.53) is

$$\rho_{nn} = \left[1 - \exp\left(-\frac{\hbar\omega}{k_B T}\right)\right]\exp\left(-\frac{n\hbar\omega}{k_B T}\right) = \frac{1}{n_{\text{th}}+1}\left(\frac{n_{\text{th}}}{n_{\text{th}}+1}\right)^n, \qquad (7.54)$$

where the normalization condition $\sum_n \rho_{nn} = \sum_n \left(\frac{n_{\text{th}}}{n_{\text{th}}+1}\right)^n \rho_{00} = 1$ is used. Detailed balance in this case gives the thermal (Bose-Einstein) distribution with an av-

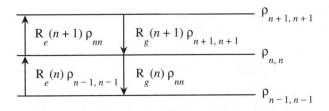

FIGURE 7-3: The flow of photon number probability.

erage photon number

$$\langle n \rangle \equiv \sum_n n \rho_{nn} = \frac{1}{\exp\left(\frac{\hbar\omega}{k_B T}\right) - 1} = n_{\text{th}}, \qquad (7.55)$$

a result we have been using all along for thermal radiation field modes. Although the field $\hat{\rho}_f(t)$ may initially be in a pure state represented by a single state vector, the process of tracing over the (unobserved) atomic states after the field-atom interaction leads to a field in a mixed state represented by a statistical mixture of photon number states $\hat{\rho}_f = \sum_n \rho_{nn} |n\rangle\langle n|$. In particular, the effect of the atomic beam is to bring the field to the same temperature as that of atoms.

The key point of the reservoir theory lies in the process of tracing over the reservoir coordinates, which leads to an irreversible dynamics for the system. This process corresponds to the lack of measurement as to whether the atom is in the upper level or in the lower level after interaction with the field. If the initial and final states of the atom are known (i.e., if information concerning the atomic beam is not discarded), the field remains in a pure state, which is a familiar problem in quantum measurements. The primary difference between the reservoir and quantum measurement theories is whether information stored in the environment (R) that interacts with system (S) is discarded or read out.

7.2.3. Atom Damping by Field Reservoirs

Next consider the reverse process; namely, a single two-level atom damped by a field reservoir in free space. The field reservoir is described by the annihilation operators $\hat{b}_k$ and creation operators $\hat{b}_k^\dagger$ at a frequency ω_k and the interaction Hamiltonian is

$$\hat{H}_{\text{int}} = \hbar \sum_k g_k\, \hat{\sigma}_-\, \hat{b}_k^\dagger\, e^{-i(\omega - \omega_k)t} + \text{adjoint}. \qquad (7.56)$$

This problem was solved in Chapter 6 using the Wigner-Weisskopf approximation; however, only the average spontaneous emission decay rate was calculated and a zero temperature reservoir was assumed. Here, we assume a reservoir density operator that corresponds to a multimode thermal field; namely,

$$\hat{\rho}_R = \prod_k \sum_n \exp\left(-\frac{\hbar\omega_k n}{k_B T}\right) \left\{1 - \exp\left(-\frac{\hbar\omega_k}{k_B T}\right)\right\} |n\rangle_{kk}\langle n|. \qquad (7.57)$$

Using (7.39), the equation of motion for $\text{Tr}_R[\hat{\rho}(t)] \equiv \hat{\rho}_a(t)$ is obtained:

$$\dot{\hat{\rho}}_a = -\frac{1}{2}\Gamma\left\{n_{\text{th}}[\hat{\sigma}_-\hat{\sigma}_+\hat{\rho}_a - \hat{\sigma}_+\hat{\rho}_a\hat{\sigma}_-] + (n_{\text{th}} + 1)[\hat{\sigma}_+\hat{\sigma}_-\hat{\rho}_a - \hat{\sigma}_-\hat{\rho}_a\hat{\sigma}_+]\right\} + \text{adjoint}, \qquad (7.58)$$

where Γ is as defined in Chapter 6. We reiterate that the validity of the Born-Markov approximation relies on the fact that the system operator does not change appreciably during the reservoir correlation time τ_c [i.e., $\tau_c \ll 1/\Gamma$]. An excellent detailed discussion of these time scales and the validity of the Born–Markov approximation is presented in Ref. [10].

7.2.4. Field Damping by Field Reservoirs

As a last example, we reconsider the system-reservoir interaction analyzed in the previous section—namely, a single electromagnetic field mode in a cavity with a finite leakage rate (Fig. 7-1). The external field reservoir once again consists of a multimode field in a statistical mixture of photon number states (Bose-Einstein distribution) given in (7.57). The time evolution of $\hat{\rho}_f(t) = \text{Tr}_R(\hat{\rho}(t))$ is given by

$$\frac{d}{dt}\hat{\rho}_f(t) = -\int_{t_0}^{t} dt' \sum_k g_k^2 \left\{ n_{\text{th}} \left[\hat{a}\hat{a}^\dagger \hat{\rho}_f(t') - \hat{a}^\dagger \hat{\rho}_f(t')\hat{a} \right] e^{-i(\omega-\omega_k)(t-t')} \right.$$
$$\left. + (n_{\text{th}}+1) \left[\hat{a}^\dagger\hat{a}\hat{\rho}_f(t') - \hat{a}\hat{\rho}_f(t')\hat{a}^\dagger \right] e^{i(\omega-\omega_k)(t-t')} \right\} + \text{adjoint}. \quad (7.59)$$

Once again, by replacing $\sum_k g_k^2$ with the integral $\int d\omega_k \, D(\omega_k) g(\omega_k)^2$ in (7.59), the following time integral is obtained:

$$\int_{t_0}^{t} dt' \int d\omega_k \, D(\omega_k) g(\omega_k)^2 e^{\pm i(\omega-\omega_k)(t-t')}$$
$$= \int_{0}^{t-t_0} d\tau \int d\omega_k \, D(\omega_k) g(\omega_k)^2 e^{\pm i(\omega-\omega_k)\tau}$$
$$\simeq \int d\omega_k \, D(\omega_k) g(\omega_k)^2 \cdot \pi \delta(\omega - \omega_k)$$
$$= \pi D(\omega) g(\omega)^2$$
$$\equiv \frac{1}{2}\left(\frac{\omega}{Q_e}\right), \quad (7.60)$$

where $\frac{\omega}{Q_e}$ is the cavity photon decay rate due to leakage (output coupling) via a partially reflecting mirror. Using (7.60) in (7.59), we have

$$\frac{d}{dt}\hat{\rho}_f(t) = -\frac{1}{2}\left(\frac{\omega}{Q_e}\right)\left\{ n_{\text{th}}(\hat{a}\hat{a}^\dagger\hat{\rho}_f - \hat{a}^\dagger\hat{\rho}_f\hat{a}) \right.$$
$$\left. + (n_{\text{th}}+1)(\hat{a}^\dagger\hat{a}\hat{\rho}_f - \hat{a}\hat{\rho}_f\hat{a}^\dagger) \right\} + \text{adjoint}. \quad (7.61)$$

Note that (7.61) has the same form as (7.50); thus, if the single-mode field is simultaneously damped by the atomic beam and external field reservoirs, we can use an effective total cavity photon decay rate

$$\frac{\omega}{Q} = \frac{\omega}{Q_0} + \frac{\omega}{Q_e}. \tag{7.62}$$

Finally, we reiterate that the master equations we have derived in these three examples account for both the dissipation/amplification and fluctuations arising from system-reservoir interactions. The master equation in all cases, however, is a deterministic linear operator differential equation for a statistical quantity.

7.3. THE FOKKER-PLANCK EQUATION

7.3.1. Glauber-Sudarshan P Representation

We have already discussed the phase-space representations in Chapter 4. These are especially useful when we are dealing with a dissipative quantum system, where they can be used to convert the density operator master equation into a c-number Fokker-Planck type of equation. The simplest phase-space representation is that of the diagonal coherent state $P(\alpha)$ expansion of Glauber and Sudarshan:

$$\hat{\rho}(t) = \int d^2\alpha \, P(\alpha) |\alpha\rangle\langle\alpha|. \tag{7.63}$$

We now apply the P-representation to the problem of field damping by field reservoirs. Substitution of (7.63) into (7.61) yields

$$\int d^2\alpha \, \dot{P} |\alpha\rangle\langle\alpha| = -\frac{1}{2} \int d^2\alpha \, P \left\{ R_e[\hat{a}\hat{a}^\dagger |\alpha\rangle\langle\alpha| - \hat{a}^\dagger |\alpha\rangle\langle\alpha|\hat{a}] \right.$$
$$\left. + R_g[\hat{a}^\dagger \hat{a} |\alpha\rangle\langle\alpha| - \hat{a}|\alpha\rangle\langle\alpha|\hat{a}^\dagger] \right\} + \text{adjoint}. \tag{7.64}$$

The expression in curly brackets can be simplified using the relation [4]

$$\hat{a}^\dagger |\alpha\rangle\langle\alpha| = \left(\frac{\partial}{\partial\alpha} + \alpha^*\right) |\alpha\rangle\langle\alpha|, \tag{7.65}$$

and its adjoint

$$|\alpha\rangle\langle\alpha|\hat{a} = \left(\frac{\partial}{\partial\alpha^*} + \alpha\right) |\alpha\rangle\langle\alpha|, \tag{7.66}$$

which follow from

$$\frac{\partial}{\partial\alpha}|\alpha\rangle\langle\alpha| = (-\alpha^* + \hat{a}^\dagger)|\alpha\rangle\langle\alpha| \quad \text{or} \quad \frac{\partial}{\partial\alpha^*}|\alpha\rangle\langle\alpha| = |\alpha\rangle\langle\alpha|(-\alpha + \hat{a}), \tag{7.67}$$

since $|\alpha\rangle\langle\alpha| = \exp(-\alpha\alpha^*)\exp(\alpha\hat{a}^\dagger)|0\rangle\langle 0|\exp(\alpha^*\hat{a})$. Using (7.65) and (7.66) in (7.64) and integrating the result by parts, we obtain

$$\int d^2\alpha\, \dot{P}|\alpha\rangle\langle\alpha| = -\int d^2\alpha \left\{ \frac{1}{2}(R_e - R_g)\left[\frac{\partial}{\partial \alpha}(\alpha P) + c.c.\right] \right.$$
$$\left. - R_e \frac{\partial^2}{\partial\alpha\partial\alpha^*} P \right\} |\alpha\rangle\langle\alpha|. \qquad (7.68)$$

Here, the fact that $P(\alpha)$ is real and vanishes in the limit $\alpha_1 \to \pm\infty$ and $\alpha_2 \to \pm\infty$ is used; that is,

$$\left[[\alpha P(\alpha)|\alpha\rangle\langle\alpha|]_{-\infty}^{\infty}\right]_{-\infty}^{\infty} = 0. \qquad (7.69)$$

Identifying the coefficients of $|\alpha\rangle\langle\alpha|$ in both sides of (7.68), the equation of motion for $P(\alpha)$ is obtained:

$$\dot{P}(\alpha) = -\frac{1}{2}(R_e - R_g)\left\{\frac{\partial}{\partial\alpha}[\alpha P(\alpha)] + c.c.\right\} + R_e \frac{\partial^2}{\partial\alpha\,\partial\alpha^*} P(\alpha), \qquad (7.70)$$

which is the Fokker-Planck equation.

In order to study the damping and amplification processes in an open dissipative system in the Schrödinger picture, either the (density operator) master (7.61) or the Fokker-Planck (7.70) equations can be used. The advantage of the Fokker-Planck equation is that it significantly simplifies the calculation process for the fields that are approximately coherent states. On the other hand, the simple diagonal $P(\alpha)$ expansion used to obtain (7.70) does not exist in certain cases. Specifically, when the field state becomes nonclassical, (7.70) is no longer a well-behaved Fokker-Planck equation due to the fact that the Glauber-Sudarshan $P(\alpha)$ distribution does not exist for nonclassical light such as squeezed and photon number states. In order to map an arbitrary nonclassical state into a classical probability density, the dimension of the phase space must at least be doubled. The resulting off-diagonal or positive-P representations have been quite popular in quantum optics [9] but are beyond the scope of this book.

7.3.2. Stochastic Differential Equations

One way of solving the Fokker-Planck equation given in (7.70) is to simulate the corresponding classical Ito stochastic differential equation [9]. This stochastic differential equation can be considered as the classical counterpart of the (quantum) Langevin equation of (7.20) and is given by

$$\frac{d}{dt}\beta = -\frac{\gamma}{2}\beta + \overline{f}_\beta, \qquad (7.71)$$

where β is a random variable (c-number) associated with $\hat{A}$. The random noise source satisfies

$$\langle \overline{f}_\beta(t)\overline{f}_\beta(t')\rangle = \langle \overline{f}_{\beta*}(t)\overline{f}_{\beta*}(t')\rangle^* = 2D_{\beta\beta}\delta(t-t') = 0, \qquad (7.72)$$

$$\langle \overline{f}_\beta(t)\overline{f}_{\beta*}(t')\rangle = 2D_{\beta\beta*}\delta(t-t') = \gamma n_{\text{th}}\delta(t-t'). \qquad (7.73)$$

We refrain from a more detailed discussion of Eq. (7.71) and its simulation as our emphasis will be on stochastic wavefunction simulations, which we will discuss in the following chapter.

7.4. QUANTUM REGRESSION THEOREM

In the previous sections, we have seen the principal methods of analyzing the dynamics of open quantum systems. Our goal in this section is to show how to utilize these techniques in the evaluation of the electromagnetic field correlation functions. Quantum regression theorem provides a framework to carry out two-time correlation functions, which are of most interest in the context of quantum optics.

The master equation for an arbitrary open quantum system can be written as

$$\frac{d\hat{\rho}(t)}{dt} = \mathcal{L}_o(\hat{\rho}) + \mathcal{L}_{\text{relax}}(\hat{\rho}), \qquad (7.74)$$

where $\mathcal{L}_o$ and $\mathcal{L}_{\text{relax}}$ denote the system Liouville and relaxation superoperators, respectively. The reduced density operator determined by (7.74) is a *one-time* operator and could be directly used to determine the expectation value of a coupled set of system operators; that is, $\langle \hat{X}(t)\rangle = \text{Tr}_s[\hat{X}\,\hat{\rho}(t)]$. If a finite set of system operators is coupled by the coherent + dissipative dynamics, then we can write the coupled equations for the expectation values of the operators in matrix form:

$$\frac{d\langle \hat{\mathbf{X}}(t)\rangle}{dt} = \underline{\underline{\mathbf{M}}}\langle \hat{\mathbf{X}}(t)\rangle + \underline{\lambda}, \qquad (7.75)$$

where $\underline{\lambda}$ is a constant vector, determined by the particular case of interest. The coupled system operator *vector* is assumed to be finite: $\hat{\mathbf{X}}(t) = [\hat{X}_1(t), \ldots, \hat{X}_n(t)]^T$. For example, (7.75) will be equivalent to the optical Bloch equations in the case of an atomic system interacting with the radiation field reservoir. We can now introduce the Quantum regression theorem.

Theorem (quantum regression): Given the equation of motion of the one-time averages of coupled system operators in (7.75), the two-time averages of the form $\langle \hat{X}_k(t+\tau)\hat{X}_i(t)\rangle$ obey

$$\frac{d\langle \hat{X}_i(t+\tau)\hat{X}_k(t)\rangle}{d\tau} = M_{ij}\langle \hat{X}_j(t+\tau)\hat{X}_k(t)\rangle + \lambda_i\langle \hat{X}_k(t)\rangle. \qquad (7.76)$$

Proof: The proof is based on the Heisenberg-Langevin equations that we have seen in the first section of this chapter [10]. We start from the equation for the operator $\hat{X}_i(t)$,

$$\frac{d\hat{X}_i(t)}{dt} = M_{ij}\hat{X}_j(t) + \lambda_i + \hat{f}_i(t), \qquad (7.77)$$

where $\hat{f}_i(t)$ is the Langevin fluctuation operator with $\langle \hat{f}_i(t) \rangle = 0$ and $\langle \hat{f}_i^\dagger(t) \hat{f}_j(t') \rangle = 2D_{ij}\delta(t-t')$. Taking the expectation value of both sides of (7.77) gives the ith component of (7.75). Now we multiply both sides of (7.77) by $\hat{X}_k(t')$ (with fixed t') and then take the expectation value. We obtain

$$\frac{d\langle \hat{X}_i(t) \hat{X}_k(t') \rangle}{dt} = M_{ij}\langle \hat{X}_j(t)\hat{X}_k(t') \rangle + \lambda_i \langle \hat{X}_k(t') \rangle + \langle \hat{f}_i(t)\hat{X}_k(t') \rangle. \qquad (7.78)$$

The direct integration of the Langevin equation (7.77) shows that $\hat{X}_k(t')$ depends linearly on the values of the Langevin fluctuation operator $\hat{f}_i(t'')$ for $t'' < t'$ [10]. Since the noise operator itself has a very short correlation time (given by the reservoir correlation time τ_c), for times $t - t' > \tau_c$, $\langle \hat{f}_i(t)\hat{X}_k(t') \rangle = 0$. Since the last term drops out, (7.78) is equal to (7.76).

We now use the quantum regression theorem to calculate the resonance fluorescence spectrum and the second-order coherence properties of light generated by a laser-driven two-level atom. The relevant coupled operator equations in this case are the optical Bloch equations [10]

$$\frac{d}{dt}\begin{bmatrix} \langle \hat{\sigma}_{eg}(t) \rangle \\ \langle \hat{\sigma}_{ee}(t) - \hat{\sigma}_{gg}(t) \rangle \\ \langle \hat{\sigma}_{ge}(t) \rangle \end{bmatrix} = \begin{bmatrix} -\gamma_{\text{tot}} + i\Delta\omega & i\frac{\Omega}{2} & 0 \\ i\Omega & -\Gamma & -i\Omega \\ 0 & -i\frac{\Omega}{2} & -\gamma_{\text{tot}} - i\Delta\omega \end{bmatrix}$$
$$\begin{bmatrix} \langle \hat{\sigma}_{eg}(t) \rangle \\ \langle \hat{\sigma}_{ee}(t) - \hat{\sigma}_{gg}(t) \rangle \\ \langle \hat{\sigma}_{ge}(t) \rangle \end{bmatrix} - \begin{bmatrix} 0 \\ \Gamma \\ 0 \end{bmatrix}, \qquad (7.79)$$

where Γ, γ_{tot}, and Ω denote the spontaneous emission rate, the coherence dephasing rate, and the coupling Rabi frequency, respectively. $\Delta\omega = \omega_{eg} - \omega_L$, where ω_{eg} and ω_L are the atomic transition and laser frequencies. Using the solution to (7.79) and the quantum regression theorem, we can evaluate the vectors

$$\mathbf{V}(t, t') = \begin{bmatrix} \langle \hat{\sigma}_{eg}(t)\hat{\sigma}_{ge}(t') \rangle, & \langle (\hat{\sigma}_{ee}(t) - \hat{\sigma}_{gg}(t))\hat{\sigma}_{ge}(t') \rangle, & \langle \hat{\sigma}_{ge}(t)\hat{\sigma}_{ge}(t') \rangle \end{bmatrix}^T$$

$$\mathbf{W}(t, t') = \begin{bmatrix} \langle \hat{\sigma}_{eg}(t')\hat{\sigma}_{eg}(t)\hat{\sigma}_{ge}(t') \rangle, & \langle \hat{\sigma}_{eg}(t')(\hat{\sigma}_{ee}(t) - \hat{\sigma}_{gg}(t))\hat{\sigma}_{ge}(t') \rangle, \\ & \langle \hat{\sigma}_{eg}(t')\hat{\sigma}_{ge}(t)\hat{\sigma}_{ge}(t') \rangle \end{bmatrix}^T.$$

Before proceeding, we note that the positive frequency component of the electric field generated by a single atom (at position $\mathbf{R}_i$) is proportional to the dipole operator $\hat{\sigma}_{eg}(t)$ [11]:

$$\hat{E}_i^+(\mathbf{r}, t) = \frac{\omega_{eg}^2 \mu_{ge}}{4\pi\epsilon_0 c^2 |\mathbf{r} - \mathbf{R}_i|} \hat{\sigma}_{eg}\left(t - \frac{|\mathbf{r} - \mathbf{R}_i|}{c}\right). \quad (7.80)$$

As a direct consequence of this relation, the two-time dipole correlation function $\langle \hat{\sigma}_{eg}(t) \hat{\sigma}_{ge}(t') \rangle$ is proportional to the first-order coherence function of the generated electric field, whose Fourier transform directly gives the desired resonance fluorescence spectrum [5], [11]. On the other hand, $\langle \hat{\sigma}_{eg}(t')[\hat{\sigma}_{ee}(t) - \hat{\sigma}_{gg}(t)]\hat{\sigma}_{ge}(t') \rangle$ gives the second-order coherence function of the electromagnetic field, which was shown to be of fundamental interest in quantum optics. The inital condition vectors satisfy

$$\mathbf{V}(t', t') = \left[\langle \hat{\sigma}_{ee}(t') \rangle, -\langle \hat{\sigma}_{ge}(t') \rangle, 0\right]^T$$
$$\mathbf{W}(t', t') = \left[0, -\langle \hat{\sigma}_{ee}(t') \rangle, 0\right]^T.$$

To evaluate the correlation functions, we first have to solve the optical Bloch equations, as this solution is required to evaluate the initial conditions.

The solution of (7.79) is lengthy but straightforward. Using Laplace transform techniques and assuming $\gamma_{\text{tot}} = \Gamma/2$ (i.e., pure spontaneous emission broadening) and $\Delta\omega = 0$, we find

$$\langle \tilde{\sigma}_{eg}(s) \rangle = \frac{(s+\Gamma)(s+\Gamma/2) + \Omega^2/2}{(s-s_1)(s-s_2)(s-s_3)} \langle \hat{\sigma}_{eg}(0) \rangle$$
$$+ \frac{\Omega^2/2}{(s-s_1)(s-s_2)(s-s_3)} \langle \hat{\sigma}_{ge}(0) \rangle$$
$$+ \frac{i\Omega/2}{(s-s_2)(s-s_3)} \langle \hat{\sigma}_{ee}(0) - \hat{\sigma}_{gg}(0) \rangle$$
$$+ \frac{i\Omega/2}{s(s-s_2)(s-s_3)}(-\Gamma), \quad (7.81)$$

$$\langle \tilde{\sigma}_{ee}(s) - \tilde{\sigma}_{gg}(s) \rangle = \frac{i\Omega}{(s-s_2)(s-s_3)} \langle \hat{\sigma}_{eg}(0) - \hat{\sigma}_{ge}(0) \rangle$$
$$+ \frac{s + \Gamma/2}{(s-s_2)(s-s_3)} \langle \hat{\sigma}_{ee}(0) - \hat{\sigma}_{gg}(0) \rangle$$
$$+ \frac{s + \Gamma/2}{s(s-s_2)(s-s_3)}(-\Gamma), \quad (7.82)$$

where s_1, s_2, s_3 are the roots of the characteristic polynomial $\det(s\underline{I} - \underline{M})$ with $s_1 = -\Gamma/2$, $s_{2,3} = -3\Gamma/4 \pm \sqrt{\Gamma^2/16 - \Omega^2} = -3\Gamma/4 \pm \kappa$. s is the transform variable and $\langle \tilde{\sigma}_{eg}(s) \rangle$, $\langle \tilde{\sigma}_{ee}(s) - \tilde{\sigma}_{gg}(s) \rangle$ denote the transformed quantities [12]. The physically interesting initial condition is that of an atom initially in the ground state $\langle\hat{\sigma}_{gg}(0)\rangle = 1$: For this choice we obtain

$$\langle\hat{\sigma}_{eg}(t)\rangle = -i\frac{\Omega\Gamma}{\Gamma^2 + 2\Omega^2}$$
$$- i\frac{\Omega}{4\kappa}e^{-3\Gamma t/4}\left[\frac{\kappa + \Gamma/4}{\kappa - 3\Gamma/4}e^{\kappa t} + \frac{-\kappa + \Gamma/4}{\kappa + 3\Gamma/4}e^{-\kappa t}\right] \quad (7.83)$$

$$\langle\hat{\sigma}_{ee}(t)\rangle = \frac{\Omega^2}{\Gamma^2 + 2\Omega^2}$$
$$- \frac{1}{4\kappa}e^{-3\Gamma t/4}\left[\frac{\kappa - \Gamma/4}{\kappa - 3\Gamma/4}(\kappa + \Gamma/4)e^{\kappa t} + \frac{\kappa + \Gamma/4}{\kappa + 3\Gamma/4}(\kappa - \Gamma/4)e^{-\kappa t}\right]. \quad (7.84)$$

We now turn to the two-time correlation functions. The power of the quantum regression theorem is that by using (7.81), we can write the solution in the transform domain as

$$\langle\tilde{\sigma}_{eg}(s,t')\tilde{\sigma}_{ge}(t')\rangle = \frac{(s+\Gamma)(s+\Gamma/2)+\Omega^2/2}{(s-s_1)(s-s_2)(s-s_3)}\langle\hat{\sigma}_{ee}(t')\rangle$$
$$- \frac{i\Omega^2/2}{(s-s_2)(s-s_3)}\langle\hat{\sigma}_{ge}(t')\rangle$$
$$+ \frac{i\Omega^2/2}{s(s-s_2)(s-s_3)}(-\Gamma)\langle\hat{\sigma}_{ge}(t')\rangle, \quad (7.85)$$

where the transform variables are $\tau = t - t'$ and s. The contribution in the second line in (7.85) comes from the last term of (7.76). Assuming that the correlation function is evaluated long after the initial turn-on, we replace the *initial values* $\langle\hat{\sigma}_{ee}(t')\rangle$ and $\langle\hat{\sigma}_{ge}(t')\rangle$ with their steady-state values obtained from (7.84). The resulting expression for the first-order coherence function $g^{(1)}(\tau) = \langle\hat{\sigma}_{eg}(\tau,t')\hat{\sigma}_{ge}(t')\rangle e^{-i\omega_L\tau}/\langle\hat{\sigma}_{ee}(t')\rangle$ discussed in Chapter 6 is

$$g^{(1)}(\tau) = \frac{\Gamma^2}{\Gamma^2 + 2\Omega^2}e^{-i\omega_L\tau} + \frac{\Omega^2/2}{(\kappa - \Gamma/4)(-\kappa - \Gamma/4)}e^{-\Gamma\tau/2 - i\omega_L\tau}$$
$$+ \left[\frac{(\kappa + \Gamma/4)(\kappa - \Gamma/4) + \Omega^2/2}{2\kappa(\kappa - \Gamma/4)} + \frac{(\kappa + \Gamma/4)\Gamma/2}{2\kappa(\kappa - 3\Gamma/4)}\right]e^{(-3\Gamma/4+\kappa)\tau - i\omega_L\tau}$$
$$+ \left[\frac{(-\kappa + \Gamma/4)(-\kappa - \Gamma/4) + \Omega^2/2}{2\kappa(\kappa + \Gamma/4)}\right.$$
$$\left. + \frac{(-\kappa + \Gamma/4)\Gamma/2}{2\kappa(\kappa + 3\Gamma/4)}\right]e^{(-3\Gamma/4-\kappa)\tau - i\omega_L\tau}. \quad (7.86)$$

Inspection of (7.86) reveals several interesting features about the fluorescence properties of two-level atoms. First, we see that only the first term in the expression for $g^{(1)}(\tau)$ is first-order coherent; this is the Rayleigh scattering contribution. The contribution of this *coherent* term to the resonance fluorescence spectrum [obtained by Fourier transforming $g^{(1)}(\tau)$] is a δ-function. If we had made the crude decorrelation approximation $\langle \hat{\sigma}_{eg}(\tau, t')\hat{\sigma}_{ge}(t') \rangle \simeq \langle \hat{\sigma}_{eg}(\tau, t') \rangle \langle \hat{\sigma}_{ge}(t') \rangle$, then a Rayleigh scattering contribution would have appeared as the only contribution to the resonance fluorescence. For the exact result of (7.86), coherent emission dominates the spectrum only in the very weak field limit ($\Gamma/4 \gg \Omega$).

For the first-order *incoherent* terms of $g^{(1)}(\tau)$, the relative magnitude of the decay rate and the Rabi frequency determine the center frequency and/or width of the fluorescence terms. In the limit, $\Gamma/4 \geq \Omega$, κ is real and the resonance fluorescence consists of three Lorentzians with different widths all centered at $\omega_L = \omega_{eg}$. For $\Gamma/4 < \Omega$, on the other hand, κ is imaginary and the corresponding spectrum consists of three Lorentzians with different center frequencies (in addition to the Rayleigh scattering contribution). Figure 7-4 shows the spectrum in the limit $\Gamma/4 \ll \Omega$; the three fluorescence peaks are referred to as *the Mollow triplets*, which we discussed earlier [5]. When $\Gamma/4 \ll \Omega$, the Rayleigh contribution vanishes; in this limit the resonance fluorescence is totally *incoherent* and originates from the fluctuations in the dipole operator. The reader is referred to Ref. [11] for an analysis of the effects of finite detuning and dephasing on the resonance fluorescence.

Next, we turn to the calculation of the second-order coherence function: Using (7.82) and the initial condition vector $\mathbf{W}(t', t')$, we obtain

$$\langle \tilde{\sigma}_{eg}(t')[\tilde{\sigma}_{ee}(s, t') - \tilde{\sigma}_{gg}(s, t')]\tilde{\sigma}_{ge}(t') \rangle = -\frac{(s+\Gamma)(s+\Gamma/2)}{s(s-s_2)(s-s_3)} \langle \hat{\sigma}_{ge}(t') \rangle. \quad (7.87)$$

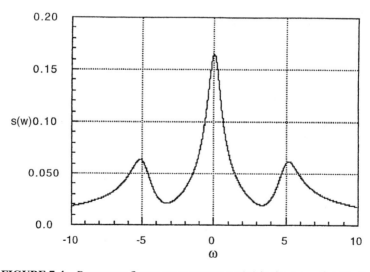

FIGURE 7-4: Resonance fluorescence spectrum s(w) in the strong field limit.

Once again, by replacing $\langle \hat{\sigma}_{ge}(t') \rangle$ with its steady-state value obtained from (7.84), we find

$$g^{(2)}(\tau) = \frac{\langle \hat{\sigma}_{eg}(t')[\hat{\sigma}_{ee}(\tau, t') - \hat{\sigma}_{gg}(\tau, t')]\hat{\sigma}_{ge}(t') \rangle + \langle \hat{\sigma}_{ee}(t') \rangle}{2\langle \hat{\sigma}_{ee}(t') \rangle^2}$$

$$= 1 + \frac{\Gamma^2 + 2\Omega^2}{4\kappa} \left[\frac{e^{\kappa\tau}}{\kappa - 3\Gamma/4} + \frac{e^{-\kappa\tau}}{\kappa + 3\Gamma/4} \right] \exp\left[\frac{-3\Gamma\tau}{4}\right]. \quad (7.88)$$

Despite its simplicity, (7.88) demonstrates that light generated by the most elementary of all systems (that is, a single two-level atom) exhibits coherence properties that cannot be explained by classical field theories. For all values of the coupling field strength, we find

$$g^{(2)}(0) = 0. \quad (7.89)$$

As we have discussed in Chapter 6, this result implies photon antibunching and it has a very simple physical explanation in the present system: If the atom emits a photon at $\tau = 0$, it is impossible for it to emit another one at $\tau = 0^+$ as it is necessarily in the ground state (due to the first emission event). The next photon can only be emitted after a *waiting time* determined by the strength of the coherent excitation and the spontaneous emission rate. For example, in the strong coupling ($\Omega \gg \Gamma/4$) regime, we find

$$g^{(2)}(\tau) = 1 - e^{-3\Gamma\tau/4} \cos(\Omega\tau). \quad (7.90)$$

For these parameters, the atom makes Rabi oscillations; therefore, at $\tau = \pi/\Omega$ the atom is (almost) guaranteed to be in the excited state and the likelihood of the second photon emission is higher than it would be in a Poissonian photon stream. Figure 7-5 shows $g^{(2)}(\tau)$ in this strong coupling limit.

So far we have discussed the two-time correlation functions of light emitted by a single atom. It is instructive to consider how the predicted coherence properties get modified for an ensemble of atoms. The total field generated by M-atoms is given by

$$\hat{E}^+(\mathbf{r}, t) = \sum_{i=1}^{M} \hat{E}_i^+(\mathbf{r}, t), \quad (7.91)$$

where $\hat{E}_i^+(\mathbf{r}, t)$ is as given in (7.80). In most of the practical cases, these M atoms will be randomly distributed in space, giving rise to random phase factors $k|\mathbf{r} - \mathbf{R}_i|$. Collisions and Doppler broadening also contribute to the randomization of the phase factors. As a result, the net electric field generated by such an ensemble vanishes.

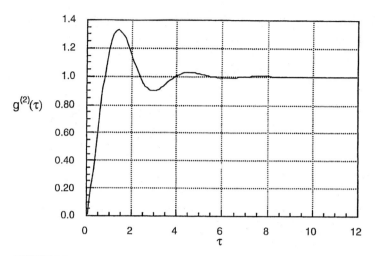

FIGURE 7-5: Second-order correlation function in the strong field limit.

The (unnormalized) first-order coherence function in this case is

$$\langle \hat{E}^-(t+\tau)\hat{E}^+(t)\rangle = \sum_{i,j}^{M}\langle \hat{E}_i^-(t+\tau)\hat{E}_j^+(t)\rangle$$

$$= \sum_{i}^{M}\langle \hat{E}_i^-(t+\tau)\hat{E}_i^+(t)\rangle$$

$$= M\langle \hat{E}_i^-(t+\tau)\hat{E}_i^+(t)\rangle, \quad (7.92)$$

where we have assumed that the atomic phases are random and that each atom has identical coherence properties. A straightforward extension of (7.92) gives

$$g_M^{(1)}(\tau) = \frac{\langle \hat{E}^-(t+\tau)\hat{E}^+(t)\rangle}{\langle \hat{E}^-(t)\hat{E}^+(t)\rangle} = g^{(1)}(\tau); \quad (7.93)$$

that is, the first-order coherence properties are independent of the number of atoms generating the fluorescence.

Similarly, for the second-order coherence we find

$$\langle \hat{E}^-(t)\hat{E}^-(t+\tau)\hat{E}^+(t+\tau)\hat{E}^+(t)\rangle = \sum_{i}^{M}\langle \hat{E}_i^-(t)\hat{E}_i^-(t+\tau)\hat{E}_i^+(t+\tau)\hat{E}_i^+(t)\rangle$$

$$+ \sum_{i\neq j}^{M}[\langle \hat{E}_i^-(t)\hat{E}_j^-(t+\tau)\hat{E}_j^+(t+\tau)\hat{E}_i^+(t)\rangle$$

$$+ \langle \hat{E}_i^-(t)\hat{E}_j^-(t+\tau)\hat{E}_i^+(t+\tau)\hat{E}_j^+(t)\rangle \,]$$
$$= M \langle \hat{E}_i^-(t)\hat{E}_i^-(t+\tau)\hat{E}_i^+(t+\tau)\hat{E}_i^+(t)\rangle$$
$$+ M(M-1)\left[|\langle \hat{E}_i^-(t)\hat{E}_i^+(t)\rangle|^2\right.$$
$$\left. + |\langle \hat{E}_i^-(t+\tau)\hat{E}_i^+(t)\rangle|^2\right], \qquad (7.94)$$

which in turn gives

$$g_M^{(2)}(\tau) = \frac{g^{(2)}(\tau)}{M} + \frac{M-1}{M}(1 + |g^{(1)}(\tau)|^2). \qquad (7.95)$$

Let's now consider the limit $M \gg 1$; the first term in (7.95) vanishes in comparison to the second, giving

$$g_M^{(2)}(\tau) = 1 + |g^{(1)}(\tau)|^2, \qquad (7.96)$$

which is the result we have given for broadband incoherent light in Chapter 6. As expected, an ensemble of randomly phased atoms generates classical thermal light, erasing all the quantum signatures of the light generation process. Intuitively, we do not expect a *dead time* between fluorescence events, since there are M uncorrelated atoms that could contribute to the fluorescence process.

REFERENCES

[1] M. Sargent III, M. O. Scully, and W. E. Lamb, Jr., *Laser Physics* (Addison-Wesley, Reading, Mass., 1974).

[2] H. Haken, *Light* (North-Holland, Amsterdam, 1985).

[3] C. W. Gardiner, *Handbook of Stochastic Methods* (Springer-Verlag, Berlin, 1985).

[4] W. H. Louisell, *Quantum Statistical Properties of Radiation* (Wiley, New York, 1973).

[5] D. F. Walls and G. J. Milburn, *Quantum Optics* (Springer-Verlag, Berlin, 1994).

[6] A. O. Caldeira and A. J. Leggett, Ann. Phys. (N.Y.) **149**, 374 (1983).

[7] S. van Smaalen and T. F. George, J. Chem. Phys. **87**, 5504 (1987).

[8] B. L. Hu, J. P. Paz, and Y. Zhang, Phys. Rev. D **45**, 2843 (1992).

[9] C. W. Gardiner, *Quantum Noise* (Springer-Verlag, Berlin, 1991).

[10] C. Cohen-Tannoudji, J. Dupont-Roc, and G. Grynberg, *Atom-Photon Interactions* (Wiley-Interscience, New York, 1992).

[11] R. Loudon, *The Quantum Theory of Light* (Oxford University Press, New York, 1985).

[12] F. W. Byron, Jr. and R. W. Fuller, *Mathematics of Classical and Quantum Physics* (Dover, New York, 1992).

8 Stochastic Wavefunction Methods

In Chapter 7, we discussed the density operator methods and derived the master equation for a *system* density operator in a number of specific cases that are of utmost importance in quantum optics. Phase space methods such as Glauber-Sudarshan *P* representation convert the operator master equation into a Fokker-Planck equation for a probability distribution, which significantly simplifies the calculation process. For example, we shall see in the discussion of the quantum theory of laser that the linewidth calculation from the master equation is relatively involved, whereas once the Fokker-Planck equation is obtained, the linewidth appears directly as the coefficient of the diffusion term. In many other cases of interest, however, direct numerical integration of the master equation is useful.

According to the standard interpretation of quantum mechanics, the (reduced) density operator provides us the maximum quantum-statistical information about the system under investigation. If no observation is being carried out, the system-reservoir interaction is equivalent to a *nonreferring* measurement, during which the system evolves into a mixed state. As we have discussed earlier, the density operator corresponding to such a mixed state provides *statistical* information about an ensemble of identical quantum systems. For example,

$$\hat{\rho}_s(t) = \sum_i P_i |\psi_i(t)\rangle\langle\psi_i(t)| \tag{8.1}$$

tells us that among the ensemble, the probability of finding a single-quantum system with wavefunction $|\psi_j(t)\rangle$ is given by P_j. Such a description is necessary when our measurements involve averages over many single-system experiments or when we cannot keep track of all the information output from the system.

Advances in quantum optics experiments, on the other hand, have made it possible to observe and manipulate single quantum systems embedded in dissipative environments [1]–[3]: The prototypical example is a cavity mode (optical or microwave) interacting with atoms and weak (classical) coherent fields. These experiments, which are detailed elsewhere, provide us the *quantum trajectory* [4] that the particular single-quantum system under observation has followed in time. Once we analyze the experimental data, we can write down a time-dependent wavefunction for the system. Naturally, this wavefunction is modified discontinuously at particular *measurement times* due to an observation of a photon at the detector.

Motivated in part by these experimental developments, a number of groups have recently developed new (related) methods for analyzing the dynamics of dissipative quantum systems [4]–[7]. The underlying idea behind these stochastic wavefunction methods is the description of the quantum systems using a stochastic wavefunction [8] that is subjected to a continuous measurement [9]. The stochasticity enters through the measurement results, which are decided by choosing pseudorandom numbers. A set of such random numbers then gives us a possible quantum trajectory for the single-quantum system. The results pertaining to the ensemble are obtained by averaging over many such trajectories. The new method is formally equivalent to the master equation from which it is derived. It has no additional predictive power but provides new insight into the problem by describing what may happen (or what is likely to happen) during the time evolution.

A significant advantage of the new stochastic wavefunction methods becomes clear when the system under consideration has a large number of variables: Assume, for example, that we need to consider N Fock states to describe the dynamics of a cavity mode. The master equation treatment would require the propagation of roughly N^2 elements in time. For the stochastic wavefunction methods, this number is N; clearly, if N is very large ($\gg 1$), as is the case in some of the quantum optics problems, the new method is significantly more favorable. In fact, if $N \simeq 10^5$, numerical simulation of master equation using modern computers is practically impossible due to memory limitations. This argument does not apply directly to computation time considerations; the accuracy requirements determine the number of averaging needed in the stochastic wavefunction treatment, which in turn determines the final computation time.

In the following two sections, we present two different realizations of the stochastic wavefunction methods: First, we consider the quantum Monte Carlo wavefunction (MCWF) approach, developed independently by Carmichael [4], Dalibard et al. [5], and Gardiner et al. [6]. The treatment we present closely follows that of Ref. [8]. In the last section, we present a brief discussion of the quantum state diffusion (QSD) model of Gisin and Percival [7].

8.1. MONTE CARLO WAVEFUNCTION APPROACH

In this section, we concentrate on the quantum MCWF approach, which describes the dissipative dynamics of an open quantum system using two complementary processes: A nonunitary time evolution by an effective non-Hermitian Hamiltonian, and quasi-continuous *gedanken measurements* (quantum jumps) whose outcomes are determined by random numbers. Following Ref. [8], we start from the general form of the master equation for the reduced system density operator $\hat{\rho}_s(t)$ in the Born-Markov approximation

$$\frac{d\hat{\rho}_s(t)}{dt} = \frac{1}{i\hbar}[\hat{H}_o, \hat{\rho}_s] - \frac{1}{2}\sum_i(\hat{C}_i^\dagger\hat{C}_i\hat{\rho}_s(t) + \hat{\rho}_s(t)\hat{C}_i^\dagger\hat{C}_i - 2\hat{C}_i\hat{\rho}_s(t)\hat{C}_i^\dagger). \quad (8.2)$$

Here $\hat{H}_o$ is the system Hamiltonian, which may include coherent interactions with classical fields. Essentially for all the fundamental systems in quantum optics, the dissipative interactions appear in the form shown in (8.2), which is referred to as the Linblad form. The *collapse operators* $\hat{C}_i$, on the other hand, vary according to the particular system that is being studied. If, for example, the system under study is a harmonic oscillator that is coupled to a bath of harmonic oscillators at zero temperature, then $\hat{C} = \sqrt{\kappa}\,\hat{a}$, where $\hat{a}$ is the annihilation operator for the system mode and κ is the golden-rule decay rate. For a two-level atom interacting with the (zero temperature) radiation field reservoir, $\hat{C} = \sqrt{\Gamma}\,\hat{\sigma}_{ge}$, with $\hat{\sigma}_{ge}$ and Γ denoting the atomic lowering operator and the spontaneous emission rate, respectively.

8.1.1. Description and Equivalence to Master Equation

Let's start by postulating mathematically the MCWF prescription described previously: First, the nonunitary evolution is described by the unique effective Hamiltonian:

$$\hat{H}_{\text{eff}} = \hat{H}_o - \frac{i\hbar}{2}\sum_i \hat{C}_i^\dagger \hat{C}_i. \tag{8.3}$$

The stochastic wavefunction that is assigned to the system $|\psi(t)\rangle$ is evolved from t to $t + \delta t$ using this effective Hamiltonian; δt is chosen to be much shorter than the system time scales. The premeasurement ket $|\psi_{\text{pm}}(t+\delta t)\rangle$ obtained as a result is not normalized due to the anti-Hermitian part of (8.3).

The second step in the algorithm is to compare the probability for a decay event in the time interval $(t, t+\delta t)$

$$\delta p = 1 - \langle \psi_{\text{pm}}(t+\delta t) | \psi_{\text{pm}}(t+\delta t) \rangle \tag{8.4}$$

to a pseudorandom number ϵ; this process is used to *decide* the outcome of the gedanken measurement and thereby determine the system wavefunction at $t + \delta t$. If $\delta p < \epsilon$, then we conclude that the measurement result was *negative* (i.e., no quantum jump or collapse process has taken place); we then set

$$|\psi(t+\delta t)\rangle = \frac{1}{\sqrt{1-\delta p}}|\psi_{\text{pm}}(t+\delta t)\rangle. \tag{8.5}$$

If $\delta p > \epsilon$, then the measurement result is positive and we decide that a quantum jump event took place. If there is only one collapse operator $\hat{C}$, then

$$|\psi(t+\delta t)\rangle = \sqrt{\frac{\delta t}{\delta p}}\,\hat{C}|\psi(t)\rangle. \tag{8.6}$$

If the dissipative interactions can be described by two or more collapse operators, we also need to decide which process has occured according to the respective probabilites of each event; that is,

$$|\psi(t+\delta t)\rangle = \sqrt{\frac{\delta t}{\delta p_i}}\,\hat{C}_i|\psi(t)\rangle, \tag{8.7}$$

with a probability $P_i = \delta p_i/\delta p$, where $\delta p = \sum_i \delta p_i$ [8]. Having determined the wavefunction at time $t + \delta t$, we can proceed with the initial step and propagate the system using $\hat{H}_{\text{eff}}$ until $t + 2\delta t$.

We now show that the equation of motion of an ensemble of these stochastic wavefunctions gives us the master equation of (8.2). Since we are assuming that δt is much smaller than the system time scales, we can write

$$|\psi_{\text{pm}}(t + \delta t)\rangle \simeq \left(1 - \frac{i\,\delta t}{\hbar}\hat{H}_{\text{eff}}\right)|\psi(t)\rangle. \tag{8.8}$$

To first order in δt, the squared norm of $|\psi_{\text{pm}}(t + \delta t)\rangle$ is given by

$$\begin{aligned}
\langle\psi_{\text{pm}}(t + \delta t)|\psi_{\text{pm}}(t + \delta t)\rangle &= \langle\psi(t)|\left(1 + \frac{i\,\delta t}{\hbar}\hat{H}_{\text{eff}}^{\dagger}\right)\left(1 - \frac{i\,\delta t}{\hbar}\hat{H}_{\text{eff}}\right)|\psi(t)\rangle \\
&\simeq 1 - \frac{i\,\delta t}{\hbar}\langle\psi(t)|\hat{H}_{\text{eff}} - \hat{H}_{\text{eff}}^{\dagger}|\psi(t)\rangle \\
&= \delta t \sum_i \langle\psi(t)|\hat{C}_i^{\dagger}\hat{C}_i|\psi(t)\rangle \\
&= 1 - \sum_i \delta p_i. \tag{8.9}
\end{aligned}$$

We therefore see that it is the anti-Hermitian part that results in the decay of the squared norm of the stochastic wavefunction, which in turn gives us the probability that a quantum jump would take place in that time interval. According to the MCWF algorithm, if we start from a specific wavefunction $|\psi(t)\rangle$ at time t, we have a probability $1 - \delta p$ of finding the system in $|\psi_{\text{pm}}(t + \delta t)\rangle/\sqrt{1 - \delta p}$, and a probability δp_i for $\sqrt{\delta t}\,\hat{C}_i|\psi(t)\rangle/\sqrt{\delta p_i}$. If we consider an ensemble of identical systems all with the same wavefunction at time t, we can write the density operator at time $t + \delta t$ according to these probabilities:

$$\begin{aligned}
\hat{\rho}_1(t + \delta t) &= (1 - \delta p)\frac{1}{1 - \delta p}|\psi_{\text{pm}}(t + \delta t)\rangle\langle\psi_{\text{pm}}(t + \delta t)| \\
&\quad + \sum_i \delta p_i \frac{\delta t}{\delta p_i}\hat{C}_i|\psi(t)\rangle\langle\psi(t)|\hat{C}_i^{\dagger} \\
&= \left(1 - \frac{i\,\delta t}{\hbar}\hat{H}_{\text{eff}}\right)\hat{\rho}_1(t)\left(1 + \frac{i\,\delta t}{\hbar}\hat{H}_{\text{eff}}^{\dagger}\right) + \delta t \sum_i \hat{C}_i\hat{\rho}_1(t)\hat{C}_i^{\dagger} \tag{8.10} \\
&\simeq \hat{\rho}_1(t) + \frac{\delta t}{i\hbar}[\hat{H}_o, \hat{\rho}_1(t)] \\
&\quad - \frac{\delta t}{2}\sum_i(\hat{C}_i^{\dagger}\hat{C}_i\hat{\rho}_1(t) + \hat{\rho}_1(t)\hat{C}_i^{\dagger}\hat{C}_i - 2\hat{C}_i\hat{\rho}_1(t)\hat{C}_i^{\dagger}).
\end{aligned}$$

Finally, we average over all possible initial stochastic wavefunctions $|\psi(t)\rangle$ and set $\overline{\hat{\rho}_1} \to \hat{\rho}_s$. When we take the limit as $\delta t \to 0$, we obtain the master equation in (8.2). This short exercise proves that for any system-reservoir interaction whose corresponding master equation can be put in the Linblad form of (8.2), the quantum MCWF formalism is equivalent to the master equation formalism. We remark that we have not made any assumption about the collapse operators $\hat{C}_i$; these could be Hermitian or non-Hermitian, and the number of such operators necessary for complete description may in some cases be infinite.

8.1.2. Two-Time Correlation Functions

One immediate application of the quantum MCWF formalism as a calculational tool is the evaluation of expectation values of system observables. However, since the formalism is equivalent to the master equation, we would expect it to be applicable to the calculation of two-time correlation functions as well. We have already seen in Chapter 7 that the quantum regression theorem allows us to carry out such calculations. We now discuss the method developed in Ref. [8] to show how to use MCWF in the calculation of $\langle \hat{A}(t+\tau)\hat{B}(t)\rangle$, where $\hat{A}$ and $\hat{B}$ are any two system operators.

The first step is to start from any initial condition at t_o and propagate the stochastic wavefunction $|\psi(t)\rangle$ until t: It is important that $t - t_o$ is long compared to the relevant time scales so that the system has reached its steady state. We then form four new wavefunctions using the projections

$$|\varphi_1(\tau = 0)\rangle = \frac{1}{\sqrt{N_1}} (1 + \hat{B})|\psi(t)\rangle \qquad (8.11)$$

$$|\varphi_2(\tau = 0)\rangle = \frac{1}{\sqrt{N_2}} (1 - i\hat{B})|\psi(t)\rangle \qquad (8.12)$$

$$|\varphi_3(\tau = 0)\rangle = \frac{1}{\sqrt{N_3}} (1 - \hat{B})|\psi(t)\rangle \qquad (8.13)$$

$$|\varphi_4(\tau = 0)\rangle = \frac{1}{\sqrt{N_4}} (1 + i\hat{B})|\psi(t)\rangle, \qquad (8.14)$$

where N_i are the normalization factors. The next step in the calculation is to propagate these wavefunctions $|\varphi_i(\tau)\rangle$ in time using the MCWF algorithm and calculate the single-time averages:

$$c_i(\tau) = \langle \varphi_i(\tau)|\hat{A}|\varphi_i(\tau)\rangle. \qquad (8.15)$$

The correlation function $\langle \hat{A}(t+\tau)\hat{B}(t)\rangle$ is then given by

$$\langle \hat{A}(t+\tau)\hat{B}(t)\rangle = \frac{1}{4} \sum_j i^{j-1} N_j \overline{c_j(\tau)}, \qquad (8.16)$$

where the averages are first taken over the trajectories of the projected wavefunctions $|\varphi_i(\tau)\rangle$, and then over the possible outcomes of the initial time evolution of $|\psi(t)\rangle$. Proof of (8.16) is carried out in Ref. [8]. Extension of this method to the calculation of second-order coherence functions such as $\langle \hat{B}^\dagger \hat{A}^\dagger(t+\tau)\hat{A}(t+\tau)\hat{B}(t)\rangle$ presents no difficulties.

We now proceed by applying the quantum MCWF formalism to specific problems. First we consider a two-level atom driven by a classical laser field. As a second and last example, we will simulate the dynamics of a single-cavity mode driven by a thermal field reservoir.

8.1.3. Two-Level Atom Driven by a Laser Field

One of the most elementary systems in (atomic) quantum optics is the coherently driven two-level atom. Early work on first- and second-order coherence properties of light generated by this system has established new concepts such as dressed states and photon antibunching. In Chapter 7, we derived the master equation for a coherently driven two-level atom, which can be written as

$$\frac{d\hat{\rho}_s(t)}{dt} = \frac{1}{i\hbar}[\hat{H}_o, \hat{\rho}_s] - \frac{\Gamma}{2}(\hat{\sigma}^\dagger_{ge}\hat{\sigma}_{ge}\hat{\rho}_s(t) + \hat{\rho}_s(t)\hat{\sigma}^\dagger_{ge}\hat{\sigma}_{ge} - 2\hat{\sigma}_{ge}\hat{\rho}_s(t)\hat{\sigma}^\dagger_{ge}), \quad (8.17)$$

where $\hat{H}_o = \hbar\omega_{eg}\hat{\sigma}_{ee} + (0.5\mu_{eg}E\exp[-i\omega_L t]\hat{\sigma}_{eg} + h.c.)$ and $\hat{\sigma}_{ee} = \hat{\sigma}_{eg}\hat{\sigma}_{ge}$. μ_{eg} and Γ denote the dipole matrix element and the atomic spontaneous emission rate, respectively. We can clearly see from (8.17) that $\hat{C} = \sqrt{\Gamma}\hat{\sigma}_{ge}$ is the collapse operator describing the lowering of the atomic excitation following the spontaneous emission event. Following the MCWF algorithm, we then obtain the effective Hamiltonian in the rotating frame:

$$\hat{H}_{\text{eff}} = \hbar(\omega_{eg} - \omega_L)\hat{\sigma}_{ee} + \hbar\frac{\Omega_{eg}}{2}(\hat{\sigma}_{eg} + \hat{\sigma}_{ge}) - \frac{i\hbar}{2}\Gamma\hat{\sigma}_{ee}, \quad (8.18)$$

where $\Omega_{eg} = \mu_{eg}E/\hbar$ is the Rabi frequency of the transition, chosen to be real for simplicity. The simulations based on this model are extremely simple as they require the expansion of the *system* wavefunction as a superposition of two eigenkets [i.e., $|\psi_s\rangle = a_g(t)|g\rangle + a_e(t)|e\rangle$]. A typical simulation result for the expectation value of the atomic population difference $\langle\hat{\sigma}_{ee} - \hat{\sigma}_{gg}\rangle$ is shown in Fig. 8-1, where the Rabi frequency is assumed to be twice as large as the spontaneous decay rate. We observe that coherent Rabi oscillations of the atoms are interrupted by discrete quantum jumps corresponding to spontaneous emission events. It should be pointed out that the quantum trajectory depicted here is qualitatively the same as those obtained in trapped single-ion experiments [10]–[12].

Figure 8-2 compares the result of an average over 10 and 100 stochastic wavefunction simulations for the atomic population difference. The MCWF method clearly does not represent any computational advantage over the master equation for this particular problem.

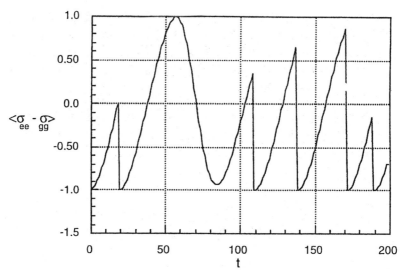

FIGURE 8-1: A typical MCWF simulation result showing the time evolution of the atomic population difference.

8.1.4. Single-Mode Cavity Driven by a Thermal Field

We now consider a single-cavity mode coupled to the radiation field reservoir at a finite temperature T_{rad} through the cavity mirrors. As we have seen before, this system can be modeled as a harmonic oscillator coupled to a bath of harmonic oscillators. The master equation has already been derived in Chapter 7:

$$\frac{d\hat{\rho}_s(t)}{dt} = \frac{1}{i\hbar}[\hbar\omega_s \hat{a}^\dagger \hat{a}, \hat{\rho}_s] - \frac{\kappa(1+\bar{n})}{2}(\hat{a}^\dagger \hat{a}\hat{\rho}_s(t) + \hat{\rho}_s(t)\hat{a}^\dagger \hat{a} - 2\hat{a}\hat{\rho}_s(t)\hat{a}^\dagger)$$

$$- \frac{\kappa \bar{n}}{2}(\hat{a}\hat{a}^\dagger \hat{\rho}_s(t) + \hat{\rho}_s(t)\hat{a}\hat{a}^\dagger - 2\hat{a}^\dagger \hat{\rho}_s(t)\hat{a}), \tag{8.19}$$

where $\hat{a}$ is the annihilation operator for the cavity mode and $\bar{n} = 1/(\exp[\hbar\omega_s/kT_{\text{rad}}]-1)$ denotes the average reservoir occupancy. κ is the cavity decay rate. There are two collapse operators in this problem; $\hat{C}_1 = \sqrt{\kappa(1+\bar{n})}\,\hat{a}$ gives the loss of cavity photons to the reservoir and $\hat{C}_2 = \sqrt{\kappa \bar{n}}\,\hat{a}^\dagger$ gives the incoherent pumping of the cavity by the thermal reservoir. The effective Hamiltonian may in turn be written as

$$\hat{H}_{\text{eff}} = \hbar\omega_s \hat{a}^\dagger \hat{a} - \frac{i\hbar}{2}\kappa(1+\bar{n})\hat{a}^\dagger \hat{a} - \frac{i\hbar}{2}\kappa \bar{n}\hat{a}\hat{a}^\dagger. \tag{8.20}$$

Figure 8-3(a) shows a typical simulation result for $\langle \hat{n} \rangle$ starting from an initial cavity occupancy of 5 and $\bar{n} = 0.1$. We observe that all the initial photons are lost into the reservoir. The reservoir injects a single photon into the cavity mode once in a while, but the average number of photons in the cavity is much less than unity. The solid line in Fig. 8-3(b) show the average over 100 stochastic wavefunctions:

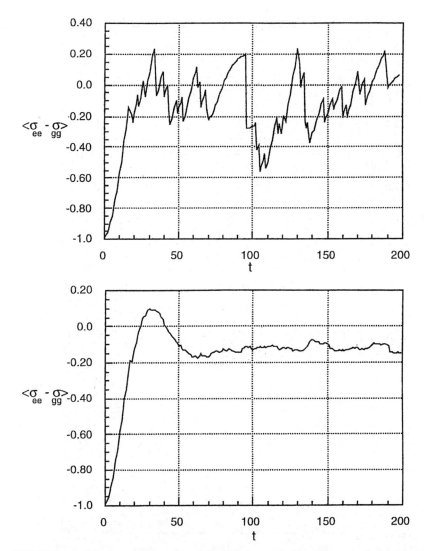

FIGURE 8-2: The expectation value of the atomic population difference obtained by an average over 10 (dashed) and 100 (solid) stochastic wavefunctions.

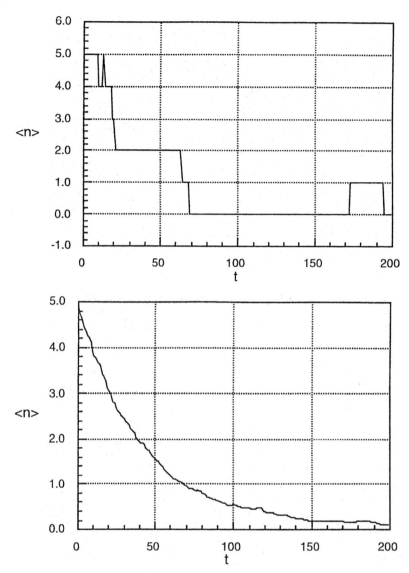

FIGURE 8-3: MCWF simulation of a cavity mode coupled to a thermal radiation field reservoir with $\bar{n} = 1$. The solid line shows a typical simulation result, whereas the dashed lines show an average over 100 trajectories.

We clearly observe that the expectation value of the cavity occupancy is 0.1. In other words, the cavity mode is driven into thermal equilibrium with the radiation field reservoir.

8.2. QUANTUM STATE DIFFUSION MODEL

The quantum MCWF approach that we have described in the previous section uses *gedanken measurements* of quantum jump events to incorporate the fluctuations introduced by the system-reservoir coupling. In experiments, if we carry out direct detection of emitted photons in either of the preceding examples, we would encounter these quantum jumps. However, we have also discussed that these quantum jumps exist only if we look for them; that is, if we do not carry out a direct detection, we cannot talk about actual quantum jumps taking place. Conversely, if we carry out other measurements on the same system, we can obtain results that would seem to contradict the quantum jump model; such is the case with homodyne or heterodyne measurements of the output field [13].

A stochastic wavefunction formalism that actually precedes the MCWF approach is the quantum state diffusion (QSD) model of Gisin and Percival [7]. In QSD, we unravel the underlying master equation to obtain a single stochastic Schrödinger equation where the reservoir-induced fluctuations are introduced directly into the equation by a complex Wiener process. Given the general form of the master equation in (8.2), the QSD stochastic equation is

$$d|\psi_s(t)\rangle = -\frac{i}{\hbar}\hat{H}_o|\psi(t)\rangle\,dt + \frac{1}{2}\sum_i[2\langle\hat{C}_i^\dagger\rangle\hat{C}_i - \hat{C}_i^\dagger\hat{C}_i - \langle\hat{C}_i^\dagger\rangle\langle\hat{C}_i\rangle]|\psi_s(t)\rangle\,dt$$
$$+ \sum_i(\hat{C}_i - \langle\hat{C}_i\rangle)|\psi_s(t)\rangle\,d\xi(t), \tag{8.21}$$

where $d\xi(t)$ denotes the complex Wiener process with $\overline{d\xi^*(t)d\xi(t)} = dt$ [7], [13]. In QSD formalism, the reservoir fluctuations are directly introduced into the stochastic Hamiltonian via $d\xi(t)$.

The stochastic wavefunction equation of (8.21) may be obtained within the MCWF formalism by assuming that the output photons are detected by heterodyne detection. In the limit where the classical local oscillator field is much stronger than the output field from the system, the collapse events become very frequent. Each collapse event, however, only slightly changes the wavefunction as the detected photon is much more likely to be a local oscillator photon. When the local oscillator field amplitude is very large, the collapse events appear as a complex Wiener process. This equivalence has been shown in Ref. [13]; we will not repeat that derivation, but instead we will show that (8.21) reproduces the master equation after averaging.

We have already seen that the reduced system density operator may be defined as an ensemble average over stochastic wavefunctions. The density operator at time

$t + dt$ may be written as

$$\hat{\rho}_s(t+dt) = \overline{[|\psi_s(t)\rangle + |d\psi_s(t)\rangle][\langle\psi_s(t)| + \langle d\psi_s(t)|]}$$
$$= \overline{|\psi_s(t)\rangle\langle\psi_s(t)| + |d\psi_s(t)\rangle\langle\psi_s(t)| + |\psi_s(t)\rangle\langle d\psi_s(t)|}$$
$$+ \overline{|d\psi_s(t)\rangle\langle d\psi_s(t)|}. \tag{8.22}$$

The first term in the last line of (8.22) simply gives $\hat{\rho}_s(t)$. After straightforward but slightly lengthy algebra, we show using (8.21) that

$$\overline{|d\psi_s(t)\rangle\langle\psi_s(t)| + |\psi_s(t)\rangle\langle d\psi_s(t)|} = -\frac{i}{\hbar}[\hat{H}_o\hat{\rho}_s(t)]\,dt$$
$$+ \sum_i [\langle\hat{C}_i^\dagger\rangle\hat{C}_i\hat{\rho}_s(t) + \langle\hat{C}_i\rangle\hat{\rho}_s(t)\hat{C}_i^\dagger]\,dt$$
$$- \frac{1}{2}\sum_i [(\hat{C}_i^\dagger\hat{C}_i\hat{\rho}_s(t) + \hat{\rho}_s(t)\hat{C}_i^\dagger\hat{C}_i)$$
$$+ 2\langle\hat{C}_i^\dagger\rangle\langle\hat{C}_i\rangle\hat{\rho}_s(t)]\,dt. \tag{8.23}$$

The terms proportional to $d\xi(t)$ have no contribution to (8.23) since $\overline{d\xi(t)} = 0$. Finally, the last term in the third line of (8.22) gives

$$\overline{|d\psi_s(t)\rangle\langle d\psi_s(t)|} = \sum_i (\hat{C}_i - \langle\hat{C}_i\rangle)\hat{\rho}_s(t)(\hat{C}_i^\dagger - \langle\hat{C}_i^\dagger\rangle)\,dt, \tag{8.24}$$

where we have used $\overline{d\xi^*(t)d\xi(t)} = dt$. Putting together the terms in (8.23) and (8.24), we recover the master equation of (8.2).

Even though the predictions of the QSD and MCWF formalisms are identical for an ensemble of systems, the physical pictures that they describe are completely different. One appealing feature of the QSD formalism is that it replaces the usual Schrödinger equation with an explicitly stochastic equation, without invoking gedanken measurements. The effective Hamiltonian that is implicit in (8.21) has terms corresponding to dissipation (drift) and fluctuation (diffusion). The complex Wiener process describing the fluctuation may be implemented into the numerical integration procedure by using Gaussian random variables with zero mean and dt variance.

The differences between the two stochastic wavefunction formalisms become more apparent when we consider the simulations of identical systems. Figure 8-4 shows the QSD simulation of the atomic population difference in a laser-driven two-level atom for the same parameters as that of Figs. 8-1 and 8-2. The single realization result shown in Fig. 8-4(a) has all the essential features as the master equation result obtained by averaging over 10 stochastic wavefunctions (Fig. 8-4(b)). Unless a

QUANTUM STATE DIFFUSION MODEL 173

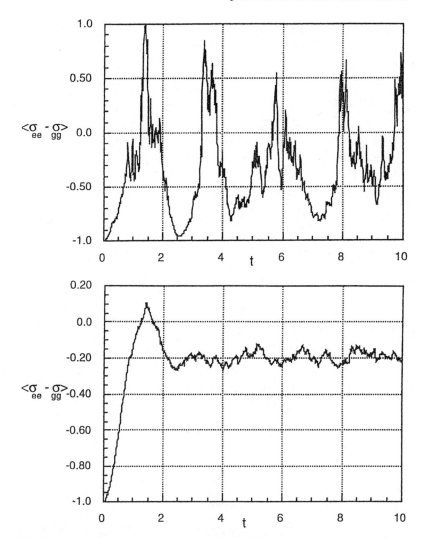

FIGURE 8-4: QSD simulation of a coherently driven two-level atom. The solid line shows a typical simulation result, whereas the dashed lines show an average over 10 trajectories.

particular measurement scheme is employed in the analyzed experiments, there is no fundamental reason to choose one stochastic wavefunction formalism with respect to another. However, computation-time requirements may favor a particular realization.

REFERENCES

[1] G. Rempe, H. Walther, and N. Klein, Phys. Rev. Lett. **58**, 353 (1987).

[2] M. G. Raizen, R. J. Thompson, R. J. Brecha, H. J. Kimble, and H. J. Carmichael, Phys. Rev. Lett. **63**, 240 (1989).

[3] M. Brune, S. Haroche, J. M. Raimond, L. Davidovich, and N. Zagury, Phys. Rev. A **45**, 5193 (1992).

[4] H. J. Carmichael, *An Open Systems Approach to Quantum Optics*, Lecture Notes in Physics—New Series m: Monographs (Springer-Verlag, Berlin, 1993).

[5] J. Dalibard, Y. Castin, and K. Molmer, Phys. Rev. Lett. **68**, 580 (1992).

[6] C. W. Gardiner, A. S. Parkins, and P. Zoller, Phys. Rev. A **46**, 4363 (1992).

[7] N. Gisin and I. C. Percival, J. Phys. A: Math. Gen. **25**, 5677 (1992).

[8] K. Molmer, Y. Castin, and J. Dalibard, J. Opt. Soc. Am. B **10**, 524 (1993).

[9] V. B. Braginsky and F. Ya. Khalili, *Quantum Measurement* (Cambridge, Cambridge, UK, 1992).

[10] F. Diedrich and H. Walther, Phys. Rev. Lett. **58**, 203 (1987).

[11] W. Nagourney, J. Sandberg, and H. G. Dehmelt, Phys. Rev. Lett. **56**, 2797 (1986).

[12] Th. Sauter, W. Neuhauser, R. Blatt, and P. Toschek, Phys. Rev. Lett. **57**, 1696 (1986).

[13] H. M. Wiseman and G. J. Milburn, Phys. Rev. A **47**, 1652 (1993).

9 Quantum Nondemolition Measurements

The sensitivity for detecting weak external forces using a mechanical harmonic oscillator has been studied extensively due to the applications in atomic force microscopy and gravitational wave detection. The fundamental limit imposed by the Heisenberg uncertainty principle is elucidated for the repeated measurements of the position of a free mass or a mechanical harmonic oscillator and is termed *standard quantum limit (SQL)*. Soon after it was proposed that a certain *nonstandard* measurement, based on a carefully chosen observable and carefully prepared measuring device, can exceed the aforementioned standard quantum limit. Such a nonstandard measurement was first pointed out by Landau and Pierles in the 1930s as the repeated measurements of a free particle momentum [1]. A rigorous mathematical model was given in the late 1970s by Braginsky and his colleagues [2]. The name *quantum nondemolition (QND) measurement* was coined at this time. A related concept, *contractive state measurement*, was also proposed to circumvent the SQL [3].

Even though the concept of a QND measurement was first investigated in connection to the experimental efforts to construct a sensitive gravitational wave detector, the idea has not been demonstrated for such a mechanical harmonic oscillator or a free mass. Instead, the concept was applied to quantum optics using advanced laser and nonlinear optics techniques. The QND measurement of photon numbers using nonlinear optical media is the central playground for experimental QND measurements at present.

The basic concept of QND measurements was presented in Chapter 1 (Sec. 1.3.5). Section 9.1 describes the QND measurement of photon number in Heisenberg and Schrödinger pictures. The experimental QND measurements using nonlinear optical media are reviewed in Sec. 9.2. In Sec. 9.3 the QND measurement of quadrature amplitudes is discussed.

9.1. QND MEASUREMENT OF PHOTON NUMBER

9.1.1. Heisenberg Picture

The photon number $\hat{n}$ of an electromagnetic field is a QND observable since the photon number operator commutes with a free-field Hamiltonian $\hat{\mathcal{H}} = \hbar\omega(\hat{n} + \frac{1}{2})$. Consider the experimental setup shown in Fig. 9-1 [4]. A signal wave at frequency ω_s

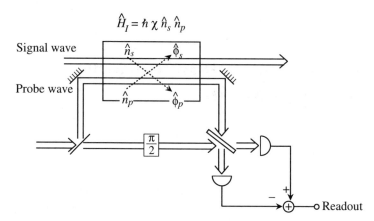

FIGURE 9-1: A QND measurement setup of photon number using the cross-phase modulation in an optical Kerr medium.

propagates in an optical Kerr medium with a probe wave at frequency ω_p. The phase $\hat{\phi}_p$ of the probe wave is modulated by the photon number $\hat{n}_s$ of the signal wave due to cross-phase modulation effect. The phase modulation of the probe wave is detected, for instance, by a Mach-Zehnder interferometric detector, and thus the information about the signal photon number can be extracted. The interaction Hamiltonian for this measurement is that of cross-phase modulation between the signal and probe waves:

$$\hat{\mathcal{H}}_I = \hbar \chi \hat{n}_s \hat{n}_p. \tag{9.1}$$

The constant χ represents the optical Kerr nonlinearity. Equation (9.1) does not commute with the readout observable $\hat{\phi}_p$ and is a function of the measured observable $\hat{n}_s$. Thus it satisfies the general requirements for an indirect measurement. Moreover, (9.1) commutes with the measured observable $\hat{n}_s$, so it is a back action evading type [see (1.84)].

The input and output Heisenberg operators of the signal and probe waves are, respectively, calculated by integrating the Heisenberg equations of motion:

$$\hat{a}_{s,\text{out}} = e^{i\chi t \hat{n}_p} \hat{a}_{s,\text{in}}, \tag{9.2}$$

$$\hat{a}_{p,\text{out}} = e^{i\chi t \hat{n}_s} \hat{a}_{p,\text{in}}. \tag{9.3}$$

Here t is the interaction time in the Kerr medium and $\hat{a}_{p,\text{in}}$ and $\hat{a}_{s,\text{in}}$ are the input probe and signal wave operators. If the Mach-Zehnder interferometer has a phase difference $\phi = \frac{\pi}{2}$ between the two arms and the two photodiode currents are subtracted as shown in Fig. 9-1, the readout observable corresponding to the difference current is the quadrature-phase amplitude [4]:

$$\hat{a}_{p,\text{out},2} \equiv \frac{1}{2i}(\hat{a}_{p,\text{out}} - \hat{a}^{\dagger}_{p,\text{out}})$$
$$= \hat{a}_{p,\text{in},1} \sin(\chi t \hat{n}_s) + \hat{a}_{p,\text{in},2} \cos(\chi t \hat{n}_s). \tag{9.4}$$

If the probe phase $\hat{a}_{p,\text{in}}$ is adjusted for satisfying $\langle \hat{a}_{p,\text{in},2} \rangle = 0$ and the cross-phase modulation is small, (9.4) is approximated by

$$\hat{a}_{p,\text{out},2} \simeq \langle \hat{a}_{p,\text{in},1} \rangle \chi t \hat{n}_s + \Delta \hat{a}_{p,\text{in},2}. \tag{9.5}$$

A properly normalized readout observable corresponding to the signal photon number is defined by

$$\hat{n}_s^{(\text{obs})} \equiv \frac{\hat{a}_{p,\text{out},2}}{\langle \hat{a}_{p,\text{in},1} \rangle \chi t} = \hat{n}_s + \frac{\Delta \hat{a}_{p,\text{in},2}}{\langle \hat{a}_{p,\text{in},1} \rangle \chi t}. \tag{9.6}$$

Taking the average and the variance of (9.6), we obtain

$$\langle \hat{n}_s^{(\text{obs})} \rangle = \langle \hat{n}_s \rangle, \tag{9.7}$$

$$\langle \Delta \hat{n}_s^{(\text{obs})2} \rangle = \langle \Delta \hat{n}_s^2 \rangle + \frac{\langle \Delta \hat{a}_{p,\text{in},2}^2 \rangle}{\langle \hat{n}_p \rangle (\chi t)^2}. \tag{9.8}$$

The first term of (9.8) is the intrinsic uncertainty of the measured observable $\hat{n}_s$, while the second term represents the finite measurement error due to the quantum noise of the quadrature-phase amplitude of the probe wave. This measurement error can be decreased arbitrarily by increasing the probe photon number $\langle \hat{n}_p \rangle$ and the nonlinear interaction strength χt. Such a QND measurement is ideal in the sense that the expectation value $\langle \hat{n}_s^{(\text{obs})} \rangle$ and the variance $\langle \Delta \hat{n}_s^{(\text{obs})2} \rangle$ of measurement results are identical to those of the signal wave.

When the probe wave is a coherent state, $\langle \Delta \hat{a}_{p,\text{in},2}^2 \rangle = \frac{1}{4}$, the measurement error is reduced to $1/4\langle \hat{n}_p \rangle (\chi t)^2$. Figure 9-2 illustrates the noise distributions of the output (phase-modulated) probe wave. We can see from the figure how the measurement error is decreased by increasing $\langle \hat{n}_p \rangle$ and $(\chi t)^2$.

In the process of this measurement, the phase noise of the signal wave is increased by the quantum uncertainty of the probe wave photon number:

$$\hat{a}_{s,\text{out},2} \equiv \frac{1}{2i}(\hat{a}_{s,\text{out}} - \hat{a}^{\dagger}_{s,\text{out}})$$
$$\simeq \langle \hat{a}_{s,\text{in},1} \rangle \chi t \hat{n}_p + \Delta \hat{a}_{s,\text{in},2}, \tag{9.9}$$

$$\langle \Delta \hat{\phi}_{s,\text{out}}^2 \rangle \equiv \frac{\langle \Delta \hat{a}_{s,\text{out},2}^2 \rangle}{\langle \hat{a}_{s,\text{in},1} \rangle^2} = \langle \Delta \hat{\phi}_{s,\text{in}}^2 \rangle + (\chi t)^2 \langle \Delta \hat{n}_p^2 \rangle. \tag{9.10}$$

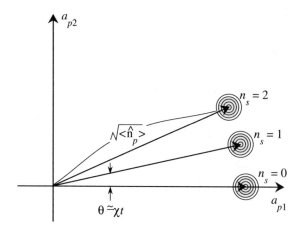

FIGURE 9-2: The noise distributions of the output (phase-modulated) probe wave.

Here the first term of (9.10), $\langle\Delta\hat{\phi}_{s,\text{in}}^2\rangle \equiv \frac{\langle\Delta\hat{a}_{s,\text{in},2}^2\rangle}{\langle\hat{a}_{s,\text{in},1}\rangle^2}$, is the initial phase noise of the signal wave and the second term, $(\chi t)^2 \langle\Delta\hat{n}_p^2\rangle$, represents the back action noise imposed on the phase of the signal wave by the photon number noise of the probe wave.

When the probe wave is a coherent state, $\langle\Delta\hat{a}_{p,\text{in},1}^2\rangle = \frac{1}{4}$, the back action noise is reduced to $(\chi t)^2 \langle\hat{n}_p\rangle$. In such a case the measurement error and the back action noise satisfy the minimum uncertainty product

$$\langle\Delta\hat{n}_s^2\rangle_{\text{meas. error}} \langle\Delta\hat{\phi}_s^2\rangle_{\text{back action}} = \frac{1}{4}. \tag{9.11}$$

This relation holds for the more general case where the probe wave is prepared in a phase squeezed state. In this case the measurement error is decreased with the given average probe photon number, while the back action noise is increased [5].

9.1.2. Schrödinger Picture

Suppose the signal and probe waves incident on the Kerr medium are in coherent states $|\alpha\rangle_s$ and $|\beta\rangle_p$. The joint-correlated density operator after the cross-phase modulation is expressed by the unitary evolution operator $\hat{U} \equiv \exp(i\chi t \hat{n}_s \hat{n}_p)$:

$$\hat{\rho}_{sp} = \hat{U}\hat{\rho}_s \otimes \hat{\rho}_p \hat{U}^\dagger = \hat{U}|\alpha\rangle_{ss}\langle\alpha| \otimes |\beta\rangle_{pp}\langle\beta|\hat{U}^\dagger. \tag{9.12}$$

The destructive measurement of the quadrature-phase amplitude of the probe wave by the Mach-Zehnder interferometer is described by the projection operator $|\beta_2\rangle_{pp}\langle\beta_2|$, where β_2 (= real number) is the measurement result and $|\beta_2\rangle_p$ is the eigenstate of the observable $\hat{a}_{p2}$ with an eigenvalue of β_2. The value of β_2 has a one-to-one correspondence to the inferred signal photon number $\tilde{n}_s$:

$$\tilde{n}_s = \frac{\beta_2}{\beta \chi t}. \qquad (9.13)$$

The probability $p(\tilde{n}_s)$ of obtaining the measurement result $\tilde{n}_s$ and the postmeasurement state $\hat{\rho}_s(\tilde{n}_s)$ for a specific measurement result $\tilde{n}_s$ are calculated by

$$p(\tilde{n}_s) = \text{Tr}_s[\hat{X}(\tilde{n}_s)\hat{\rho}_s], \qquad (9.14)$$

$$\hat{\rho}(\tilde{n}_s) = \frac{1}{p(\tilde{n}_s)} \hat{Y}(\tilde{n}_s)\hat{\rho}_s \hat{Y}^\dagger(\tilde{n}_s). \qquad (9.15)$$

The generalized projection operator $\hat{X}(\tilde{n}_s)$ is given by

$$\hat{X}(\tilde{n}_s) = \hat{Y}^\dagger(\tilde{n}_s)\hat{Y}(\tilde{n}_s), \qquad (9.16)$$

where

$$\hat{Y}(\tilde{n}_s) = {}_p\langle \tilde{n}_s | \hat{U} | \beta \rangle_p. \qquad (9.17)$$

The variances of the photon number and the sine operator are evaluated for the postmeasurement state [6]:

$$\langle \Delta \hat{n}_s^2 \rangle \equiv \text{Tr}_s[(\hat{n}_s - \langle \hat{n}_s \rangle)^2 \hat{\rho}_s(\tilde{n}_s)] \simeq \left[\frac{1}{|\alpha|^2} + 4(\chi t)^2 |\beta|^2 \right]^{-1}, \qquad (9.18)$$

$$\langle \Delta \hat{S}^2 \rangle \equiv \text{Tr}_s[(\hat{S} - \langle \hat{S} \rangle)^2 \hat{\rho}_s(\tilde{n}_s)] \simeq \frac{1}{4|\alpha|^2} + \frac{1}{2}[1 - \exp(-2(\chi t)^2 |\beta|^2)]. \qquad (9.19)$$

The first term of (9.18) is the inverse of the initial photon number noise for the coherent state $|\alpha\rangle_s$, while the second term represents the inverse of the measurement error for the photon number. Equation (9.18) indicates that the photon number noise of the postmeasurement state is decreased by decreasing measurement error. The first term of (9.19) is the initial sine operator (phase) noise for the coherent state $|\alpha\rangle_s$, while the second term represents the additive back action noise imposed on the sine operator. Equation (9.19) indicates that the sine operator noise of the postmeasurement state is increased with decreasing measurement error.

Figure 9-3 shows that the number-phase uncertainty product $P_{ns} \equiv \langle \Delta \hat{n}_s^2 \rangle \langle \Delta \hat{S}^2 \rangle / \langle \hat{C} \rangle^2$ of the postmeasurement state is equal to the minimum value $(=\frac{1}{4})$ as long as the nonlinear interaction $\chi t (= \sqrt{F})$ is not too strong [6]. When χt exceeds a certain critical value, the photon number noise increases rather than decreases. This is because the periodic nature of the cross-phase modulation by a multiple of 2π results in a case where different signal photon number values $\tilde{n}_s$ are simultaneously identified by the same phase measurement readout β_2. If the nonlinear interaction

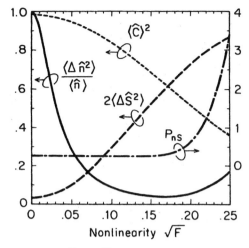

FIGURE 9-3: Uncertainties $\langle \Delta \hat{n}^2 \rangle$, $\langle \Delta \hat{S}^2 \rangle$ and uncertainty product $P_{ns} = \langle \Delta \hat{n}^2 \rangle \langle \Delta \hat{S}^2 \rangle / \langle \hat{C} \rangle^2$ of the state after the measurement; ($|\alpha_0| = |\beta_0| = 4$, $\beta_2 = 0$).

is not too strong, the initial coherent state $|\alpha\rangle_s$ is transformed to a number-phase squeezed state with reduced photon number noise.

9.2. EXPERIMENTAL QND MEASUREMENTS

Various Kerr nonlinear media have been tested for realizing efficient cross-phase modulation between signal and probe waves. If there is a loss in the Kerr nonlinear medium, uncorrelated vacuum fluctuations are introduced into both the signal and probe waves. This would degrade the quantum correlation between the two waves and also deform the system quantum state, thus preventing the realization of ideal QND measurements. If there is an extra noise source in the Kerr nonlinear medium, it would also prevent the realization of ideal QND measurements.

A fused silica fiber has an ultralow loss ($\simeq$ 0.2 dB/km) and almost zero group velocity dispersion at a wavelength of 1.55 μm. Thus the signal and idler waves are confined in a small core region and co-propagate over a long distance, which is expected to result in an efficient cross-phase modulation between the two waves. The QND measurements of photon number using an optical fiber were demonstrated in a normal linear propagation mode [7] and also in a nonlinear soliton propagation mode [8]. An optical soliton is a stable photonic wavepacket bound together by the balance between positive Kerr nonlinearity and negative group velocity dispersion of a fiber. A single soliton pulse at $\lambda = 1.5$ μm with pulse duration of a few picoseconds contains roughly 10^8 photons. Two optical solitons with slightly different center frequencies propagate with different group velocities. If the separation between two center frequencies and the initial positions of the two solitons (signal and probe) are carefully chosen, they collide inside a fiber and cross-phase modulate with each other

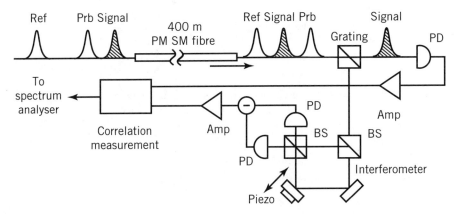

FIGURE 9-4: (a) A schematic diagram of the quantum nondemolition (QND) measurement. A signal soliton (Signal) at 1460.7 nm and identical probe (Prb) and reference (Ref) solitons at 1455 nm propagate in a 400-m polarization-maintaining (PM), single-mode (SM) fiber having anomalous group-velocity dispersion. The signal and probe solitons collide in the fiber, causing the phase of the probe soliton to shift in response to the photon number of the signal soliton. After leaving the fiber, the solitons are separated by a dispersing grating. A photodiode (PD) records the photon-number fluctuations of the signal soliton and an interferometer (with one arm slightly longer than the other so that the probe and reference pulses overlap) measures the phase shift fluctuations of the probe solitons. Correlation measurements with a radio-frequency spectrum analyzer show the fluctuations to be correlated, indicating a partial QND measurement.

(Fig. 9-4). We can measure the photon number of the signal soliton nondestructively by measuring the phase shift of the probe soliton by constructing an interferometer for the probe soliton [9].

However, the third-order nonlinearity of a fused silica is very small so that the cross-phase modulation is not large enough even though the interaction length is long and the peak intensity is enhanced by the temporal (soliton) and spatial (fiber core) confinement. Moreover, excess noise is generated in a fiber due to stimulated Brillouin scattering and Raman scattering. To circumvent these drawbacks, another optical soliton formed by the resonant coupling between light pulses and assembly of two level atoms, self-induced transparency soliton, was studied as a means to realize efficient cross-phase modulation [10].

It is well known that Kerr nonlinearity is enhanced by using an atomic resonance line. The interactive cross-phase modulation using a beam of Na atoms with a three-level cascade system was employed in the realization of QND measurement of photon numbers (Fig. 9-5) [11]. However, such resonant Kerr nonlinearity is associated with linear absorption and spontaneous emission noise. The residual Doppler broadening of the transition lines and the atom number fluctuation also introduce excess noise. The use of cold trapped atoms instead of an atomic beam could suppress this excess noise and improve efficiency [12].

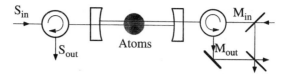

FIGURE 9-5: The QND measurement of photon number using atoms in an optical cavity. $S_{in, out}$ and $M_{in, out}$ are the inputs and outputs of the signal and meter waves.

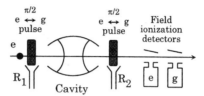

FIGURE 9-6: The Ramsey atom interferometer for QND measurement of photon number.

The resonant Kerr nonlinearity in an atomic four-level system has the advantage of enhancing the third-order nonlinearity and simultaneously eliminating linear (one photon) absorption completely based on the principle of electromagnetically induced transparency [13]. It is predicted that the cross-phase modulation even by one photon is observable by using such a destructive quantum interference effect [14].

The dispersive atom-field interaction inside a high-Q cavity modulates the phase of the atom due to the photon-number-dependent anticrossing behavior of the two dressed states [15]. Therefore, if the phase of a throughput atom is measured by constructing an atom interferometer of Ramsey type, the QND measurement of the photon number inside the cavity can be realized (Fig. 9-6). The detection of the atomic phase shift already reported the difference between a vacuum state and one photon state [15].

9.3. QND MEASUREMENT OF QUADRATURE AMPLITUDES

The slowly varying quadrature amplitudes $\hat{a}_1$ and $\hat{a}_2$ of the electromagnetic field are also QND observables. A two-way degenerate parametric amplifier with a single mirror, shown in Fig. 9-7, realizes the QND measurement of the quadrature amplitude [5]. We will describe the operation of this apparatus in a Heisenberg picture. The right propagating input and output waves in a degenerate parametric amplifier are related by

$$\hat{c} = \sqrt{G}\,\hat{b} + \sqrt{G-1}\,\hat{b}^\dagger, \tag{9.20}$$

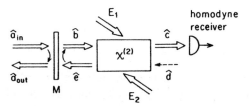

FIGURE 9-7: Configuration for QND measurement of quadrature amplitude.

where G is the parametric amplifier gain. The left propagating input and output waves are related by

$$\hat{e} = \sqrt{G}\,\hat{d} - \sqrt{G-1}\,\hat{d}^{\dagger}, \qquad (9.21)$$

where $\hat{d}$ represents the vacuum field emitted from the detector surface. $\hat{d}$ is interpreted as the meter input wave in this configuration. The difference of the signs in (9.20) and (9.21) is due to π phase difference of the two pump waves E_1 and E_2. The incident and output waves of the mirror M with the transmission coefficient T are related by

$$\begin{pmatrix}\hat{a}_{\text{out}}\\ \hat{b}\end{pmatrix} = \begin{pmatrix}\sqrt{1-T} & \sqrt{T}\\ \sqrt{T} & -\sqrt{1-T}\end{pmatrix}\begin{pmatrix}\hat{a}_{\text{in}}\\ \hat{e}\end{pmatrix}. \qquad (9.22)$$

By combining (9.20), (9.21), and (9.22), the meter output wave is given by

$$\begin{aligned}\hat{c} &= \sqrt{T}\,(\sqrt{G}\,\hat{a}_{\text{in}} + \sqrt{G-1}\,\hat{a}_{\text{in}}^{\dagger}) + \sqrt{1-T}\,\hat{d}\\ &= \sqrt{T}\,(e^{r}\hat{a}_{\text{in},1} + ie^{-r}\hat{a}_{\text{in},2}) + \sqrt{1-T}\,\hat{d},\end{aligned} \qquad (9.23)$$

where $e^r = \sqrt{G} + \sqrt{G-1}$ and $e^{-r} = \sqrt{G} - \sqrt{G-1}$. Note that the meter output wave carries the (amplified) information about the quadrature amplitude $\hat{a}_{\text{in},1}$ with the background noise $\sqrt{1-T}\,\hat{d}$. The quadrature amplitude $\hat{c}_1$ of the meter output wave is measured by an optical homodyne detector. The normalized output of the homodyne detector is expressed by

$$\hat{a}_{\text{in},1}^{(\text{obs})} \equiv \frac{\hat{c}_1}{\sqrt{T}\,e^r} = \hat{a}_{\text{in},1} + \sqrt{\frac{1-T}{T}}\,\frac{1}{e^r}\,\hat{d}_1. \qquad (9.24)$$

Since $\langle \hat{d}_1 \rangle = 0$, this measurement is free from signal offset [i.e., $\langle \hat{a}_{\text{in},1}^{(\text{obs})} \rangle = \langle \hat{a}_{\text{in},1} \rangle$]. The measurement error for $T \ll 1$ is given by

$$\langle \Delta \hat{a}_1 \rangle_{\text{meas. error}} \simeq \frac{1}{T} e^{-2r} \langle \Delta \hat{d}_1^2 \rangle, \qquad (9.25)$$

which can be reduced to far below the fluctuation of the vacuum state if $e^{-2r}/T < 1$.

The signal output wave is given by

$$\hat{a}_{\text{out}} = \sqrt{1-T}\,\hat{a}_{\text{in}} + \sqrt{T}\,(e^{-r}\hat{d}_1 + ie^r\hat{d}_2). \quad (9.26)$$

The measured quadrature amplitude is untouched, $\hat{a}_{\text{out},1} \simeq \hat{a}_{\text{in},1}$, if $T \ll 1$ and $Te^{-r} \ll 1$. The other quadrature amplitude has the back action noise

$$\langle \Delta \hat{a}_2^2 \rangle_{\text{back action}} = Te^{2r}\langle \Delta \hat{d}_2^2 \rangle. \quad (9.27)$$

We remark once again that the measurement error (9.25) and the back action noise (9.27) satisfy the minimum uncertainty product.

The essential part of this QND measurement of the quadrature amplitude is the combination of a high-reflection mirror and a squeezed vacuum state to simultaneously preserve the quadrature amplitude information in the signal output wave and realize the good signal-to-noise ratio in the meter output wave [16]–[18].

The experimental implementation of this QND measurement scheme was realized by using a type II phase matched degenerate parametric amplifier, which amplifies one polarization and deamplifies the other polarization. The mirror is replaced by a polarizing cube and a half-wavelength plate [19]. Two cascade QND measurements using a cavity parametric amplifier have been demonstrated [20].

REFERENCES

[1] L. Landau and R. Peierls, Z. Phys. **69**, 56 (1931).

[2] V. B. Braginsky and F. Y. Khalili, *Quantum Measurement* (Cambridge University Press, Cambridge, UK, 1992).

[3] H. P. Yuen, Phys. Rev. Lett. **51**, 719 (1983).

[4] N. Imoto, H. A. Haus, and Y. Yamamoto, Phys. Rev. A **32**, 2287 (1985).

[5] Y. Yamamoto et al., in *Progress in Optics XXVIII* (North-Holland, Amsterdam, 1990), p. 87.

[6] M. Kitagawa, N. Imoto, and Y. Yamamoto, Phys. Rev. A **35**, 5270 (1987).

[7] M. D. Levenson, R. M. Shelby, M. Reid, and D. F. Walls, Phys. Rev. Lett. **57**, 2473 (1986).

[8] S. R. Friberg, S. Machida, and Y. Yamamoto, Phys. Rev. Lett. **69**, 3165 (1992).

[9] H. A. Haus, K. Watanabe, and Y. Yamamoto, J. Opt. Soc. Am. B **6**, 1138 (1989).

[10] K. Watanabe, H. Nakano, A. Honold, and Y. Yamamoto, Phys. Rev. Lett. **62**, 2257 (1989).

[11] P. Grangier, J. F. Roch, and G. Royer, Phys. Rev. Lett. **66**, 1418 (1991).

[12] J. F. Roch, K. Vigneron, Ph. Grelu, A. Sinatra, J. Ph. Poizat, and P. Grangier, Phys. Rev. Lett. **78**, 634 (1997).

[13] A. Imamoğlu and S. E. Harris, Opt. Lett. **14**, 1344 (1989).

[14] H. Schmidt and A. Imamoğlu, Opt. Lett. **21**, 1936 (1996).

[15] M. Brune, P. Nussenzveig, F. Schmidt-Kaler, F. Bernardot, A. Maali, J. M. Raimond, and S. Haroche, Phys. Rev. Lett. **72**, 3339 (1994).

[16] J. H. Shapiro, Opt. Lett. **5**, 351 (1980).
[17] B. Yurke, Phys. Rev. A **29**, 408 (1984).
[18] R. M. Shelby and M. D. Levenson, Opt. Commun. **64**, 553 (1987).
[19] A. LaPorta, R. E. Slusher, and B. Yurke, Phys. Rev. Lett. **62**, 28 (1989).
[20] K. Bencheikh, J. A. Levenson, P. Grangier, and O. Lopez, Phys. Rev. Lett. **75**, 3422 (1995).

10 Semiconductor Bloch Equations

In Chapter 6, we presented the theory of atom-field interactions and discussed some of the fascinating quantum optical phenomena observed in this fundamental system. Even though *atomic quantum optics* continues to be the field in which the majority of research activity is concentrated, semiconductor radiation-field systems provide a challenging paradigm for investigating quantum and nonlinear effects in optical physics. This relatively new field of *semiconductor quantum optics* has the added advantage of possible practical applications in optoelectronics technology. Our goal in this and the following two chapters is to present the background necessary for discussing quantum optical phenomena in semiconductors.

In this chapter, we will concentrate on developing the basic theory of semiconductor-light interactions starting from the first principles. Our main goal here is to provide a relatively simple derivation of the so-called semiconductor Bloch equations and to emphasize the differences with the familiar optical Bloch equations that govern the ensemble-averaged dynamics of atoms interacting with the radiation field. Since we use a field-theoretic technique, our derivation does not assume familiarity with semiconductor physics. Most of the material covered in the first part of this chapter is discussed (albeit from a different perspective) in texts by H. Haken [1] and H. Haug and S. W. Koch [2].

This chapter is organized as follows: In the first section, we present an overview of the field-theoretic approach to semiconductors. Semiconductor Bloch equations (SBEs) are derived in Sec. 10.2, where we also present a comparison with the corresponding equations in atomic systems. In Sec. 10.3, we introduce the concept of excitons by solving SBE in the limit of weak excitation (i.e., Wannier equation). The nonlinear optical properties arising from the quantum statistics and residual Coulomb interactions are discussed in Sec. 10.4 using a Gross-Pitaevskii type of equation [3].

10.1. FIELD THEORY OF SEMICONDUCTORS

In Chapter 2 we have seen the canonical quantization of the Schrödinger matter field. With the exception of the burgeoning fields of laser-cooled atoms and trapped Bose-Einstein condensates, a field-theoretic treatment of atoms is not necessary in atomic quantum optics. In contrast, when dealing with semiconductor-field interactions, the quantum field theory of semiconductors is an indispensable tool. First, one is necessarily dealing with a many-body system consisting of $\simeq 10^{23}$ particles. The problem greatly simplifies if one concentrates on quasi-particles, which are commonly referred to as *conduction-band electrons* and *valence-band holes*. Absorption or emission of a photon creates or annihilates an electron-hole pair, and therefore the quasi-particle number is not conserved in radiation field interactions. Naturally, this is easily treated in the quantum field-theoretic formalism.

Second, unlike a low-density atomic vapor, the elementary excitations in a semiconductor cannot be treated as independent electron-hole pairs under any circumstance. Both strong Coulomb interactions and quantum statistics via the Pauli-exclusion principle or bosonic stimulation are essential in determining the response. Once again, the field theory is essential since it allows for straightforward treatment of quantum statistical effects.

The starting point of our analysis is the second-quantized Hamiltonian of the semiconductor obtained using the canonical quantization method described in Chapter 2:

$$\hat{H}_0 = \int d^3 r\, \hat{\Psi}^\dagger(\mathbf{r}) \left(-\frac{\hbar^2}{2m}\Delta + V_L(\mathbf{r}) \right) \hat{\Psi}(\mathbf{r})$$
$$+ \frac{1}{2} \int d^3 r \int d^3 r'\, \hat{\Psi}^\dagger(\mathbf{r})\hat{\Psi}^\dagger(\mathbf{r}') \frac{e^2}{|\mathbf{r}-\mathbf{r}'|} \hat{\Psi}(\mathbf{r}')\hat{\Psi}(\mathbf{r}), \quad (10.1)$$

where $\hat{\Psi}(\mathbf{r})$ denotes the field operator that annihilates an electron at position $\mathbf{r}$. Just as in the electromagnetic field case, it is convenient to rewrite the Eq. (10.1) in terms of annihilation and creation operators of the eigenmodes of the semiconductor; that is, using the Ansatz

$$\hat{\Psi}(r) = \sum_{i,k} \varphi_{i,k}(r) \hat{a}_{i,k}. \quad (10.2)$$

Here $\varphi_{i,k}(r)$ and $\hat{a}_{i,k}$ denote the single-particle wavefunction and the annihilation operator of eigenstate k of the ith band of the semiconductor. We dropped the vector notation for r and k and suppressed the spin index for simplicity. In contrast to the Maxwell field, the determination of the single-particle wavefunction is an extremely difficult task and it is one of the principal problems of solid-state theory. Since we are interested in the optical properties, all we need to do is to consider the quantum field theory of elementary excitations, assuming that determination of the single-particle wavefunctions $\varphi_{i,k}(r)$ has been carried out. We do, however, start by discussing the steps and approximations involved in determining $\varphi_{i,k}(r)$. The fact that we do not need to consider the band calculations is not surprising when we recall that the details of the atomic-state wavefunctions were, to a large extent, irrelevant for the description of quantum optical processes in atoms.

The underlying assumption in the determination of the bands is the Hartree-Fock (HF) approximation, which states that the many-body system may be described by an effective single-body problem [4], [6]. The ground state of the semiconductor at

$T = 0$ K in the HF approximation is

$$|\Phi\rangle = \hat{a}^\dagger_{v_1,k_1} \hat{a}^\dagger_{v_1,k_2} \cdots \hat{a}^\dagger_{v_1,k_N} \hat{a}^\dagger_{v_2,k_1} \cdots \hat{a}^\dagger_{v_m,k_N} |0\rangle, \quad (10.3)$$

where we have explicitly assumed that only the states of the m valence bands are occupied. N is determined by the size of the crystal and is on the order of 10^{23}. The conduction band creation operators do not appear in (10.3) as all of them are unoccupied at $T = 0$ K in the absence of an external perturbation. We will further simplify the problem by assuming that only a single valence band and a single conduction band are relevant for optical transitions; strictly speaking, this approximation does not hold for most semiconductors of interest. However, qualitative and in many cases quantitative understanding of the optical properties may be obtained with this simplified model.

The HF equations that we obtain for the single-particle wavefunctions starting from the Hamiltonian of (10.1) are coupled nonlinear integrodifferential equations. If we write these equations in the form of a standard time-independent Schrödinger equation for a single-particle wavefunction, we obtain [1]

$$H_{\text{eff}} \varphi_{i,k}(r) = \left[-\frac{\hbar^2}{2m} \Delta + V_{\text{eff}}(r) \right] \varphi_{i,k}(r) = E \varphi_{i,k}(r). \quad (10.4)$$

It is essential to note that V_{eff} appearing in (10.4) is not an ordinary potential; in fact it depends on the wavefunction $\varphi_{i,k'}(r)$ of all the other electrons as well as on $\varphi_{i,k}(r)$, making (10.4) nonlinear. In addition, V_{eff} contains an exchange term [1]. Despite these complications, we can assume that V_{eff} is a periodic potential with the same periodicity as the lattice. Much insight into the properties of single-electron wavefunctions can then be obtained through Bloch's theorem, which states that whatever the detailed structure of V_{eff} is, we can always write

$$\varphi_{i,k}(r) = \frac{1}{\sqrt{V_c}} e^{ikr} u_{i,k}(r), \quad (10.5)$$

where $u_{i,k}(r) = u_{i,k}(r + a)$ with a denoting a lattice vector. $\varphi_{i,k}(r)$ is referred to as the Bloch wavefunction and V_c is the quantization volume.

The determination of the periodic part $u_{i,k}(r)$ of $\varphi_{i,k}(r)$ is still a difficult many-body problem. Further simplification, however, is provided by the *effective-mass theorem*, which states that upon *diagonalization* of the Hamiltonian of (10.4), we can describe the single-electron states as *free-particle* states with a different, renormalized effective mass. Provided that the momentum exchange in an interaction process is small compared to π/a, we can assume that interacting particles are free *quasi-particles* with a new effective mass. In other words, the effect of the complicated effective potential V_{eff} is reduced to a simple change of mass.

Assuming that determination of $\varphi_{i,k}(r)$, the corresponding eigenenergies $(\varepsilon_{i,k})$, and effective masses (m_i^*) has been carried out, we can proceed with the field theory of semiconductors. To this end, we use the expansion for the field operators given in (10.2) to express the Hamiltonian of (10.1) in terms of mode operators. We obtain

$$\hat{H}_0 = \sum_k \left[\varepsilon_{c,k} \hat{a}_{c,k}^\dagger \hat{a}_{c,k} + \varepsilon_{v,k} \hat{a}_{v,k}^\dagger \hat{a}_{v,k} \right]$$
$$+ \frac{1}{2} \sum_{\substack{k,k' \\ q \neq 0}} V(q) \left[\hat{a}_{c,k+q}^\dagger \hat{a}_{c,k'-q}^\dagger \hat{a}_{c,k'} \hat{a}_{c,k} \right.$$
$$\left. + \hat{a}_{v,k+q}^\dagger \hat{a}_{v,k'-q}^\dagger \hat{a}_{v,k'} \hat{a}_{v,k} + 2\hat{a}_{c,k+q}^\dagger \hat{a}_{v,k'-q}^\dagger \hat{a}_{v,k'} \hat{a}_{c,k} \right], \quad (10.6)$$

where $V(q)$ denotes the Fourier transform of the Coulomb interaction; in a three-dimensional system $V(q) = e^2/(\epsilon q^2)$, with ϵ denoting the dielectric constant in the medium. We note that the crystal momentum given by the index k of the Bloch wavefunction (or the electron operators) is a conserved quantity in (10.6). Actually, there is nontrivial physics and mathematics involved in obtaining (10.6) from (10.1). First, we note that the summation (or integral in the continuum limit) over q is restricted to $q \neq 0$; the reason for this is that the contributions from the $q = 0$ terms of the electron-electron interactions exactly cancel the electron-lattice interaction term, proportional to V_L in (10.1).

Second, if we classify the interaction terms according to the band indices, there are, in general, 2^4 terms. In the HF approximation, the only nonzero terms are those with the identical number of annihilation and creation operators belonging to any given band i. Equivalently, the number of electrons in any given band is conserved in the electron-electron interaction. This assumption can alternatively be justified as an analog of the rotating wave approximation we have previously encountered: Typically, the bandgap energies in semiconductors are much larger than the electron-electron interaction strength, implying that energy-nonconserving terms have negligible contribution.

Introduction of HF approximation leaves six possible interaction terms; two of these are referred to as electron-hole exchange terms. The effective interaction strength of these terms is much smaller than the other four that we have kept in (10.6), due to the orthogonality of the Bloch wavefunctions. A detailed discussion on these exchange terms, which are important in exciton spin dynamics, can be found elsewhere [1]. Finally, the interaction terms with indices $cvvc$ (i.e., terms such as $\hat{a}_{c,k+q}^\dagger \hat{a}_{v,k'-q}^\dagger \hat{a}_{v,k'} \hat{a}_{c,k}$) and $vccv$ are equivalent, resulting in the factor of 2 in the last term of (10.6).

For the majority of the problems in semiconductor quantum optics, the ground state of the semiconductor in the absence of an applied electromagnetic field is a completely filled valence band and an empty conduction band. This is true even at room temperature since $k_B T$ ($\simeq 0.026$ eV) $\ll E_g \simeq 1 - 3$ eV, where $E_g = \varepsilon_{c,k}^{\min} - \varepsilon_{v,k}^{\max}$ is the bandgap energy. It is therefore logical to choose the full valence band as the *vacuum state* $|0\rangle_V$ of the semiconductor; that is,

$$|0\rangle_V = \hat{a}_{v,k_1}^\dagger \hat{a}_{v,k_2}^\dagger \cdots \hat{a}_{v,k_N}^\dagger |0\rangle, \quad (10.7)$$

where $|0\rangle$ denotes the (physically irrelevant) *true vacuum*. When we deal with the new vacuum $|0\rangle_V$, which will be denoted simply by $|0\rangle$ from now on, the properties of the electron annihilation and creation operators will depend on the band index. For example,

$$\hat{a}_{c,k}|0\rangle = 0, \hat{a}^\dagger_{c,k}|0\rangle \neq 0, (\hat{a}^\dagger_{c,k})^2|0\rangle = 0$$
$$\hat{a}_{v,k}|0\rangle \neq 0, \hat{a}^\dagger_{v,k}|0\rangle = 0, (\hat{a}_{v,k})^2|0\rangle = 0. \tag{10.8}$$

We can therefore identify the conduction band annihilation and creation operators as standard fermionic operators. The action of the valence band operators on the new vacuum state, on the other hand, is reversed. Another complication associated with the valence-band electronic states comes from the band theory; in all cases of interest, the (non-interacting) valence band electron energy can be written as $\varepsilon_{v,k} = \tilde{E}_v - \hbar^2 k^2/(2|m_v|)$, which in turn implies that valence band electrons behave as particles with negative effective mass.

Both of these complications are resolved by introducing the concept of a hole. Mathematically, we define a new annihilation operator

$$\hat{h}_{k,\sigma_k} = \hat{a}^\dagger_{v,-k,-\sigma_k}, \tag{10.9}$$

together with the corresponding equation for the hole creation operator $\hat{h}^\dagger_k$. σ_k in (10.9) denotes the spin degree of freedom that we normally suppress for simplicity. It is straightforward to show that the hole operators satisfy the standard anticommutation relations. In order to understand the properties of quasi-particles generated by hole creation operators, we substitute the transformation of (10.9) into (10.6). In addition, if we let $\hat{a}_{c,k} \to \hat{e}_k$, and use $\hat{a}^\dagger_{v,-k}\hat{a}_{v,-k} = \hat{h}_k \hat{h}^\dagger_k = 1 - \hat{h}^\dagger_k \hat{h}_k$ in the (bilinear) energy term, we obtain

$$\hat{H}_0 = \sum_k \left[\varepsilon_{c,k} \hat{e}^\dagger_k \hat{e}_k - \varepsilon_{v,-k} \hat{h}^\dagger_k \hat{h}_k \right] + \sum_k \varepsilon_{v,-k}$$
$$+ \frac{1}{2} \sum_{\substack{k,k' \\ q \neq 0}} V(q) \left[\hat{e}^\dagger_{k+q} \hat{e}^\dagger_{k'-q} \hat{e}_{k'} \hat{e}_k + \hat{h}_{-k+q} \hat{h}_{-k'-q} \hat{h}^\dagger_{-k'} \hat{h}^\dagger_{-k} \right.$$
$$\left. + 2 \hat{e}^\dagger_{k+q} \hat{h}_{-k'-q} \hat{h}^\dagger_{-k'} \hat{e}_k \right]. \tag{10.10}$$

The first line of (10.10) conveys a simple and convenient result; the (bare) energy of a hole with wavevector k is $\tilde{\varepsilon}_{h,k} = -\varepsilon_{v,-k} = -\tilde{E}_v + \hbar^2 k^2/(2|m_v|)$, which implies that holes behave as particles with positive effective mass $m_h = |m_v| = -m_v$.

Further insight into the properties of holes can be obtained by putting the interaction term of (10.10) in standard normally ordered form. By carrying out the anticommutation of the hole operators, we obtain

$$\hat{H}_0 = \sum_k \left[\varepsilon_{e,k} \hat{e}_k^\dagger \hat{e}_k + \varepsilon_{h,k} \hat{h}_k^\dagger \hat{h}_k \right] + E_0$$

$$+ \frac{1}{2} \sum_{\substack{k,k' \\ q \neq 0}} V(q) \left[\hat{e}_{k+q}^\dagger \hat{e}_{k'-q}^\dagger \hat{e}_{k'} \hat{e}_k \right.$$

$$\left. + \hat{h}_{k+q}^\dagger \hat{h}_{k'-q}^\dagger \hat{h}_{k'} \hat{h}_k - 2 \hat{e}_{k+q}^\dagger \hat{h}_{k'-q}^\dagger \hat{h}_{k'} \hat{e}_k \right]. \quad (10.11)$$

Equation (10.11) is the standard electron-hole Hamiltonian of a semiconductor. We note that the last term, which describes the interaction of an electron and a hole, now has a negative sign, indicating that electron and hole have opposite charges [1]. The Hamiltonian of (10.11) therefore describes interactions between negatively charged electrons with mass m_e and positively charged holes with mass m_h. The new hole energy $\varepsilon_{h,k}$ includes direct Coulomb interaction with the valence band electrons; this contribution arises from the anticommutation relations employed in the derivation of (10.11). For simplicity, we will assume from now on that $E_0 = 0$ and

$$\varepsilon_{e,k} = E_g + \frac{\hbar^2 k^2}{2m_e^*}$$

$$\varepsilon_{h,k} = \frac{\hbar^2 k^2}{2m_h^*}. \quad (10.12)$$

To complete the derivation of the elementary Hamiltonian of semiconductor quantum optics, we consider semiconductor-electromagnetic field interactions. In the electric-dipole approximation whose validity we shall discuss shortly, the second quantized interaction Hamiltonian is

$$\hat{H}_{\text{int-rad}} = \int d^3 r \, \hat{\Psi}^\dagger(r) [-e\mathbf{r} \cdot \hat{\mathbf{E}}(r)] \hat{\Psi}(r). \quad (10.13)$$

To express (10.13) in terms of mode operators, we recall that in the two-band approximation we can write the field operator as

$$\hat{\Psi}(r) = \sum_k \varphi_{c,k}(r) \hat{e}_k + \sum_k \varphi_{v,k}(r) \hat{h}_{-k}^\dagger, \quad (10.14)$$

which directly follows (10.2) and (10.9). This expansion states that the field operator $\hat{\Psi}(r)$, which annihilates an electron at position **r**, can be decomposed using mode operators that annihilate a (conduction band) electron with wavevector k or generates a (valence band) hole with wavevector $-k$. As we have seen in Chapter 2, the quantized electric field operator is given by

$$\hat{E}(r) = i \sum_q E_q \hat{a}_q e^{i\mathbf{q}\cdot\mathbf{r}} - \text{h.c.}, \tag{10.15}$$

where $\mathbf{E}_q = \left(\frac{\hbar\omega_q}{2\epsilon V_{\text{opt}}}\right)^{1/2} \mathbf{u}_q = E_q \mathbf{u}_q$, with $V_{\text{opt}} = V_c$ denoting the electromagnetic field quantization volume. $\mathbf{u}_q$ is the polarization vector.

Substituting (10.14) and (10.15) into (10.13), we obtain

$$\hat{H}_{\text{int-rad}} = -i \sum_{k,k',q} eE_q \int d^3r\, \mathbf{u}_q \cdot \mathbf{r} \left[\varphi^*_{c,k}(r)\hat{e}^\dagger_k + \varphi^*_{v,k}(r)\hat{h}_{-k}\right]$$

$$\left[\hat{a}_q e^{i\mathbf{q}\cdot\mathbf{r}} - \hat{a}^\dagger_q e^{-i\mathbf{q}\cdot\mathbf{r}}\right]\left[\varphi_{c,k'}(r)\hat{e}_{k'} + \varphi_{v,k'}(r)\hat{h}^\dagger_{-k'}\right], \tag{10.16}$$

where we have assumed the polarization vector to be real for simplicity. Equation (10.16) contains eight terms; two of these can be eliminated using the rotating wave approximation (RWA). These are the terms where a photon and an electron-hole pair are simultaneously annihilated or created. As we shall see shortly, the carrier-photon coupling strength in semiconductors has the same order of magnitude as that in atoms; therefore, RWA is an excellent approximation for all field strengths of interest.

The terms that correspond to annihilation of an electron (hole) with wavevector k_1 and creation of another electron (hole) with wavevector k_2 along with either creation or annihilation of a photon correspond to *intraband transitions*. In an ideal three-dimensional (3D) semiconductor, the contribution of these four terms vanishes due to crystal momentum conservation. In a quantum-well (QW) structure, on the other hand, momentum conservation along the growth direction is relaxed due to the absence of translational symmetry, and *subbands* within both conduction and valence bands form. The corresponding *inter-subband transitions* have giant dipole moments and are of significant interest in infrared optics. Due to the order-of-magnitude difference in transition energies, however, we can discard these transitions when we are dealing with (visible) *interband transitions* in QWs. The calculation of the interaction Hamiltonian and the dipole moment of inter-subband transitions can be found elsewhere [7]. After discarding the corresponding four terms, we end up with the interaction Hamiltonian

$$\hat{H}_{\text{int-rad}} = -\sum_{k,k',q} ieE_q \hat{e}^\dagger_k \hat{h}^\dagger_{-k'} \hat{a}_q \int d^3r\, \mathbf{u}_q \cdot \mathbf{r}\varphi^*_{c,k}(r)e^{i\mathbf{q}\cdot\mathbf{r}}\varphi_{v,k'}(r) + \text{h.c.}$$

$$= -\sum_{k,q} i\hbar g_{k,q} \hat{e}^\dagger_{k+q}\hat{h}^\dagger_{-k}\hat{a}_q + \text{h.c.} \tag{10.17}$$

We note that the first line of (10.17) is valid in any dimension, whereas the second line only applies to a 3D semiconductor. In 3D, the calculation of the coupling strength $g_{k,q}$ is straightforward:

$$g_{k,q} = \sum_{k'} \frac{eE_q}{\hbar} \int d^3r\, \mathbf{u}_q \cdot \mathbf{r}\varphi^*_{c,k}(r) e^{i\mathbf{q}\cdot\mathbf{r}} \varphi_{v,k'}(r)$$

$$= \sum_{k'} \frac{eE_q}{\hbar} \int d^3r\, \mathbf{u}_q \cdot \mathbf{r} u^*_{c,k}(r) u_{v,k'}(r) e^{i(\mathbf{k'}+\mathbf{q}-\mathbf{k})\mathbf{r}}. \tag{10.18}$$

In all cases of interest, we deal with transitions between conduction and valence band states with $k, k' \ll \pi/a_L$, where a_L denotes the lattice constant of the semiconductor. Band structure calculations show that in this range of crystal momenta, $u_{c,k}(r)$ and $u_{v,k'}(r)$ are essentially independent of k, k'. This allows us to set

$$u_{c,k}(r) \rightarrow u_{c,k=0}(r)$$

$$u_{v,k}(r) \rightarrow u_{v,k=0}(r)$$

in (10.18). We can understand this approximation by noting that $u_{c,k}(r)$ and $u_{v,k}(r)$ are periodic functions with period a_L (i.e., they vary significantly within a unit cell). $\exp[ikr]$ factors for $k \ll \pi/a_L$, on the other hand, are essentially constant within a unit cell. We can therefore rewrite the integral in (10.18) as a sum over integrals in unit cells: $\int d^3r \rightarrow \sum_i \int d^3r_i$, where $\mathbf{r} = \mathbf{R}_i + \mathbf{r}_i$ and $\int d^3r_i$ runs over the ith unit cell. We obtain

$$g_{k,q} \simeq \sum_{k'} \frac{E_q}{\hbar} \int d^3r_i\, e\mathbf{u}_q \cdot \mathbf{r}_i u^*_{c,0}(r) u_{v,0}(r) \sum_i e^{i(\mathbf{k'}+\mathbf{q}-\mathbf{k})\mathbf{R}_i}$$

$$\cong \sum_{k'} \frac{\mu_{cv} E_q}{\hbar} \sum_i e^{i(\mathbf{k'}+\mathbf{q}-\mathbf{k})\mathbf{R}_i} = \frac{\mu_{cv} E_q}{\hbar} \cdot \delta_{\mathbf{k'}+\mathbf{q},\mathbf{k}} \tag{10.19}$$

where we have introduced the dipole matrix element μ_{cv} between the conduction and valence bands. Typically, $\mu_{cv} \sim 3-5$ Å, which a posteriori justifies the electric-dipole approximation that we have used since $\lambda_{\text{optical}} = 2\pi\hbar c/E_g \sim 2000$ Å. We also remark that μ_{cv} is larger than but comparable to atomic dipole matrix elements, indicating the validity of RWA for essentially all electric field strengths of interest. The derivation of g_q for a 2D semiconductor interacting with a 3D radiation field reservoir explicitly shows that only the in-plane momentum is conserved.

In most practical problems, we can make two further simplifications: First, the photon momentum q is approximately two orders of magnitude smaller than typical electron and hole momenta. We can therefore discard the momentum carried by the photon and set $q = 0$ (i.e., assume that optical transitions are *vertical* and do not change the momentum). Second, the typical field occupancies in experiments that do not involve a microcavity are high enough that a classical field approximation is valid. We can then set $E_q < \hat{a}_q >\rightarrow iE_c^{(+)}(t) = iE_c \exp[-i\omega_c t]$. This assumption will be lifted in Chapter 11, where we analyze cavity polaritons. The semiclassical

Hamiltonian we use in this chapter is then

$$\hat{H} = \sum_k \left[\varepsilon_{e,k} \hat{e}_k^\dagger \hat{e}_k + \varepsilon_{h,k} \hat{h}_k^\dagger \hat{h}_k \right] + \sum_k \mu_k (E_c^{(-)} \hat{\sigma}_k + \text{h.c.})$$
$$+ \frac{1}{2} \sum_{\substack{k,k' \\ q \neq 0}} V(q) \left[\hat{e}_{k+q}^\dagger \hat{e}_{k'-q}^\dagger \hat{e}_{k'} \hat{e}_k + \hat{h}_{k+q}^\dagger \hat{h}_{k'-q}^\dagger \hat{h}_{k'} \hat{h}_k - 2 \hat{e}_{k+q}^\dagger \hat{h}_{k'-q}^\dagger \hat{h}_{k'} \hat{e}_k \right],$$
(10.20)

where $\mu_k = \hbar g_k / E_q = \mu_{cv}$ and $\hat{\sigma}_k = \hat{h}_{-k} \hat{e}_k$ is the polarization operator. Introduction of quantized electromagnetic fields may be achieved by replacing $E_c^{(-)}$ and $E_c^{(+)}$ with the corresponding operators from (10.15).

10.2. SEMICONDUCTOR BLOCH EQUATIONS

Semiconductor Bloch equations (SBEs) are the semiconductor counterpart of optical Bloch equations that we have discussed in the context of atom-field interactions [2]: They describe the coupled time evolution of the expectation values of the electron and hole number operators, and the polarization operator. The novelty in the case of SBE arises from the presence of interactions and the band dispersion. The starting point for our derivation is the Heisenberg equation for $\hat{\sigma}_k$, using the Hamiltonian of (10.20):

$$\frac{d\hat{\sigma}_k}{dt} = \frac{1}{i\hbar} \left[\hat{\sigma}_k, \hat{H} \right]$$
$$= -\frac{i}{\hbar}(\varepsilon_{e,k} + \varepsilon_{h,k}) \hat{\sigma}_k + \frac{i \mu_k E_c^{(+)}}{\hbar} \left[1 - \hat{e}_k^\dagger \hat{e}_k - \hat{h}_{-k}^\dagger \hat{h}_{-k} \right]$$
$$+ \frac{i}{\hbar} \sum_{k',q \neq 0} V(q) \left[\delta_{-k,k'-q} \hat{h}_{-k+q} \hat{e}_{k-q} + \hat{h}_{k'-q}^\dagger \hat{h}_{k'} \hat{h}_{-k} \hat{e}_{k-q} \right.$$
$$\left. + \hat{e}_{k'+q}^\dagger \hat{e}_{k'} \hat{h}_{-k+q} \hat{e}_k \right]$$
$$+ \frac{i}{\hbar} \sum_{k',q \neq 0} V(q) \left[\hat{e}_{k'-q}^\dagger \hat{e}_{k'} \hat{e}_{k-q} \hat{h}_{-k} + \hat{h}_{k'-q}^\dagger \hat{h}_{-k-q} \hat{h}_{k'} \hat{e}_k \right]. \quad (10.21)$$

Next, we let $\sigma_k = <\hat{\sigma}_k>$, $n_{ek} = <\hat{e}_k^\dagger \hat{e}_k>$, and $n_{hk} = <\hat{h}_{-k}^\dagger \hat{h}_{-k}>$, and then take the ensemble average of both sides of (10.21) to obtain

$$\frac{d\sigma_k}{dt} = -\frac{i}{\hbar}(\varepsilon_{e,k} + \varepsilon_{h,k}) \sigma_k + \frac{i \mu_k E_c^{(+)}}{\hbar} [1 - n_{ek} - n_{hk}]$$
$$+ \frac{i}{\hbar} \sum_{k',q \neq 0} V(q) \left[\delta_{-k,k'-q} \sigma_{k-q} + <\hat{h}_{k'-q}^\dagger \hat{h}_{k'} \hat{h}_{-k} \hat{e}_{k-q}> \right.$$

$$+ <\hat{e}^\dagger_{k'+q}\hat{e}_{k'}\hat{h}_{-k+q}\hat{e}_k>\Big]$$

$$+ \frac{i}{\hbar}\sum_{k',q\neq 0} V(q)\Big[<\hat{e}^\dagger_{k'-q}\hat{e}_{k'}\hat{e}_{k-q}\hat{h}_{-k}> + <\hat{h}^\dagger_{k'-q}\hat{h}_{-k-q}\hat{h}_{k'}\hat{e}_k>\Big]. \quad (10.22)$$

We can see from (10.22) that unlike the atomic case, σ_k does not couple only to n_{ek} and n_{hk}; due to the interactions, there is coupling to all other polarization terms (σ_{k-q}). More important, σ_k couples to expectation values of four-operator terms, which will in turn couple to six-operator terms. Therefore, the exact ensemble-averaged polarization equations will not form a closed set. This is also true for the corresponding equations for n_{ek} and n_{hk}. To obtain a closed set of equations in the presence of interactions, we will need to introduce time dependent Hartree-Fock approximation. However, we first consider the illustrative case of a non-interacting semiconductor.

10.2.1. Semiconductor Bloch Equations in the Non-interacting Limit

The Bloch equations for the electron-hole occupancies and polarization simplify significantly in this case:

$$\frac{d\sigma_k}{dt} = -\frac{i}{\hbar}(\varepsilon_{e,k} + \varepsilon_{h,k})\sigma_k + \frac{i\mu_k E_c^{(+)}}{\hbar}[1 - n_{ek} - n_{hk}] \quad (10.23)$$

$$\frac{dn_{ek}}{dt} = -\frac{i\mu_k}{\hbar}\left(\sigma_k E_c^{(-)} - \sigma_k^* E_c^{(+)}\right) \quad (10.24)$$

$$\frac{dn_{hk}}{dt} = -\frac{i\mu_k}{\hbar}\left(\sigma_k E_c^{(-)} - \sigma_k^* E_c^{(+)}\right). \quad (10.25)$$

It is worth mentioning that there is a one-to-one correspondence between these equations and the optical Bloch equations, provided that we set $\sigma_k \to \sigma_{ge}$ and $n_{ek} + n_{hk} \to 2\sigma_{ee}$. Since these equations do not include external pumping via carrier injection, we have $n_{ek} = n_{hk}$. Finally, we remark that (10.23) through (10.25) form a closed set [i.e., (σ_k, n_{ek}, n_{hk}) do not couple to ($\sigma_{k'}', n_{ek'}, n_{hk'}$) for $k \neq k'$, implying that electrons with different crystal momenta respond to an external radiation field as independent oscillators with resonance frequency $\tilde{E}_{res} = E_g + \hbar k^2/(2m_r)$, where $m_r = m_e^* m_h^*/(m_e^* + m_h^*)$]. The resulting absorption profile would have an inhomogeneous lineshape with a non-Gaussian lineshape [2].

We conclude this section by noting an inconsistency in the interaction-free model: Even though electron-electron and hole-hole interactions can in principle be neglected by assuming a very low (conduction band) electron density, the same procedure cannot be justified for electron-hole interactions. A photon absorbed at position r, creates an interacting electron-hole pair at that position. The electron-hole interaction is therefore important even for a single pair, as we shall see quantitatively in the next section.

10.2.2. Semiconductor Bloch Equations with Interactions

In order to obtain a tractable set of equations describing the carrier-light interactions in the presence of interactions, we need to introduce time dependent HF approximation, where we write the four operator expectation values in terms of all possible combinations of two operator expectation values [2]. As a result, all terms on the RHS of (10.22) can be written in terms of polarization and population of modes with different momenta. We obtain

$$\frac{d\sigma_k}{dt} = -\frac{i}{\hbar}\left[\varepsilon_{e,k} + \varepsilon_{h,k}) - \sum_{q\neq 0} V(q)(n_{e,k-q} + n_{h,k-q})\right]\sigma_k$$

$$+ \left[\frac{i\mu_k E_c^{(+)}}{\hbar} + \frac{i}{\hbar}\sum_{q\neq 0} V(q)\sigma_{k-q}\right][1 - n_{ek} - n_{hk}]. \quad (10.26)$$

Equation (10.26) has two important features: First, it predicts that the interactions change the resonance energy of each mode

$$\tilde{E}_{res} \to E_k = E_g + \frac{\hbar^2 k^2}{2m_r^*} - \sum_{q\neq 0} V(q)(n_{e,k-q} + n_{h,k-q}), \quad (10.27)$$

which represents a contribution to the well-known bandgap renormalization in semiconductor physics, originating from (repulsive) electron-electron and hole-hole interactions [2]. We can understand this term by noting that the presence of a significant hole density means that there are many available valence band states and the valence electrons can adjust their wavefunctions so as to screen out the repulsive interactions.

Second, the Rabi frequency is enhanced as compared to the non-interacting case:

$$\mu_k E_c^{(+)} \to \hbar\Omega_k = \mu_k E_c^{(+)} + \sum_{q\neq 0} V(q)\sigma_{k-q}. \quad (10.28)$$

This enhancement is commonly referred to as *excitonic* or *Coulomb enhancement*; it originates from the electron-hole (attractive) interactions. Due to attraction, the electron and hole adjust their relative motion so as to spend more time in each other's vicinity, as compared to non-interacting particles. Since photon generation/annihilation requires the simultaneous presence of an electron and a hole at the same position, this positive correlation or *bunching* enhances the transition probability.

When these replacements are made, the equations for the electron and hole occupancies retain their original form [i.e., (10.24) and (10.25); their implications, however, change completely as we now have coupling between different momenta and the equations are nonlinear]. The set of SBE is then [2]

$$\frac{d\sigma_k}{dt} = -\frac{i}{\hbar}E_k\sigma_k + i\Omega_k[1 - n_{ek} - n_{hk}] \quad (10.29)$$

$$\frac{dn_{ek}}{dt} = -i(\Omega_k^* \sigma_k - \Omega_k \sigma_k^*), \qquad (10.30)$$

with $\dot{n}_{hk} = \dot{n}_{ek}$. The numerical solution of full (nonlinear) SBE is useful in making accurate predictions about the gain spectrum of semiconductor lasers [2]. However, the simple form of SBE given in (10.29) and (10.30) cannot be used to calculate two-time correlation functions. Due to the time dependent HF approximation introduced to obtain a closed set of equations, quantum regression theorem cannot be utilized. In this chapter, however, we are interested in understanding the elementary optical excitation spectrum obtained under weak excitation conditions.

10.3. EXCITONS

For low excitation densities, we can assume that the electron and hole occupancies remain much smaller than unity; in this limit, we can set $n_{ek} \simeq n_{hk} \simeq 0$, $\forall k$ to obtain

$$\frac{d\sigma_k}{dt} = -\frac{i}{\hbar}\left(E_g + \frac{\hbar^2 k^2}{2m_r^*}\right)\sigma_k + \frac{i}{\hbar}\left(\mu_k E_c^{(+)} + \sum_{q \neq 0} V(q)\sigma_{k+q}\right), \qquad (10.31)$$

which is a linear equation. We see that the effect of electron-hole interactions is to couple various σ_k. To understand the implications of this equation, we take the inverse-Fourier transform of both sides of (10.31); we obtain

$$\frac{d\sigma(r)}{dt} = -\frac{i}{\hbar}\left(E_g - \frac{\hbar^2}{2m_r^*}\nabla^2 - \frac{e^2}{4\pi\epsilon r}\right)\sigma(r) + \frac{i}{\hbar}\mu_k E_c^{(+)} \delta^{(3)}(r) V. \qquad (10.32)$$

The solution to this linear inhomogeneous equation can be written in terms of a superposition of the solutions to the homogeneous time-independent eigenvalue equation:

$$E_n \varphi_n(r) = \left(-\frac{\hbar^2}{2m_r^*}\nabla^2 - \frac{e^2}{4\pi\epsilon r}\right)\varphi_n(r), \qquad (10.33)$$

which is referred to as the Wannier equation in solid-state physics [2]. The direct analogy to the Schrödinger equation for the hydrogen atom tells us that the solutions describing the relative motion of an electron and a hole will be hydrogenic. The spectrum consists of a discrete (with $E_n < 0$) and a continuous (with $E_n > 0$) part.

The total optical polarization created by the applied classical field $E_c^{(+)}$ can be found after straightforward algebra:

$$\wp(\omega_c) = \epsilon \chi^{(1)} E_c = \sum_n \frac{|\mu_{cv}|^2 |\varphi_n(r=0)|^2}{\epsilon(E_n + E_g - \hbar\omega_c - i\Gamma_n/2)} \epsilon E_c, \qquad (10.34)$$

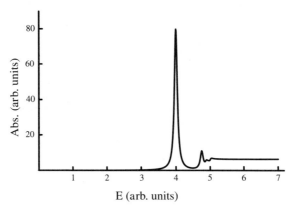

FIGURE 10-1: The absorption spectra of a two-band semiconductor.

where ω_c denotes the laser frequency. Γ_n is the imaginary part of the self-energy arising from coupling to all the other radiation field modes. We use the same variable to denote the real and momentum space wavefunctions for simplicity; unless k dependence is noted explicitly, φ_n should be taken to be the wavefunction in real space. The total polarization exhibits resonances at energies $E_g + E_n$; since $E_n < 0$ for the discrete part of the spectrum, these resonances appear below the bandgap energy. We identify these resonances as exciton lines descibing a bound electron-hole pair state with a hydrogenic wavefunction describing the relative motion. The continuum absorption, which is the only type of absorption present in the absence of interactions, starts at the *bare bandgap energy*. The absorption spectra derived from $\wp$ is shown in Fig. 10-1. We will consider the properties of excitons and their interaction with light in depth in the next chapter.

We finish this section by noting that the contribution of each exciton line to total polarization is proportional to $|\varphi_n(0)|^2$ (i.e., the probability of finding the electron and the hole at the same position, as we predicted earlier). We immediately see that only the s-states contribute to optical absorption. We also note that for the $n = 1s$ solution, $|\varphi_n(0)|^2 \propto 1/a_B^d$, implying that the optical coupling strength will increase with increasing Coulomb attraction [here a_B is the Bohr radius determined by (10.33) and d is the dimensionality].

10.4. NONLINEAR RESPONSE OF EXCITONS

The optical response predicted by (10.31) is purely linear, as we have already calculated. This result is a direct consequence of the fact that we have neglected the electron and hole populations generated by the applied laser field. In this section, we include the nonlinear response of the semiconductor by taking into account the first-order changes in the occupancies induced by the laser field. For $|\sigma_k| \ll 1, \forall k$, we can show that

$$n_{ek} = n_{hk} \simeq |\sigma_k|^2, \qquad (10.35)$$

provided that decoherence can be neglected. Substituting this expression in (10.26) and rearranging the terms, we obtain

$$\frac{d\sigma_k}{dt} = -\frac{i}{\hbar}\left[\left(E_g + \frac{\hbar^2 k^2}{2m_r^*}\right)\sigma_k - \sum_{q\neq 0} V(q)\sigma_{k-q}\right]$$
$$+ \frac{i}{\hbar}\mu_k E_c^{(+)}(1 - 2|\sigma_k|^2)$$
$$+ 2\sum_q V(q)\sigma_{k-q}(\sigma_{k-q}^* - \sigma_k^*)\sigma_k. \qquad (10.36)$$

Equation (10.36) is explicitly nonlinear in σ_k, and the optical response that one would obtain based on this equation will also be nonlinear. By confining ourselves to the low excitation regime, we were able to obtain a nonlinear equation with a single set of variables (σ_k).

The first line of (10.36) descibes the Wannier equation in Fourier space, which has the eigenstates $\varphi_n(k)$, as we have seen in the previous section. The second line descibes the coupling to the external field; We notice that the coupling coefficient is reduced in proportion to the magnitude of the occupancies. This is the *phase-space filling* (PSF) effect that introduces the first kind of optical nonlinearity. The third line has a product of three polarization terms and describes the *exciton-exciton* interactions. We therefore find that within the time dependent HF approximation and in the low excitation regime where $n_{ek} = n_{hk} \ll 1, \forall k$, PSF and exciton-exciton interactions provide the two physically distinct sources of nonlinearity in a semiconductor. A novelty in the case of semiconductors is the nonlinearity due to exciton-exciton interactions, which is absent in the case of nondegenerate atoms interacting with radiation fields. We also point out that recent experiments in bulk GaAs under high magnetic fields indicate that exciton-exciton correlations that are neglected in the HF approximation can play a dominant role in determining the nonlinear response [8].

We can further simplify (10.36) by assuming that only the 1s excitonic state contributes to the polarization: $\sigma_k \simeq \varphi_{1s}(k)\psi(t)$. This assumption is valid provided that the laser frequency satisfies near-resonance conditions with the 1s exciton resonance and remains far off resonant from the higher-energy resonances. Multiplying both sides of (10.36) by $\varphi_{1s}^*(k)$ and summing over all k, we obtain

$$i\hbar\frac{d\psi}{dt} = (E_g - E_{1s})\psi - \varphi_{1s}^*(0)\mu_{cv}E_c^{(+)} + 2\mu_{cv}E_c^{(+)}\sum_k |\varphi_{1s}(k)|^2\varphi_{1s}^*(k)|\psi|^2$$
$$+ 2\sum_k V(q)|\varphi_{1s}(k)|^2(\varphi_{1s}^*(k) - \varphi_{1s}^*(k-q))\varphi_{1s}(k-q)|\psi|^2\psi. \qquad (10.37)$$

It is easy to recognize that (10.37) has the form of a nonlinear Schödinger equation. We conclude this section by evaluating the third-order excitonic nonlinear response using (10.37). The optical polarization is related to the exciton amplitude $\tilde{\psi} = \psi \exp[i\omega_c t]$ via

$$\wp(\omega_c) = \mu_{cv}\varphi_{1s}(0)\tilde{\psi}. \tag{10.38}$$

If we let $\psi = \psi^{(1)} + \psi^{(3)}$, where $|\psi^{(1)}| \gg |\psi^{(3)}|$, the contribution of $1s$ resonance to linear susceptibility ($\psi^{(1)}$) is given by the steady-state solution of (10.37), where only the first two terms on the RHS are taken into account:

$$\tilde{\psi}^{(1)} = \frac{\mu_{cv}\varphi_{1s}^*(0)}{\Delta} E_c. \tag{10.39}$$

Here, $\Delta = E_g + E_{1s} - \hbar\omega_c$. The nonlinear correction can then be evaluated by replacing ψ by $\psi^{(1)}$ in the remaining terms on the RHS. This gives

$$\begin{aligned}\tilde{\psi}^{(3)} &= \frac{\mu_{cv}^3|\varphi_{1s}(0)|^2}{\Delta^3} 2\sum_k |\varphi_{1s}(k)|^2 \varphi_{1s}^*(k) E_c^3 \\ &+ \frac{\mu_{cv}^3|\varphi_{1s}(0)|^2 \varphi_{1s}^*(0)}{\Delta^4} 2\sum_k V(q)|\varphi_{1s}(k)|^2 (\varphi_{1s}^*(k) \\ &- \varphi_{1s}^*(k-q))\varphi_{1s}(k-q) E_c^3. \end{aligned} \tag{10.40}$$

Calculation of various nonlinear optical processes, such as the Kerr effect, can be done using the third-order susceptibility that immediately follows (10.40). An analogous calculation can be used to derive the excitonic linear and nonlinear susceptibility in a quasi-two-dimensional quantum well structure.

We conclude this chapter by noting that we have not introduced reservoir coupling into our SBE. The inclusion of dissipation due to radiation field coupling, for instance, can be done in a straightforward way that is completely analogous to the procedure carried out for atoms. The inclusion of electron-phonon and electron-electron scattering induced dephasing, however, is much more delicate. The calculation of the rate of the latter process, for example, should be carried out self-consistently by going beyond the HF approximation [9].

REFERENCES

[1] H. Haken, *Quantum Field Theory of Solids* (North-Holland, Amsterdam, 1976).
[2] H. Haug and S. W. Koch, *Quantum Theory of the Optical and Electronic Properties of Semiconductors* (World Scientific, Singapore, 1993).
[3] S. Schmitt-Rink, D. S. Chemla, and H. Haug, Phys. Rev. A **37**, 941 (1996).
[4] C. Kittel, *Introduction to Solid State Physics* (Wiley, New York, 1996).

[5] N. W. Ashcroft and N. D. Mermin, *Solid State Physics* (Saunders College, Philadelphia, 1976).

[6] C. Kittel, *Quantum Theory of Solids* (Wiley, New York, 1987).

[7] C. Weisbuch and B. Vinter, *Quantum Semiconductor Structures* (Academic Press, San Diego, 1991).

[8] P. Kner, S. Bar-Ad, M. V. Marquezini, D. S. Chemla, and W. Schafer, Phys. Rev. Lett. **78**, 1319 (1997).

[9] F. Jahnke, M. Kira, S. W. Koch, G. Khitrova, E. K. Lindmark, T. R. Nelson, Jr., D. V. Wick, J. D. Berger, O. Lyngnes, H. M. Gibbs, and K. Tai, Phys. Rev. Lett. **77**, 5257 (1996).

11 Excitons and Polaritons

In Chapter 10, we developed the basic theory of semiconductor-light interactions starting from the first principles. Using the time dependent Hartree–Fock (HF) approximation, we arrived at the semiconductor Bloch equations (SBEs), which provide the starting point for the analysis of optical processes in semiconductors. In the ultralow excitation density limit, we have seen that the equation for the polarization term admits discrete eigenstate solutions, which are termed excitons.

Even though SBEs give a consistent and complete treatment of semiconductor-light interactions in the HF approximation, their usefulness is limited by the sheer size of the dynamical quantities one must keep track of. Except for the simple cases we have considered, SBEs can only be solved numerically, which in turn makes it difficult to obtain further insight into the physical nature of the problem. The strength of SBEs is the fact that they can be generalized to include correlation terms consistently [2], [3]; however, this procedure increases the complexity significantly, requiring ultrafast computers to carry out the calculations. It is therefore highly desirable to use a simplified basis of optical excitations such as excitons to analyze a relatively narrow but interesting class of problems in semiconductor quantum optics. Analyzing the light-matter interactions using exciton operators is tractable only in the limit of low temperatures and low excitation densities, which is the natural domain for quantum optics problems.

In this chapter, we will concentrate on the physics of excitons and their interactions with the radiation field (Sec. 11.1). Of particular interest in the context of quantum optics is the microcavity exciton polaritons, or cavity polaritons, that we analyze in Sec. 11.3. We will conclude the chapter with a discussion of physics that arise from the composite-boson character of excitons (Sec. 11.4).

11.1. NON-INTERACTING EXCITONS

We have seen in Chapter 10 that in the *ultralow density* regime, the electron and hole occupancies generated by the applied laser field can be neglected. The only important interaction term in this limit is that of electron-hole interaction [1]. Leaving the quantitative description of what constitutes the ultralow density regime for the moment, we recall the Hamiltonian (in the RWA) that we used to derive the equation of motion for the polarization:

$$\hat{H} = \sum_k \hbar\omega_p(k)\hat{a}_k^\dagger \hat{a}_k + \sum_k [\varepsilon_{e,k}\hat{e}_k^\dagger \hat{e}_k + \varepsilon_{h,k}\hat{h}_k^\dagger \hat{h}_k] - \sum_{\substack{k,k' \\ q\neq 0}} V(q)\hat{e}_{k+q}^\dagger \hat{h}_{k'-q}^\dagger \hat{h}_{k'}\hat{e}_k$$

$$- \sum_{k,q} i\hbar g_{k,q}\hat{e}_{k+q}^\dagger \hat{h}_{-k}^\dagger \hat{a}_q + h.c. \tag{11.1}$$

In this case, we have assumed coupling to a general radiation field reservoir. We observe that the first line of this equation can be diagonalized for a single electron-hole pair by introducing the electron-hole pair operator

$$\hat{C}_{\nu,k} = \sum_p \varphi_\nu(p)\hat{h}_{\frac{k}{2}-p}\hat{e}_{\frac{k}{2}+p}, \tag{11.2}$$

provided that $\varphi_\nu(p)$ is the spatial Fourier transform of the νth eigenstate solution of the Wannier equation with energy E_ν. The complete Hamiltonian can be put in a simple form using this pair operator:

$$\hat{H} = \sum_k \hbar\omega_p(k)\hat{a}_k^\dagger \hat{a}_k + \sum_{k,\nu} \hbar\omega_{\text{exc}}^\nu(k)\hat{C}_{\nu,k}^\dagger \hat{C}_{\nu,k} + \sum_{k,\nu} i\hbar g_\nu(k)\left(\hat{C}_{\nu,k}^\dagger \hat{a}_k - \hat{C}_{\nu k}\hat{a}_k^\dagger\right), \tag{11.3}$$

where $\hbar g_\nu(k) = \mu_{cv}\varphi_\nu(r=0)\sqrt{\hbar\omega_p(k)/(2\epsilon)}$ and $\hbar\omega_{\text{exc}}^\nu = \epsilon_{e,o} + \epsilon_{h,o} + E_\nu$.

11.2. BULK EXCITON POLARITONS

The complete Hamiltonian of (11.3) can be diagonalized using a simple form of Bogoluibov transformation, provided that the operator $\hat{C}_{\nu,k}$ satisfies Bose-commutation relations. A straightforward algebra shows that [2]

$$\left[\hat{C}_{\nu,k}, \hat{C}_{\mu,k'}^\dagger\right] \simeq \delta_{\nu,\mu}\delta_{k,k'} - \sum_p |\varphi_\nu(p)|^2(n_{ep} + n_{hp}), \tag{11.4}$$

where n_{ep} and n_{hp} are as defined in Chapter 10. From (11.4) we observe that we can treat the *exciton annihilation* operators $\hat{C}_{\nu,k}$ as *bosonic* provided that the (electron and hole) carrier density N satisfies

$$Na_B^d \ll 1. \tag{11.5}$$

Equation (11.5) thereby defines the ultralow density limit for which the use of the approximate Hamiltonian of (11.3) is justified. With the exception of the last section of this chapter, we will assume that this condition holds in our discussion of excitons and exciton polaritons.

We can now introduce the Bogoluibov transformation that diagonalizes the complete Hamiltonian of (11.3) in the bosonic limit:

$$\hat{p}_k = u(k)\hat{C}_k + v(k)\hat{a}_k, \qquad (11.6)$$

where $u(k)$ and $v(k)$ satisfy $|u(k)|^2 + |v(k)|^2 = 1$. In writing (11.6), we have assumed that only one internal excitonic state (i.e., $v = 1s$) is important, and we dropped the v index. By definition, $\hat{p}_k^\dagger |0\rangle$ is an eigenstate of $\hat{H}$ with energy $\Omega(k)$ and we have

$$\hat{H}(u(k)^* \hat{C}_k^\dagger + v(k)^* \hat{a}_k^\dagger)|0\rangle = \hbar\Omega(k)(u(k)^* \hat{C}_k^\dagger + v(k)^* \hat{a}_k^\dagger)|0\rangle. \qquad (11.7)$$

Equating the coefficients of $\hat{C}_k^\dagger |0\rangle$ and $\hat{a}_k^\dagger |0\rangle$ on each side, we obtain

$$u^*(k)\Omega(k) = u^*(k)\omega_{\text{exc}}(k) + v^*(k)g(k)$$
$$v^*(k)\Omega(k) = v^*(k)\omega_p(k) + u^*(k)g(k). \qquad (11.8)$$

Solution of these algebraic equations gives the two (upper and lower) polariton dispersion relations

$$\Omega_{u\atop l}(k) = \frac{1}{2}[\omega_{\text{exc}}(k) + \omega_p(k)] \pm \frac{1}{2}\sqrt{(\omega_{\text{exc}}(k) - \omega_p(k))^2 + 4|g(k)|^2}. \qquad (11.9)$$

The coefficients $u(k)$ and $v(k)$ are in turn given by

$$u_{u\atop l}(k) = \frac{g(k)}{\sqrt{(\Omega_{u\atop l}(k) - \omega_{\text{exc}}(k))^2 + |g(k)|^2}}$$

$$v_{u\atop l}(k) = \frac{\Omega_{u\atop l}(k) - \omega_{\text{exc}}(k)}{\sqrt{(\Omega_{u\atop l}(k) - \omega_{\text{exc}}(k))^2 + |g(k)|^2}}. \qquad (11.10)$$

The expressions that we obtained are clearly very reminiscent of the atom-field dressed-state expressions we have derived previously. The primary difference in the present case stems from the fact that we are now dealing with the coupling of two sets of continuum of states parametrized by their momenta. In the end, we obtain new dispersion relations describing the dependence of energy on momentum.

Figure 11-1 shows the calculated dispersion curve for polaritons in a 3D semiconductor. For $k \to 0$, we have an approximately exciton-like dispersion for the upper branch. At $k = 0$ the upper polariton approaches the longitudinal exciton eigenen-

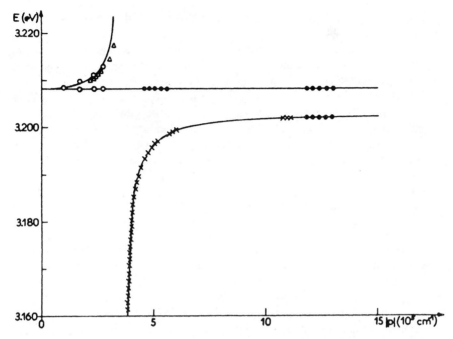

FIGURE 11-1: Bulk exciton polariton dispersion in CuCl, obtained by two-photon absorption (circles, D. Frölich et al., Phys. Rev. Lett. **26**, 554 (1971)) and hyper-Raman scattering (dots and crosses, B. Hönerlage et al., Phys. Rev. Lett. **41**, 49 (1978)).

ergy, which is shifted from the transverse exciton eigenenergy by

$$\hbar \Delta_{LT} = \hbar \frac{|g(0)|^2}{\omega_{\text{exc}}(0)} = \frac{\mu_{cv}^2}{\epsilon \pi a_B^3}, \qquad (11.11)$$

where Δ_{LT} is normally referred to as the *longitudinal-transverse splitting*. For the lower polariton branch in this limit, we have a photonlike linear dispersion.

In the high-momentum limit $k \gg E_g/(\hbar c)$, the lower branch dispersion is given by $\Omega_l(k) \simeq \omega_{\text{exc}}(k)$. We observe that the effect of the radiation field coupling is negligible in the high momentum limit.

Finally, we note that even though we have assumed coupling to a reservoir of radiation field modes, we have obtained real eigenvalues for the polariton states; this is a result of the momentum conservation that indicates that any given exciton state with momentum k can only couple to a single radiation field state with the same momentum in an ideal semiconductor.

11.3. QUANTUM-WELL EXCITONS AND CAVITY POLARITONS

The optical properties of excitons change considerably at lower dimensions. Due to the advances in molecular beam epitaxy (MBE) in the last 20 years, it is possible to grow uniform two-dimensional quantum-well (QW) structures where the thickness L_z of the QW layer can be as small as 10 Å with monolayer accuracy in growth. In such QW structures, excitons are confined in the so-called growth direction (chosen to be z) but are free to move in the x-y-plane. The solutions of the Wannier equation that describes the electron-hole pair wavefunction are slightly modified in this case; of particular interest is the 1s wavefunction in two dimensions given by

$$\varphi_{1s}^{2D}(k) = \sqrt{2\pi} \frac{a_B^2}{(1 + \frac{k^2 a_B^2}{4})^{3/2}}. \tag{11.12}$$

Due to the enhanced overlap between the electron and hole wavefunctions along the growth direction, the Bohr radius in an ideal 2D structure is reduced by a factor of 2 as compared to the 3D semiconductor ($a_B^{2D} = a_B/2$). For the same reason, the binding energy and the oscillator strength are increased by a factor of 4. In typical QW structures with $L_z \simeq a_B \simeq 100$ Å, the corresponding enhancement factors would be in the range of ~ 2 to 3.

If the QW structure is not embedded inside a cavity, the 2D excitons interact with a 3D radiation field reservoir. The interaction Hamiltonian in this case is given by

$$\hat{H}_{int} = \sum_{k=(k_\perp, k_z), \nu} i\hbar g_\nu(k) \left(\hat{C}_{\nu, k_\perp}^\dagger \hat{a}_{k_\perp, k_z} - \hat{a}_{k_\perp, k_z}^\dagger \hat{C}_{\nu, k_\perp} \right), \tag{11.13}$$

where

$$g_\nu(k) = \frac{\mu_{cv}}{2\pi a_B} \sqrt{\frac{\omega_p(k)}{\hbar \epsilon L_{opt}}}. \tag{11.14}$$

Here L_{opt} denotes the quantization length of the radiation field modes along the z-direction. We immediately see that the physics of exciton-vacuum coupling has changed completely in this case: Due to the relaxation of exciton momentum conservation along the z-direction, any given exciton state now couples to a continuum of vacuum modes with varying k_z. As a result, we expect to see irreversible coupling leading to spontaneous radiative recombination. The time scale for radiative relaxation can be computed using Fermi's golden rule:

$$\Gamma_{sp} = 2\pi \hbar |g_\nu(k)|^2 \rho(E), \tag{11.15}$$

where $\rho(E, k_\perp \simeq 0) \simeq L_{opt}/(\hbar c)$. For GaAs, this formula gives a spontaneous emission time of $\tau_{sp} = \Gamma_{sp}^{-1} = 3 \times 10^{-11}$ s: This value is about 40 times shorter than the radiative lifetime of a free electron-hole pair.

This short radiative lifetime has the same physical origin as the supperadiance for an assembly of two-level atoms discussed in Sec. 6.6.2. If only one atom is excited in the N atom system, the spontaneous emission rate is enhanced by N compared to a single-atom case. This enhancement stems from the fact that the excitation exists simultaneously in all N atoms as indicated by a linear superposition state (6.44). The dipole moments of all N atoms are coherently added to produce a *giant dipole moment*. For the Wannier exciton in semiconductors, the electron-hole spatial correlation is not perfect. The electron and hole may exist in different lattice cites. In such a case there is no optical matrix element. Nevertheless, if the sample size exceeds the exciton Bohr radius, approximately $N \simeq A/\pi(a_B^{2D})^2$ lattice dipole moments coherently contribute to form a *giant dipole moment*, where A is the effective sample area. What determines the effective sample area A in most cases is the quality of the QW wafer. If the QW structure is perfect and free from defects and scattering, the so-called geometric factor, which accounts for narrowing the radiation pattern for a wider aperture radiator and reducing the final density of radiation states, determines the ultimate value N, which is ~40 for a GaAs QW system [4].

In the absence of optical mode confinement (i.e., $L_{opt} \gg \lambda = hc/E_g$), the exciton-vacuum coupling is irreversible, as discussed previously. If, on the other hand, the optical mode is confined, we can recover the reversible coupling. In fact, the large exciton-field coupling strength $g(k)$ responsible for the fast excitonic spontaneous emission rate would, in principle, make it easier to go into the regime of reversible spontaneous emission we have discussed earlier in the context of atom-field interactions. Due to large dissipation and dephasing rates, however, it is still essential to have microcavities that can confine the optical mode within a few λ [5].

During the last decade, there has been significant activity in the MBE growth of one-dimensional microcavity structures using distributed Bragg reflectors (DBRs). The DBR layers consist of alternating layers of quarter wavelength ($\lambda/4$) semiconductor materials with different indices of refraction. For example, using approximately 20 stacks of AlAs($n_r = 2.96$) and Al$_{0.3}$Ga$_{0.7}$As ($n_r \simeq 3.4$) DBR layers for each mirror, it is possible to get quality factors Q exceeding 10^4 at wavelengths corresponding to GaAs exciton emission. The effective cavity length is normally on the order of λ. Figure 11-2 shows a typical microcavity DBR structure.

When a 100-ÅGaAs QW structure is sandwiched in between two DBR mirrors, it is possible to obtain exciton-cavity coupling strengths [$\hbar g(k)$] on the order of 4 meV; the exact value of the coupling strength can be obtained using the expression for $g(k)$ from (11.14) after replacing L_{opt} with the effective cavity length L_{cav}. Since typical high-quality exciton and cavity linewidths are on the order of 0.5 meV, such a coupling strength results in well-resolved *cavity-polariton* branches. Since $g(k) \propto 1/a_B$, large bandgap semiconductors such as GaN and ZnSe could have normal mode splittings exceeding 12 meV. We also note that by introducing M QW structures inside the microcavity, it is possible to enhance the normal-mode coupling by a factor $\sqrt{M}$, just as in the case of many-atom-field interaction.

The cavity polaritons resulting from the reversible exciton-photon coupling have a characteristic dispersion that is qualitatively different from those of bulk polaritons (Fig. 11-3). The exact dispersion relation can be calculated using (11.9), by taking

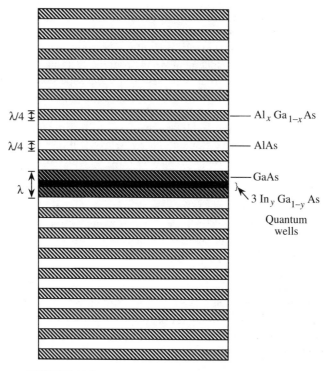

FIGURE 11-2: Semiconductor microcavity DBR structure.

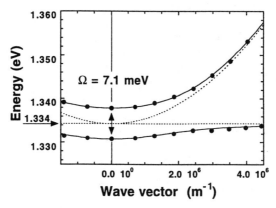

FIGURE 11-3: Cavity exciton polariton dispersion in a GaAs quantum well microcavity (R. P. Stanley et. al., in Microcavities and Photonic Bandgaps: Physics and Applications, eds. J. Rarity and C. Weisbuch (Kluwer, Dordrecht, 1996)).

into account the bare 2D exciton and cavity dispersion relations

$$\omega_{\text{exc}}(k) = E_g + E_{1s}^{2D} + \frac{\hbar k^2}{2m_{\text{exc}}^*}$$

$$\omega_{\text{cav}}(k) = \hbar c \sqrt{k^2 + \frac{\pi^2}{L_{\text{cav}}^2}}, \quad (11.16)$$

where $m_{\text{exc}}^* = m_e^* + m_h^*$ and $k = k_\perp$. If exact resonance condition is satisfied, $\omega_{\text{exc}}(k=0) = \omega_{\text{cav}}(k=0)$, then the upper and lower polariton wavefunctions are superpositions of the exciton and cavity mode wavefunctions with equal weight. We can define the mass of polaritons (for $k \sim 0$) by

$$m_{u \atop l}^* = \frac{1}{\partial^2 \Omega_{u \atop l} / \partial k^2}. \quad (11.17)$$

It is straightforward to show using (11.17) that at resonance $m_u^* = m_l^* \simeq 2 \times m_{ph}^*$, where $m_{ph}^* \simeq \pi^2 \hbar/(cL_{\text{cav}}) \simeq 10^{-4} m_{\text{exc}}^*$. Therefore, cavity polaritons are quasi-particles with extremely small mass.

Figure 11-4 shows the theoretical and experimental spatial dispersion of the exciton polaritons [6]. In this sample the optical cavity layer is tapered along one direction of the sample so that the cavity frequency varies with sample position. When the cavity frequency is tuned to the QW heavy-hole (HH)-exciton frequency (position A in Fig. 11-4), the two HH-exciton polariton states are produced, as shown in Fig. 11-5(a). Similarly, when the cavity frequency is tuned to the QW light hole (LH)-exciton frequency (position C in Fig. 11-4), the two LH-exciton polaritons are produced, as shown in Fig. 11-5(b).

The presence of such a large normal-mode splitting in the frequency domain naturally implies fast energy exchange between excitons and the cavity mode; the resulting cavity polariton oscillations in time domain have also been observed. The temporal evolution of the GaAs microcavity polariton emission is shown in Fig. 11-6 [6]. Trace A was taken at HH-exciton resonance (position A in Fig. 11-4). The excitation oscillates back and forth between the QW HH-excitons and the microcavity field, leading to the exciton polariton oscillation. The irregular oscillation traces observed in sample positions B, C, and D are due to the beating of the three polariton states, including the LH-exciton mixing.

One of the interesting features of cavity polaritons is that typical polariton linewidths are actually narrower than what we would predict naively based on the bare exciton and cavity-mode linewidths. This is due to the so-called *motional narrowing* effect: Due to their ultrasmall mass, polaritons average out the random potential arising from growth irregularities (i.e., interface roughness) and thus have smaller linewidths. As a result, the number of cavity polariton oscillations observed in the preceding experiments could easily exceed 10.

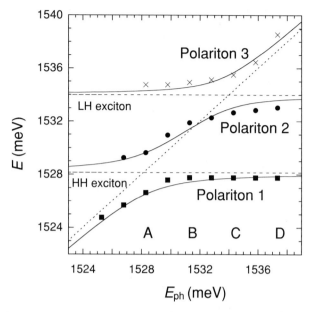

FIGURE 11-4: Exciton-polariton spatial dispersion curves deduced from absorption measurement. The solid lines are the theoretically fitted exciton-polariton dispersion curves. The dashed (dotted) lines are the uncoupled exciton (photon) dispersion curves.

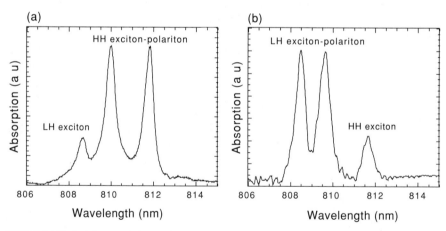

FIGURE 11-5: Measured absorption spectrum at (a) HH-exciton resonance, (b) LH-exciton resonance.

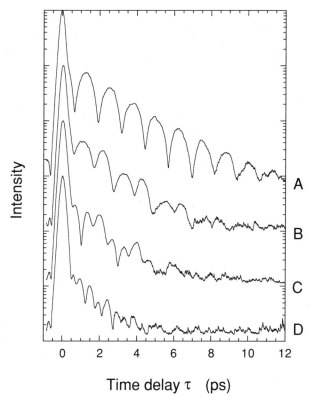

FIGURE 11-6: The measured microcavity emission as a function of the time delay τ at positions A, B, C, D of the dispersion curves in figure 11.4.

11.4. EXCITONS AS BOSONS

11.4.1. Bosonization

As we have seen in (11.4), excitons or polaritons in the ultralow density limit ($Na_B^d \ll 1$) can be considered as composite bosons, obeying standard Bose commutation relations. In the first two sections of this chapter, we have confined ourselves to this very low excitation density limit, where we can describe the semiconductor-radiation field interactions using a linearly coupled set of harmonic oscillator modes. The resulting Hamiltonian of (11.3) is bilinear and can be diagonalized using Bogoluibov transformations, as we have already seen. On the other hand, we know that the observation of quantum optical phenomena relies on the existence of nonlinearities, and we therefore need to go beyond this simple model in a self-consistent fashion.

A natural choice to investigate the (moderately) low density limit, where the first-order nonlinear corrections become important, is to use a transformation from a complete set of fermion states to a set of boson states. Such a transformation has been proposed by Marumori [7] and later modified by Steyn-Ross and Gardiner [8]. In this transformation, the transformed boson space is a subspace of the complete boson space and contains only the properly antisymmetrized boson states that guarantee the correspondance to the original fermion states and hence are *physical*. From a practical perspective, the drawback of this transformation is that it results in an infinite series of boson-operator expansions. The new boson basis therefore becomes intractable as the density approaches the so-called Mott density ($Na_B^d \sim 1$). However, it does allow us to obtain the corrections to the bilinear terms in a self-consistent manner, in the low density limit. We refrain from giving the details of the transformation, which can be found elsewhere [7], [8]. In the low-density limit, the resulting Hamiltonian is

$$\hat{H} = \hat{H}_0 + \hat{H}_{\text{exc-exc}} + \hat{H}_{\text{int}}, \quad (11.18)$$

where

$$\hat{H}_0 = \sum_{\nu,k} \hbar \omega_{\nu,k} \hat{C}^\dagger_{\nu,k} \hat{C}_{\nu,k} \quad (11.19)$$

$$\hat{H}_{\text{int}} \simeq -i\hbar \sum_{k,\nu} g_\nu(k) \hat{a}^\dagger_k \hat{C}_{\nu,k} \left[1 - \gamma^* \sum_{k,\mu} \hat{C}^\dagger_{k,\mu} \hat{C}_{k,\mu} \right] + \text{h.c.} \quad (11.20)$$

$$\hat{H}_{\text{exc-exc}} \simeq \sum_{\substack{\mu,\mu',\nu,\nu' \\ k,k',q}} \hbar g_{\text{exc-exc}}(k,k',q) \hat{C}^\dagger_{\nu',k+q} \hat{C}^\dagger_{\mu',k'-q} \hat{C}_{\mu,k'} \hat{C}_{\nu,k}. \quad (11.21)$$

All operators appearing in (11.21) are bosonic. $\hat{H}_{\text{int}}$ denotes the exciton-photon interaction Hamiltonian with γ^* giving the PSF factor. The explicit forms of the interaction coefficients are given in Ref. [8]. The Hamiltonian of (11.18) could serve as a starting point to determine the quantum and nonlinear optical properties of exciton-light interaction.

11.4.2. Bose Condensation of Excitons

The validity of composite-boson treatment of electron-hole pair excitations naturally raises the possibility of Bose-Einstein condensation (BEC) of excitons [9]. There has been considerable interest in the realization of exciton BEC over the last 20 years, and so far two groups have reported experimental evidence [10], [11]. If we assume an ideal exciton gas with no dynamic screening, we can show that the ground state of the optically excited semiconductor is approximately given by

$$|\Psi_g\rangle = |\Psi_{\text{BCS}}\rangle = \prod_k \left[u(k) + v(k) \hat{e}^\dagger_k \hat{h}^\dagger_{-k} \right] |0\rangle, \quad (11.22)$$

where $u(k)$ and $v(k)$ are subject to the normalization condition $|u(k)|^2 + |v(k)|^2 = 1$ and are determined by variational techniques [12]. In the low-density limit, we can show that $u(k) \simeq 1$, $\forall k$ and $v(k) = \sqrt{N_{\text{exc}}}\, \varphi_{1s}(k)$, where N_{exc} is the total number of electron-hole pairs in the system. We can also show that in this limit

$$|\Psi_g\rangle \simeq \exp\left[\alpha \hat{C}^\dagger\right]|0\rangle, \qquad (11.23)$$

which describes a coherent state of ground-state excitons. One of the fascinating aspects of the exciton condensation problem is that the composite-boson density is tunable by the driving field strength, which in principle allows us to explore both the low-density limit, where we should obtain a BEC of excitons, and the high-density limit, where under special circumstances a BCS-type state can form. Both of these limits are governed by the same ground-state wavefunction given in (11.24).

Of particular interest in the context of quantum optics is exciton condensation in a QW structure with weak lateral confinement. The signature of BEC in such a structure is the generation of bidirectional and all-orders coherent light. Since the ground state of the system is the $k_\perp = 0$ state, photons generated by annihilating excitons should conserve the transverse momentum and propagate along the $\pm z$-direction. The statistics of the spontaneously generated light also follow those of excitons, at least in the ultralow density limit, where anharmonic coupling (γ^*) can be neglected. In the low-density limit, we would expect to see quantum features due to the nonlinearities.

The process of condensation would require a cooling process that can be stimulated by the occupancy of the final state. In the case of excitons, electron-hole-phonon coupling provides a natural mechanism for the system to cool down to its ground state. In the ultralow density limit, quantum Boltzman equations predict that a macroscopic fraction of excitons indeed accumulate in the ground state. The role of bosonic enhancement of scattering, however, strongly depends on the density of excitons. Finally, we note that a finite radiative lifetime strongly restricts the observability of exciton condensation as it produces a heating mechanism. In the presence of such dissipation, a more appropriate description of the condensation process would be a laser or *boser*, as excitons in this case will remain out of equilibrium [13].

11.4.3. Stimulated Scattering of Excitons

In this section, we derive an expression for the statistical enhancement factor of composite boson scattering that is valid for all densities. Even though we concentrate on electron-hole-pair phonon interaction in the Born-Markov limit, our basic result can be used for a large class of composite bosons and for arbitrary dimensionality, provided that the Bardeen-Cooper-Schrieffer (BCS) type of ground state of the system is predetermined [13].

The starting point of this analysis is the interaction of a specific many-body electron-hole state with the phonon reservoir, in the fermion basis. The corresponding interaction Hamiltonian is [1]

$$\hat{H}_{\text{int}} = \hbar \sum_{k,q} \left[g_{\text{e-ph}}(q) \hat{e}^\dagger_{k+q} \hat{e}_k + g_{\text{h-ph}}(q) \hat{h}^\dagger_{k+q} \hat{h}_k \right] \left(\hat{b}_q + \hat{b}^\dagger_{-q} \right), \qquad (11.24)$$

where $g_{e-ph}(q)$ and $g_{h-ph}(q)$ are the corresponding electron-phonon and hole-phonon interaction coefficients. The spin index and the vector nature of the momenta are once again suppressed for simplicity. A similar set of variables could also be used to describe electron-hole-phonon system under large magnetic fields.

We proceed by assuming that the initial many-body electron-hole state can be written as [13]

$$|\Psi_{in}\rangle = \hat{C}^\dagger_{v,K}|\tilde{\Psi}_{BCS}\rangle = \hat{C}^\dagger_{v,K}\prod_k\left[\tilde{u}(k) + \tilde{v}(k)\hat{e}^\dagger_k\hat{h}^\dagger_{-k}\right]|0\rangle. \quad (11.25)$$

The final (electronic) state of the scattering process is simply $|\Psi_{BCS}\rangle = \prod_k[u(k) + v(k)\hat{e}^\dagger_k\hat{h}^\dagger_{-k}]|0\rangle$, where $\sum_p |v(p)|^2 = N + 1$ is the mean number of electron-hole pairs. In the following discussion, we will assume that $N \gg 1$, so that the difference between $v(p)(u(p))$ and $\tilde{v}(p)(\tilde{u}(p))$ is negligible $\forall p$.

The next step is to obtain the scattering rate in the Born-Markov approximation using Fermi's golden rule. For simplicity, we will assume that the lattice is at zero temperature and only consider the spontaneous phonon processes. We then obtain the scattering rate:

$$W_s = 2\pi \sum_q |g_{e-ph}(q)M(q) + g_{h-ph}(q)M(-q)|^2 \delta(\omega_{in} - \omega_{fin} - \omega_q), \quad (11.26)$$

where

$$M(q) = \sum_k \varphi^*_v\left(k + \frac{q}{2}\right)u^*(k)v(k)\left[1 - |v(k+q)|^2\right]. \quad (11.27)$$

Here, $\hbar\omega_{in}$ and $\hbar\omega_{fin}$ denote the energies of the initial and final many-body eigenstates, respectively. ω_q is the frequency of the emitted phonon. We reiterate that (11.28) is derived using the interaction Hamiltonian of (11.26) and no assumptions regarding the *bosonic character* of electron-hole pairs were made. The product $u^*(k)v(k)$ is proportional to the electron-hole pair wavefunction. In the Hartree-Fock approximation, $u^*(k)v(k)$ can be determined from the semiconductor Bloch equations (SBEs) [13]. As we have already seen, for ultralow density excitons, the polarization equation reduces to the Wannier equation in real space and yields the 1s exciton wavefunction $\varphi_{1s}(p)$. In the high-density limit ($Na^d_B/L^d > 1$), we obtain the counterpart of superconducting gap equation [14]. In many cases, however, Hartree-Fock approximation is not valid and the calculation of the pair wavefunction would require the inclusion of screening and scattering terms.

The factor $[1 - |v(k+q)|^2]$ in (11.29) gives the correction to the electron-hole-phonon scattering arising from the fact that the presence of a BCS ground state with a large number of composite bosons modifies the commutation relation of high-momentum electron-hole pairs as well. To the extent that $K \gg k_F$, where k_F is the Fermi wavevector, the contribution of this term is negligible. In addition, if we assume $K \sim q \gg \pi/a_B$, we can also neglect the k-dependence of $\varphi^*_v(k + \frac{q}{2})$, provided that φ_ν is a Hydrogenic wavefunction. In this limit, we obtain

$$M(q) \simeq \varphi_\nu^*\left(\frac{q}{2}\right) \sum_k u^*(k)v(k), \qquad (11.28)$$

and

$$W_s \simeq 2\pi \sum_q \left| g_{e-ph}(q)\varphi_\nu^*\left(\frac{q}{2}\right) + g_{h-ph}(q)\varphi_\nu^*\left(\frac{-q}{2}\right) \right|^2 . I . \delta(\omega_{in} - \omega_{fin} - \omega_q) \qquad (11.29)$$

where

$$I = \left| \sum_k u^*(k)v(k) \right|^2. \qquad (11.30)$$

The expression for I given in (11.32) contains the statistical enhancement factor for the scattering of a phonon by a many-body composite boson (electron-hole pair) system. In the low-density limit, where $v(k) \simeq \sqrt{N}\,\varphi_{1s}(k)$, we obtain $I \propto N$ for all composite bosons, as expected. In the high-density limit, the qualitative nature of saturation and Pauli blocking of statistical enhancement factor strongly depends on the particular BCS state [i.e., the coherence factors $u(k)$, $v(k)$]. In the low-density limit ($u(k) \sim 1, \forall k$), bosonic enhancement arises from a constructive interference of the contributions from all partially occupied pair states. In the opposite high-density limit, only the states around Fermi level for which $u^*(k)v(k) \neq 0$ contribute to I. Equivalently for this latter case, the electron-hole pairs with $k \ll k_F$ have exhausted the phase space available for them ($v(k) \simeq 1$) and can no longer participate in stimulated scattering. Physically, this is due to the fact that the mean separation of the electron-hole pairs is less than their size, which makes the Pauli exclusion dominant [16].

Equation (11.32) shows that the stimulated scattering explicitly depends on the overlap $u^*(k)v(k)$. Therefore, it is the coherence between the electron-hole pair states that results in bosonic enhancement. Conversely, if the ground state of the many-body system is an electron-hole plasma state, where $u^*(k)v(k) = 0, \forall k$, there is no final-state stimulation at any electron-hole pair density. We remark that even though we assume a BCS state in our analysis, the assumption of a well-defined condensate phase should not be relevant for the bosonic enhancement factor.

Next, we consider the special case of two-dimensional (2D) magnetoexcitons. It has been shown by several authors that in the strong magnetic field limit, where the magnetic length $a_0 = \sqrt{\hbar/eB}$ is much smaller than a_B, the magnetoexcitons become ideal non-interacting bosons. More specifically, Paquet et al. [15] have shown that the single-particle wavefunction remains unchanged for all occupancies of the lowest exciton band. For this system we have

$$v(k) = v = \sqrt{2\pi a_o^2 N/L^2}, \qquad (11.31)$$

where L is the transverse size of the 2D structure. The evaluation of the stimulated scattering contribution is then straightforward:

$$I = N \left(1 - \frac{2\pi a_0^2 N}{L^2}\right) \frac{L^2}{2\pi a_0^2}. \qquad (11.32)$$

The analytical expression given in (11.32) is valid for $1 \leq N \leq L^2/2\pi a_0^2$. In the low-density limit ($N \ll L^2/2\pi a_0^2$), magnetoexcitons behave as ideal bosons ($I \propto N$). The total scattering rate peaks at $N_{\max} = N_M = L^2/4\pi a_0^2$, where only half of the magnetoexcitons contribute to stimulation. For $N > N_M$, stimulated scattering rate into the ground state starts to decrease and goes to zero as $N \to 2N_M$; at this occupancy, all the underlying electron and hole fermionic phase space is exhausted and it is not possible to create another ground-state magnetoexciton [16].

We therefore see that stimulated scattering of excitons can indeed survive exciton densities far exceeding the more restrictive low-density limit where our bosonic Hamiltonian of (11.18) is valid. This stimulated scattering of excitons plays the key role of stimulated emission in an exciton laser [12].

11.5. STIMULATED SCATTERING EXPERIMENT WITH EXCITONS

As we have discussed, a general consequence of quantum statistics of bosons is the stimulation of the scattering rate into a final state whose population is larger than one, which holds irrespective of dimensionality, mass, and equilibrium conditions. For massless photons, this is routinely observed in the laser. Final state stimulation into a largely populated ground state of a Bose-Einstein condensate [17], [18] has been recently claimed for the massive excitons in cuprous oxide [11]. The system, consisting of highly nonequilibrium exciton reservoir and large polariton populations produced by external laser action in a semiconductor microcavity, can also lead to observation of final state stimulation of the scattering of high-energy excitons into low-energy cavity polaritons [19].

We now consider elastic scattering of two bosons into other two bosons. For a scattering event from states 1 and 2 to states 3 and 4, the initial state is $|i\rangle = |n_1\rangle_1 |n_2\rangle_2 |n_3\rangle_3 |n_4\rangle_4$ and the final state is $|f\rangle = |n_1-1\rangle_1 |n_2-1\rangle_2 |n_3+1\rangle_3 |n_4+1\rangle_4$. The transition rate is proportional to

$$|\langle i|\mathcal{H}_{\text{int}}|f\rangle|^2 = H_{fi} n_1 n_2 (1+n_3)(1+n_4) \qquad (11.33)$$

Here $\mathcal{H}_{\text{int}}$ is the interaction Hamiltonian, and H_{fi} its transition matrix element. This scattering rate is enhanced by the populations in either state 3 or 4.

The massive bosonic states 3, 4 in the experiment we describe are the exciton polaritons of a planar semiconductor microcavity structure (Fig. 11-7). The photon mass and exciton mass differ by almost four orders of magnitude: their energies are largely different already at relatively small $\mathbf{k}$, where they become essentially uncoupled (the upper polariton becomes photonlike, the lower becomes exciton-

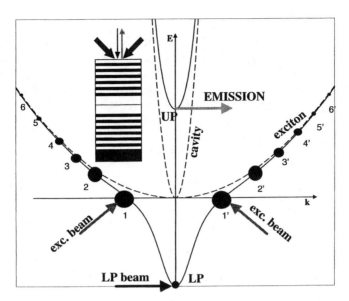

FIGURE 11-7: Dispersion of the exciton-polaritons. The dispersion of the excitons (not to scale) and cavity photons (dashed lines) is also shown. In the same figure, the exciton population is depicted by solid circles of size proportional to it, also labeled with numerals. Typical scatterings are $(c_1) 6 \to \text{UP} \pm \text{photon}$; $(c_2) 1 + 6' \to \text{UP} + 4$; $(c_2') 1 + 1' \to \text{UP} + \text{LP}$. In the inset: the cross section of the microcavity structure. Quarter wavelength dielectric stacks (distributed Bragg reflectors) confine the photon inside the cavity, into which a quantum well (QW), confining the exciton, is embedded. The microcavity is grown on a substrate and mounted in a cryostat. In the experiment, two pump beams at large angles and a probe beam at the normal direction are incident upon the top facet. The emission is also collected in the normal direction, and analyzed with a CCD spectrometer.

like). In the following, we use upper and lower polaritons to refer to the strongly coupled modes at $\mathbf{k} \simeq 0$, and excitons to refer to the lower polaritons at larger $\mathbf{k}$. Polaritons and excitons at different $\mathbf{k}$ can be excited by external laser beams, because the DBR mirrors have finite transmission. $\mathbf{k}$ is related to the external incidence angle θ through $k = (\omega/c) \sin \theta$, where $\hbar \omega$ is the excitation photon energy and c is the velocity of light. Because of this coupling to the external world, polaritons and excitons at small k can decay into external photons and have finite radiative lifetimes. In this cavity, the photon can bounce on a mirror not more than few thousand times, and the resulting polariton lifetime at $\mathbf{k} = 0$ is typically $\tau_{\text{pol}} \sim 2$ ps. For the exciton, the radiative lifetime is much longer, and excitons having $k > n\omega/c$ (n is the index of refraction of the material) have, in fact, infinite radiative lifetime as they cannot decay into external photons, due to energy and momentum conservation in the decay process. The thermalization process of these excitations involves scattering with the lattice vibrations (phonons). As a result of the largely different masses of excitons and polaritons, the latter have a much smaller scattering rate with phonons than the former. It turns out that the externally injected polaritons do not have time to scatter with phonons, and do not thermalize to the lattice temperature, before they decay into external photons, whereas the excitons do approximately thermalize. Injected excitons are prevented from further cooling down and transforming into lower polaritons by a relaxation bottleneck, related to the aforementioned slowdown of phonon scattering. Thus, the injected excitons are not in thermal equilibrium with the polaritons. The rate of emission of upper polaritons in this process is given by (11.33), where $n_1 = n_2 = n_{exc}$, $n_3 = N_{low}$, and $n_4 = N_{up}$. Then final state stimulation can be in-

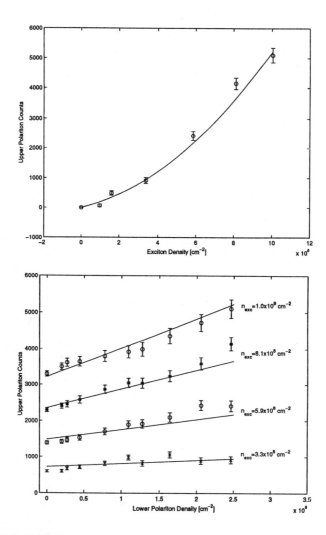

FIGURE 11-8: Experimental results and theoretical fits. a: measured upper polariton emission rate vs. the exciton density. Solid line: fits according to (11.33). b: the upper polariton emission rate, as a function of the lower polariton density. Solid lines: fit with (11.33) [R. Huang, F. Tassone and Y. Yamamoto, to be published].

duced by injection of lower polaritons, using an external probe beam. The population in the upper polariton states can instead be neglected, due to its short lifetime.

In the top graph of Fig. 11-8a, the upper polariton emission is plotted as a function of exciton density, for a fixed LP density of 2.5×10^8 cm^{-2}. The upper polariton emission linearly increases at low pump powers, and then clearly shows a quadratic dependence at larger pump intensities. The linear dependence at low powers is roughly independent of the probe intensity, whereas a larger quadratic component is found at larger probe powers. The upper polariton emission intensity as a function of LP density, for varying exciton density, is shown in Fig. 11-8b. As explained previously, the exciton-exciton scattering into lower and upper polariton is expected to behave quadratically with exciton density, and thus quadratically with P_{pump}. Then the linear dependence of the upper polariton emission on P_{probe} is itself a direct proof of final state stimulation of the scattering process.

REFERENCES

[1] H. Haug and S. W. Koch, *Quantum Theory of the Optical and Electronic Properties of Semiconductors* (World Scientific, Singapore, 1993).
[2] H. Haken, *Quantum Field Theory of Solids* (North-Holland, Amsterdam, 1976).
[3] M. Kira, F. Jahnke, and S. W. Koch, Phys. Rev. Lett. **81**, 3263 (1998).
[4] G. Björk, J. Jacobson, S. Pau, and Y. Yamamoto, Phys. Rev. B **50**, 17336 (1994).
[5] C. Weisbuch, M. Nishioka, A. Ishikawa, and Y. Arakawa, Phys. Rev. Lett. **69**, 3314 (1992).
[6] H. Cao, S. Jiang, S. Machida, Y. Takiguchi, and Y. Yamamoto, Appl. Phys. Lett. **71**, 1461 (1997).
[7] T. Marumori, M. Yamamura, and A. Tokunaja, Prog. Theor. Phys. **31**, 10096 (1964).
[8] M. L. Steyn-Ross and C. W. Gardiner, Phys. Rev. A **27**, 310 (1983).
[9] For a review of condensation effects, see A. Griffin, D. W. Snoke, and S. Stringari, eds., *Bose-Einstein Condensation* (Cambridge University Press, New York, 1995).
[10] J.-L. Lin and J. P. Wolfe, Phys. Rev. Lett. **71**, 1223 (1993).
[11] A. Mysyrowicz, E. Benson, and E. Fortin, Phys. Rev. Lett. **77**, 896 (1996).
[12] C. Comte and P. Nozieres, J. de Physique **43**, 1069 (1982).
[13] A. Imamoğlu and R. J. Ram, Phys. Lett. A **214**, 193 (1996).
[14] M. Tinkham, *Introduction to Superconductivity* (McGraw-Hill, New York, 1996).
[15] D. Paquet, T. M. Rice, and K. Ueda, Phys. Rev. B **32**, 5208 (1985).
[16] A. Imamoğlu, Phys. Rev B, **57**, R4195 (1998).
[17] S. N. Bose, Z. Phys. **26**, 178 (1924).
[18] A. Einstein, Sitzber. Kgl. Preuss. Akad. Wiss, **261**, (1924).
[19] F. Tassone, and Y. Yamamoto, Phys. Rev. B59, 10830 (1999).

12 Coulomb Blockade and Squeezing

In Chapters 10 and 11, we discussed the basic theory of semiconductor-light interactions in two- or three-dimensional structures using a quantum field theoretic approach. The semiconductor Bloch equations (SBE) that we derived using the time dependent Hartree-Fock (HF) approximation enabled us to identify excitons as the relevant optical excitations below the bandgap in the low-density, low-temperature regime. We have also seen that there are two key physical processes that lead to excitonic nonlinearities: phase-space filling (PSF) and interaction effects. PSF is a direct result of the Pauli exclusion principle, whereas interactions arise from the Coulomb repulsion between electrons.

When the size of the semiconductor becomes small in all three dimensions, both of these effects become more prominent. When the size is on the order of 100 Å, size quantization or equivalently PSF is the dominant physical mechanism determining the nonlinear response of the material. For semiconductors with dimensions $0.1\,\mu$–$1\,\mu$, it is the Coulomb interactions that dominate the transport and even optical properties; this is the mesoscopic limit or the *Coulomb blockade* regime [1].

In this chapter, we will consider the optical signatures of Coulomb interactions in a mesoscopic semiconductor. More specifically, we shall see that Coulomb blockade is at the heart of photon-number squeezing observed in semiconductor light sources. Due to the complexity of the problem, however, our approach will be semiclassical and based on phenomenological models. In Sec. 12.1, we will present a brief review of Coulomb blockade phenomenon in both n-i-n and p-i-n semiconductor junctions. The principal result of this section is the strong Coulomb blockade induced anticorrelation between individual electron or photon emission events. We will then discuss implications of this anticorrelation for large p-i-n junctions and explain photon number squeezing as a macroscopic Coulomb blockade effect (Sec. 12.2). In Sec. 12.3, we discuss a single-photon turnstile device that generates heralded single-photon states.

12.1. COULOMB BLOCKADE OF TUNNELING

We consider a metal-insulator-metal (m-I-m) junction shown in Fig. 12-1; when a bias V_0 is applied between the leads, the electrons will tunnel from the emitter side to the collector. The junction can be modeled as having a tunneling resistor R_t and a

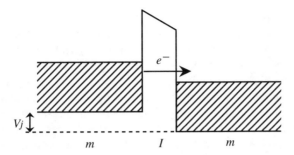

FIGURE 12-1: A mesoscopic metal-insulator-metal (m-I-m) junction.

capacitance C_d. The capacitance is given by

$$C_d = \varepsilon \frac{A_{\text{eff}}}{L_i}, \tag{12.1}$$

where ε and L_i are the dielectric constant and the length of the insulator layer, respectively. A_{eff} is the effective area of the junction. The resistance R_t can be written in terms of C_d and the tunneling rate of electrons Γ_t:

$$R_t = \frac{1}{\Gamma_t C_d}. \tag{12.2}$$

The Coulomb blockade regime can be roughly defined as the parameter range where the single-electron charging energy, given by e^2/C_d, exceeds all the other relevant energy scales, such as the characteristic energy scale for thermal fluctuations (kT) and the broadening arising from the tunneling process itself ($\hbar \Gamma_t$). The latter requirement can be stated as

$$R_t > R_Q = \frac{h}{2e^2}, \tag{12.3}$$

where R_Q is commonly referred to as the *quantum unit of resistance* [1]. By using semiclassical arguments, we can deduce that in the limit

$$\frac{e^2}{C_d} \gg kT, \hbar \Gamma_t, \tag{12.4}$$

single-electron charging can induce large correlations between successive electron injection events. More specifically, electron injection through the barrier by tunneling or across the barrier by thermionic emission can be strongly inhibited by an earlier injection event provided that the circuit recovery time is long compared to the tunneling time [1]. The last condition is satisfied only if

$$R_s \gg R_t, \tag{12.5}$$

where R_s is the (frequency independent) source resistance seen by the junction. When (12.5) is valid, the junction is said to be driven by a *constant current source* with $I \simeq V_{dc}/R_s$, where V_{dc} is the applied voltage. Physically, restoring the missing electron (and hence the preinjection junction voltage) requires a time $\tau = e/I$, leading to a dead time between the injection events as both tunneling (in the WKB limit) and thermionic emission rates depend exponentially on the junction voltage.

The functional dependence of the junction energy on the junction voltage is $E = 0.5 C_d V_j^2$. For $V_j = 0$, a tunneling event is strictly prohibited as the energy change as a result of the tunneling is $\Delta E = e^2/C$, which cannot be provided by any other source. Only when $V_j = e/2C_d$, the initial and final states of injection event have identical energy; therefore, we expect to see a tunneling event for $V_j \geq e/2C_d$. Following the initial tunneling, a second injection event will be forbidden for a time window given by $\tau = e/I$. As a result, single electrons are injected at regular time intervals separated by τ.

These simple and interesting predictions of the Coulomb blockade theory for a single constant-current-driven mesoscopic junction have not been observed experimentally due to the effects of the electromagnetic environment, which effectively shunt out the source resistance [2]. Mathematically, we need to satisfy $R_s(f) \gg R_Q$ for $f \leq f_0 = I/e$, where $R_s(f)$ is the magnitude of the real part of effective source resistance. In the case of mesoscopic junctions, $f_0 \simeq 1$ GHz, and it is well known that simple lumped circuit approximation breaks down at such frequencies.

Due to these restrictions, experimental demonstration of the Coulomb blockade phenomenon was carried out in double-barrier junctions [3]. Figure 12-2 shows a constant-voltage-driven *n-I-i-I-n* type of double-barrier semiconductor junction, where I and i denote undoped semiconductor layers with large and small conduction band energies, respectively. The electrons in the n-type regions are assumed to be degenerate, and the capacitance C_d of each junction is taken to be equal for simplicity. Provided that $e^2/C_d \gg kT$, the electron injection will only occur if the voltage drop across the first junction exceeds $e/2C_d$. Following the electron injection event, the energy of the *Coulomb island* is increased by e^2/C_d, while the emitter-collector voltage remains fixed due to very short circuit recovery time (constant voltage operation). The injected electron can only leave by tunneling out into the collector region. However, before this happens, injection of another electron from the emitter is inhibited: The electron injection events across the junction are therefore antibunched but not regulated [4].

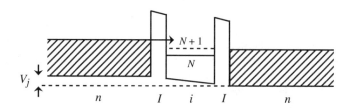

FIGURE 12-2: An *n-i-n* double-barrier tunnel junction.

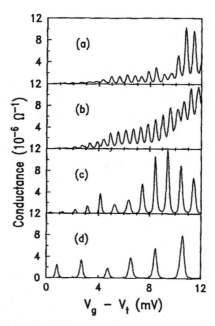

FIGURE 12-3: Conductance-gate voltage characteristics of a double-barrier mesoscopic junction with decreasing capacitance (from a to d).

The current-voltage characteristics of the double-barrier structure of Fig. 12-2 exhibit plateaus separated by e/C_d; this is due to the fact that at each step, we allow enough energy to open up a new channel for electron injection across the junction. Typically, a gate electrode independently controls the charge of the Coulomb island (i-layer). Figure 12-3 shows a typical experimental conductance-gate voltage curve exhibiting conductance oscillations that are one of the hallmarks of the Coulomb blockade effect [3].

In order to investigate the signatures of Coulomb blockade in photon statistics, one should consider p-n or p-i-n type semiconductor junctions [4]–[6]. In the case of a mesoscopic p-N junction (with bandgap energies satisfying $E_{g,p} < E_{g,N}$), for example, the electron injection events across the depletion layer would be regulated in exactly the same way as that in m-i-m junctions if constant-current operation were possible. These regulated injection events occurring at time intervals $\tau = e/I$ will be followed by electron-hole recombination, resulting in the generation of a photon. If the average waiting time for photon emission (given by the spontaneous radiative recombination time) satisfies $\tau_{sp} \ll \tau$, then generated photon statistics will be identical to that of electrons. We can therefore generate a stream of single-photon states with a well-defined clock; this is a highly nonclassical *multimode number state*. The second-order correlation function of this quantum state exhibits strong antibunching, as we shall discuss in Sec. 12.3 [7].

12.2. MACROSCOPIC COULOMB BLOCKADE OF ELECTRON INJECTION

Suppression of shot noise in semiconductor *p-n* junction lasers driven by a high-impedance constant-current source provides us with the simplest source of nonclassical light. We will leave the detailed discussion of this phenomenon to Chapter 15. The value of the second-order correlation function for the light generated by such a constant-current-driven macroscopic *p-n* junction is below but very close to the Poisson limit, indicating that anticorrelation between successive photon emission events is very small and that practically no information on the photon emission times exists: The anticorrelations only become important for a large number of photons.

The suppression of the driving current shot noise in a macroscopic conductor connected to a large resistor is an important factor in the observation of the described effects; this will be discussed in the next chapter (13). However, a constant-current source alone does not dictate the correlations between successive injection events. In this section, we will analyze the transition from a *macroscopic regulation* of many electrons in the *squeezing* regime to the strict regulation of individual electron injection events in the Coulomb blockade regime that we already discussed. More specifically, we will show that the junction capacitance (C_{dep}) and the operating temperature (T) determine the transition from the macroscopic to mesoscopic regime through the ratio $r = e^2/(kTC_{\text{dep}})$ of the single-electron charging energy to the characteristic energy of the thermal fluctuations. In the macroscopic and high-temperature limit ($r \ll 1$), the electron injection process is sub-Poissonian with variance given by $1/r$. On the other hand, for $r > 1$ (mesoscopic and low-temperature limit), the individual injection process is regulated, so that a nonstochastic spike appears in the noise spectrum at the single-electron charging frequency I/e with a squeezed background noise. We also note that for measurement times (T_{meas}) short compared to the thermionic emission time τ_{te}, the injection process is Poissonian, even with an ideal constant-current source. Once again, provided that the radiative recombination process is fast compared to these time scales, the same noise properties will be transcribed to the generated light field. The squeezed-state generation in macroscopic *p-n* junctions [8], [9] and the recently proposed Coulomb-blockade regulated single-electron injection (and single-photon stream generation) from low-capacitance mesoscopic junctions [4], [5] appear as special cases of this discussion.

To illustrate this connection, we consider a *p-i-n* AlGaAs–GaAs heterojunction driven by an ideal constant-current source. The carrier transport in such a junction occurs by thermionic emission of electrons from the *n*-type AlGaAs layer into the *p*-type GaAs layer, across an undoped (*i*) AlGaAs section. The rate of thermionic emission of electrons is given by

$$\kappa_{te}(t) = \frac{A_{\text{eff}} T^2 A^*}{e} \exp\left[\frac{e}{kT}\left(V_j(t) - V_{\text{bi}} - \frac{e}{2C_d}\right)\right]. \quad (12.6)$$

Here A^* is the Richardson's constant [5]. Under ideal constant-current operation ($dI/dt = 0$), the time dependence of the thermionic emission is exponential:

$$\kappa_{te}(t) = \kappa_{te}(0) \exp\left[\frac{t}{\tau_{te}} - rn_e(t)\right], \quad (12.7)$$

where

$$\tau_{te} = \frac{kTC_{\text{dep}}}{e^2}\frac{e}{I} = \frac{1}{r}\frac{e}{I} = \frac{1}{r}\tau. \quad (12.8)$$

The time constant τ_{te} as defined in (12.8) gives the time scale over which the thermionic emission rate changes appreciably and will be termed *thermionic emission time*. In (12.8), $n_e(t)$ denotes the number of thermionically emitted electrons in time interval $(0, t)$. The $\exp[-rn_e(t)]$ term can be regarded as providing a *feedback mechanism*: Emission of an electron results in a decrease in the thermionic emission rate and makes a second emission event less likely. This feedback is at the heart of Coulomb blockade regulation, where the decrease in the emission rate is strong enough to block a second emission event for another single electron charging time $\tau = e/I$. It is also the very physical origin of macroscopic squeezing where the decrease in the emission rate is small so that only the regulation of many electrons is possible.

We can show analytically that in the limit of $r \ll 1$ (i.e., the macroscopic regime), the probability for observing n_e electron injection events in an observation time T_{meas} is

$$P(n_e, T_{\text{meas}}) \simeq \frac{1}{N(r, \bar{n}_e)}\frac{1}{n_e!}\bar{n}_e^{n_e} \exp[-\bar{n}_e] \exp\left[-\frac{r}{2}(n_e - \bar{n}_e)^2\right], \quad (12.9)$$

where $\bar{n}_e = T_{\text{meas}}I/e$ and $N(r, \bar{n}_e)$ is the normalization factor. If $\bar{n}_e r \ll 1$ (i.e. $\tau_{te} \gg T_{\text{meas}}$), then the last term in (12.9) can be neglected and we obtain the usual Poisson distribution with $\Delta n_e = \sqrt{\bar{n}_e}$; since the observation time is much shorter than τ_{te}, we do not expect to see the effects of the feedback resulting from a decreased thermionic emission rate. If, on the other hand, $\bar{n}_e r \gg 1$ (i.e., $\tau_{te} \ll T_{\text{meas}}$), then the n_e dependence is predominantly determined by the last term (i.e., n_e is Gaussian distributed). The standard deviation in this limit is

$$\Delta n_e = \frac{1}{\sqrt{r}} = \sqrt{\frac{kTC_{\text{dep}}}{e^2}}. \quad (12.10)$$

Δn_e given in (12.10) can be considered a *fundamental noise limit for macroscopic squeezing* in constant-current-driven p-i-n heterojunctions. Even in the limit $r \ll 1$, where emission or tunneling of a single electron creates a very small voltage drop, the combined effect of many electrons is sufficient to control and regulate the electron emission to within several Δn_e. Finally, note that Δn_e obtained from (12.10) does not depend on T_{meas}, even though $\bar{n}_e$ does.

To understand the transition from the mesoscopic ($r > 1$) to macroscopic ($r \ll 1$) regime, we consider the dependence of Δn_e on the effective junction area. Figure 12-4 shows the results of a classical Monte Carlo simulation, where the normalization is such that $A_{\text{eff}} = 1$ corresponds to a junction capacitance that satisfies $r = 1$

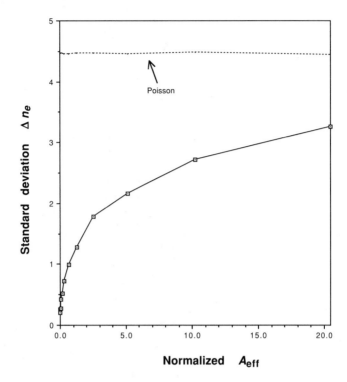

FIGURE 12-4: Δn_e as a function of the effective junction area.

[6]. For the chosen (fixed) measurement time $T_{\text{meas}} = 20/I$, the expectation value $\bar{n}_e$ of the thermionically emitted electrons is 20. We observe that for very large junction areas ($\tau_{te} > T_{\text{meas}}$, or $A_{\text{eff}} > 20$), the value of Δn_e is very close to the Poisson value of 4.47: In this limit the electron injection events have practically no effect on the thermionic emission rate. As a result, for the chosen observation time, we have *a random point process* with a constant rate, even though the junction is driven by a *perfect constant-current source*. For junction areas that give $e/I < \tau_{te} < T_{\text{meas}}$ ($1 < A_{\text{eff}} < 20$), in Fig. 12-4, Δn_e is approximately proportional to the square root of the area (or C_{dep}) and is clearly below the Poisson limit. Finally, for $e/I > \tau_{te}$ (i.e., $A_{\text{eff}} < 1$), Δn_e decreases very sharply: This is the Coulomb blockade regime where the individual electron injection events become deterministic as the single-electron charging energy e^2/C_{dep} exceeds kT.

The ratio r of the single-electron charging energy to the characteristic energy of the thermal fluctuations determines the extent of the correlations between the injected electrons: In the mesoscopic limit ($kTC_{\text{dep}}/e^2 \ll 1$), each electron is aware of the previous one due to the significant change in the junction voltage that the last thermionic emission event created: This is the Coulomb blockade regime discussed previously. In the macroscopic limit ($kTC_{\text{dep}}/e^2 \gg 1$), individual thermionic emission events practically have no effect on the expected injection time of the next electron. A large number of emission events, however, do have a combined affect on κ_{te}, and it is this feedback that keeps the standard deviation below the Poisson limit.

The dependence of the squeezing bandwidth (i.e., $1/\tau_{te}$) on current I, temperature T, and junction capacitance C due to the collective Coulomb blockade effect was confirmed experimentally using GaAs light-emitting diodes (LEDs) [10]. The light intensity fluctuation from the LEDs is decreased to below the shot noise value within the pump squeezing bandwidth $\omega_c = \frac{eI}{kTC}$ if the radiative recombination lifetime is much shorter than the inverse of ω_c. Figure 12-5 shows the measured squeezing bandwidths of four LEDs with different junction capacitances C versus the junction current at room temperature. The dotted lines are the theoretical squeezing bandwidths ω_c due to the macroscopic Coulomb blockade effect and the dashed lines are the squeezing bandwidths determined by a finite radiative recombination lifetime. The experimental results agree well with the theoretical predictions (solid

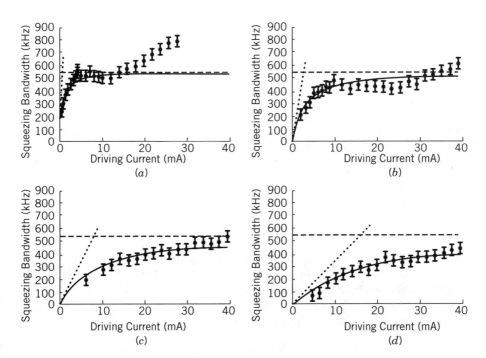

FIGURE 12-5: The measured squeezing bandwidth as a function of a driving current for various light emitting diodes (LEDs) at room temperature. Broken lines show the radiative recombination lifetime limitation of about 560 kHz (carrier lifetime of 290 ns). Dotted lines are the squeezing bandwidths due to the collective Coulomb blockade effect. The solid lines are the expected overall squeezing bandwidth. Areas of the LEDs (capacitance values to fit the data) are (a) 0.073 mm^2 (6.5 nF), (b) 0.423 mm^2 (30 nF), (c) 1.00 mm^2 (90 nF), and (d) 2.10 mm^2 (180 nF).

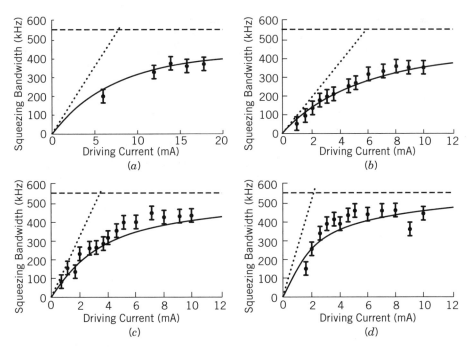

FIGURE 12-6: The measured squeezing bandwidth as a function of a driving current for various temperatures. The area of the LED was 1.00 mm^2. The temperatures are (a) 295 K [identical to Fig. 7(c)], (b) 220 K, (c) 120 K, and (d) 78 K. Broken lines show the radiative recombination lifetime limitation of about 560 kHz. Dotted lines are the squeezing bandwidths due to the collective Coulomb blockade effect and the solid lines are the overall bandwidths. A junction capacitance of 90 nF and carrier lifetime of 290 ns, estimated from room temperature measurement (Fig. 13-13), were used.

lines), including the two effects of the macroscopic Coulomb blockade and the finite radiative recombination lifetime. The measured squeezing bandwidths at various temperatures are plotted in Fig. 12-6 as a function of the junction current. The squeezing bandwidths at small junction currents are determined by the macroscopic Coulomb blockade effect (i.e., they are proportional to the junction current I and inversely proportional to the junction capacitance C and the temperature T).

12.3. AC-VOLTAGE-DRIVEN MESOSCOPIC p-i-n JUNCTIONS

As we have discussed in Sec. 12.1, there are fundamental problems in the realization of a constant-current-driven mesoscopic single-barrier p-i-n junction. Several groups have therefore investigated the possibility of obtaining similar signatures in double-barrier junctions using ac voltage sources [7], [11]. In this section, we will consider the proposal of Ref. [7], which generates heralded single photons that are regulated in essentially the same way as the photons that would be generated by single-barrier constant-current-driven junctions.

The energy-band diagram of the mesoscopic p-i_p-i-i_n-n AlGaAs–GaAs heterojunction that we analyze is illustrated in Fig. 12-7. If the junction voltage $V_j(t)$ is well below the built-in potential V_{bi} ($V_{bi} - V_j(t) \gg kT$), the carrier transport in such a structure takes place by resonant tunneling of electrons and holes through the undoped i_p and i_n–AlGaAs *barrier* layers, respectively. The injected electron-hole pairs then recombine radiatively in the i–GaAs layer. We assume that the width of the i–GaAs quantum well is small enough that the energy separation of the quantized subbands well exceed the single electron (hole) charging energy of the GaAs *Coulomb island* and that resonant tunneling into a single conduction (valence) subband need to be considered. The resonant tunneling of an electron or a hole is allowed only when the junction voltage is such that

$$E_{fn} - e^2/2C_{ni} \geq E_{\text{res},e} \geq E_{nc} - e^2/2C_{ni} \text{ (electrons)}, \quad (12.11)$$

and

$$E_{fp} + e^2/2C_{pi} \leq E_{\text{res},h} \leq E_{pv} + e^2/2C_{pi} \text{ (holes)}. \quad (12.12)$$

Here, $E_{\text{res},e}$ ($E_{\text{res},h}$) is the energy of the electron (hole) resonant subband of the i–GaAs quantum well (or dot); E_{nc} and E_{pv} are the energies of the conduction and valence bands in the n- and p-type layers, respectively; and E_{fn} and E_{fp} are the quasi-Fermi energies in the corresponding layers. C_{ni} and C_{pi} are the capacitances of the n-i_n-i and p-i_p-i regions, respectively. The energies in (12.1) are determined by the applied junction voltage $V_j(t) = V_o + v(t)$, where $v(t) = 0 (0 \leq t < T_{ac}/2)$; and $v(t) = \Delta V (T_{ac}/2 \leq t < T_{ac} = f_{ac}^{-1})$. The impurity concentrations on both n and p

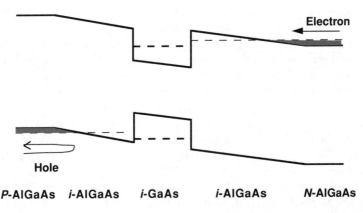

FIGURE 12-7: Energy-band diagram of the mesoscopic double-barrier p-i-n junction.

sides should be small enough that $E_{fn} - E_{nc} \simeq e^2/C_{ni}$ and $E_{pv} - E_{fp} \simeq e^2/C_{pi}$, since we want to be able to turn the tunneling of a particular carrier on and off by applying a voltage pulse whose magnitude is on the order of (but larger than) the single-charge charging energy. Finally, we assume that the Al concentrations in the two barrier regions are chosen independently so as to guarantee that peak electron and hole resonant tunneling occur at (significantly) different values of the applied junction voltage.

We choose the dc-bias voltage (V_o) so that the electron tunneling is resonantly enhanced when $v(t) = 0$. The applied square voltage pulses [$v(t) = \Delta V$] enable the resonant hole tunneling while blocking electron tunneling due to the second inequality in (12.11) (i.e., by quantum confinement). A second tunneling event of the same carrier during the time interval where the junction voltage remains unchanged is blocked by Coulomb blockade. Therefore, only one electron and one hole can tunnel into the i–GaAs layer within a single cycle of the applied ac voltage. Assuming that the radiative recombination occurs in a time scale short compared to the period of the ac voltage, a single photon is generated in each cycle with a probability approaching unity. If, in addition, the heterostructure is embedded in a micropost cavity [12] or photonic bandgap structure [13], then the photons are spontaneously emitted into a single mode of the radiation field.

Figure 12-8 shows the junction dynamics obtained using the classical Monte Carlo method. The period of the ac voltage is such that both the electron and hole tunneling occur with very high probability during the time intervals in which they are allowed. The photon emission events follow the hole tunneling in a very short time interval. We can consider the ratio of the period of the photoemission events to the jitter in the single-photon generation time as a *quality factor* Q for the generated single-photon stream. For a source with a time-independent generation rate, this ratio is unity (Poisson limit). In the proposed device, the quality factor is given by the ratio of T_{ac} to τ_{rad}, provided that the peak hole tunneling rate satisfies $\Gamma_{tunn,h} > \Gamma_{rad}$. For the parameters of Fig. 12-8, $Q = 30$. If the temperature is kept low enough to avoid secondary tunneling events in a single cycle, we can increase the quality factor by increasing T_{ac}.

The output light field generated by such a junction consists of optical pulses that contain one and only one photon. These heralded single photons in a given observation time window form a special class of *multimode number states*: They have well-defined number and emission time information, which is achieved at the expense of increased phase and energy uncertainty. This is a nonclassical state of light with a characteristic second-order coherence function, which in the ideal case exhibits peaks at $\tau = nT_{ac}, n = 1, 2, \ldots$ and vanishes elsewhere.

Recently, the first experimental evidence for the Coulomb blockade phenomenon and single-photon turnstile operation at cryogenic temperatures in double-barrier mesoscopic *p-i-n* junctions has been reported [14]: Figure 12-9(a) shows the measured current of a GaAs single-photon turnstile device as a function of ac modulation frequency, with a fixed ac amplitude at three different dc bias voltages. The measured current was in close agreement with the relation $I = ef$, $I = 2ef$, and $I = 3ef$ (solid lines). In Fig. 12-9(b) the slope I/f from the current versus frequency curves

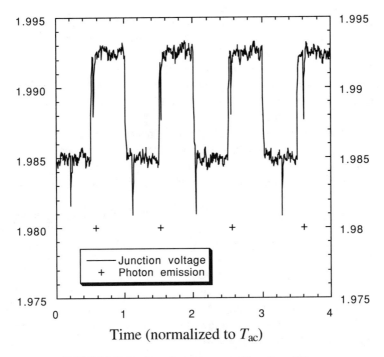

FIGURE 12-8: Junction voltage as a function of time.

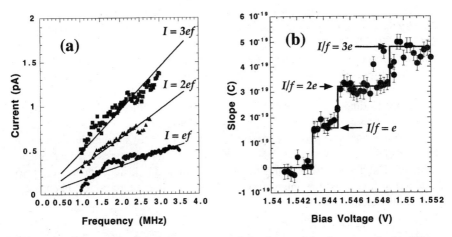

FIGURE 12-9: (a) The modulation frequency dependence of the dc current in the 600-nm turnstile device. A fixed ac amplitude of 72 mV is superposed on the three different dc voltages, $V = 1.545$, 1.547, and 1.550 V. The background dc (leakage) current, which is independent of the modulation frequency, has been subtracted. The measured current agrees with the relation $I = nef$ (solid lines), where e is the electron charge and $n = 1$, 2, and 3. (b) The slopes I/f in the current-frequency curve versus dc bias voltage. The slopes I/f are quantized and form plateaux at multiples of e, indicating the operations at $I = ef$, $2ef$, and $3ef$.

is plotted as a function of the dc bias voltage. It is found that the slope increases discretely, creating plateaus at $I/f = ne$, where $e = 1.6 \times 10^{-19}$ C is the charge of an electron and $n = 1, 2$, and 3.

The locking of the current at multiples of the modulation frequency ($I = nef$) suggests that the charge transfer through the device is strongly correlated with the external modulation signal. At the first current plateau at $I = ef$, a single electron and single hole are injected into the central QW per modulation period, resulting in single-photon emission. At the second current plateau at $I = 2ef$, two electrons and two holes are injected into the central QW per modulation period, resulting in two photon emission. Similarly, at the third current plateau at $I = 3ef$, three electrons and three holes are injected per modulation period, resulting in three-photon emission.

To observe the time correlation between the modulation input and photon emission, the time delay from the rising edge of the modulation input to the photon detection event at first current plateau ($I = ef$) and second current plateau ($I = 2ef$) is measured. The histograms of the measured time delay with 10-MHz modulation frequency is shown in Fig. 12-10(a) (for $I = ef$) and in Fig. 12-10(c) (for $I = 2ef$). The photon emission probabilities have peaks near the rising edge of the modulation input. The rapid increase of the photon emission probability is associated with the hole tunneling time ($\tau_h \simeq 4ns$), and the slow decay of the photon emission probability corresponds to the radiative recombination lifetime ($\tau_{ph} \simeq 25$ ns). The second and faster decay for $I = 2ef$ [Fig. 12-10(c)] stands for the decay of the hole via backward tunneling and radiative recombination. The associated lifetime for this decay is $(\tau_h^{-1} + \tau_{ph}^{-1})^{-1}$. The dashed lines show the analytical solutions using these parameters. The experimental results as well as the analytical traces are well reproduced by the Monte Carlo numerical simulation, as shown in Fig. 12-10(b) ($I = ef$) and Fig. 12-10(d) ($I = 2ef$). The fact that the photon emission probability decreases during the on-pulse duration is a unique signature that the number of holes injected during an on-pulse is restricted to either one or two due to the Coulomb blockade effect. In the absence of the Coulomb blockade effect, an arbitrary number of holes are allowed to tunnel into the central QW during an on-pulse, and so the resulting photon emission probability should increase monotonically with a time constant τ_{ph} to the steady-state value [experimental result in Fig. 12-10(e) and simulation result in Fig. 12-10(f)]. Such a device can be operated at higher temperatures if a quantum dot is utilized. The possibility of controlling the electron and hole injection makes mesoscopic p-i-n junctions very interesting for the study of quantum optical phenomena in semiconductors.

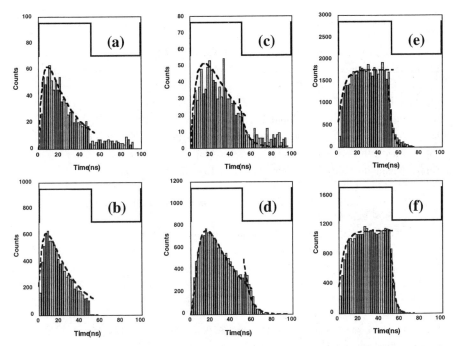

FIGURE 12-10: The photon emission characteristics of the turnstile device. (a) Measured histogram of a time delay between the rising edge of the modulation input and the photon detection event at the first plateau ($I = ef$). The solid line on top indicates the external ac modulation voltage. Since a single electron is injected at V_0 and a single hole is injected at $V_0 + \Delta V$, a photon is emitted after the rising edge of the pulse. The time delay for the photon emission probability to reach the maximum value is determined by the hole tunneling time ($\tau_h \simeq 4$ ns), and the subsequent exponential decay of the photon emission probability is determined by the radiative recombination time ($\tau_{ph} \simeq 25$ ns). The dashed line is an analytical solution with these time constants. The finite photon counts during the off-pulse period are due to the dark counts of the detector (Si SSPM). (b) Monte Carlo numerical simulation result for the photon emission probability versus the time delay at the first plateau ($I = ef$). The photon emission probability well reproduces the measurement result. (c) Measured histogram of a time delay at the second plateau ($I = 2ef$), where two electrons and two holes are injected per modulation period. The two dashed lines are the theoretical solutions considering the finite hole tunneling times, radiative recombination time and the backward tunneling time. The sharp cutoff of photon emission after the falling edge of the modulation input is due to the decay of the hole population due to simultaneous radiative recombination and reverse hole tunneling. (d) Monte Carlo numerical simulation result for the photon emission probability versus the time delay at the second plateau ($I = 2ef$). The distribution is broader since two electrons and two holes are injected, and two photons are emitted. (e) Measured histogram of a time delay for a large area device (diameter of 1.4 μm) at a temperature of 4K, where Coulomb blockade effect is absent. (f) Monte Carlo numerical simulation result for the photon emission probability versus the time delay in the absence of Coulomb blockade effect, where the number of electrons and holes injected per modulation period is not restricted. The photon emission probability increases monotonically through the duration of the on-pulse with the time constant τ_{ph}, as one would expect from a turn-on of a classical photon source.

REFERENCES

[1] D. V. Averin and K. K. Likharev, J. Low Temp. Phys. **62**, 345 (1986).

[2] M. H. Devoret, D. Esteve, H. Grabert, G. -L. Ingold, H. Pothier, and C. Urbina, Phys. Rev. Lett. **64**, 1824 (1990).

[3] T. A. Fulton and G. J. Dolan, Phys. Rev. Lett. **59**, 109 (1987).

[4] A. Imamoğlu and Y. Yamamoto, Phys. Rev. B **46**, 15982 (1992).

[5] A. Imamoğlu, Y. Yamamoto, and P. Solomon, Phys. Rev. B **46**, 9555 (1992).

[6] A. Imamoğlu and Y. Yamamoto, Phys. Rev. Lett. **70**, 3327 (1993).

[7] A. Imamoğlu and Y. Yamamoto, Phys. Rev. Lett. **72**, 210 (1994).

[8] S. Machida, Y. Yamamoto, and Y. Itaya, Phys. Rev. Lett. **58**, 1000 (1987); S. Machida and Y. Yamamoto, Phys. Rev. Lett. **60**, 792 (1988); W. H. Richardson and R. M. Shelby, Phys. Rev. Lett. **64**, 400 (1990).

[9] P. R. Tapster, J. G. Rarity, and S. Satchell, Europhys. Lett. **4**, 293 (1987).

[10] J. Kim, H. Kan, and Y. Yamamoto, Phys. Rev. B **52**, 2008 (1995).

[11] L. J. Geerlings, V. F. Anderegg, P. A. M. Holweg, J. E. Mooij, H. Pothier, D. Esteve, C. Urbina, and M. H. Devoret, Phys. Rev. Lett. **64**, 2691 (1990); L. P. Kouwenhoven, A. T. Johnson, N. C. van der Vaart, C. J. P. M. Harmans, and C. T. Foxon, Phys. Rev. Lett. **67**, 1626 (1991).

[12] G. Bjork, S. Machida, Y. Yamamoto, and K. Igeta, Phys. Rev. A **44**, 669 (1991).

[13] E. Yablonovitch, J. Opt. Soc. Am. B **10**, 283 (1993).

[14] J. Kim, O. Benson, H. Kan, and Y. Yamamoto, Nature **397**, 500 (1999).

13 Current Noise in Mesoscopic and Macroscopic Conductors

The pump process of a semiconductor p-n junction laser is the process of carrier injection across a depletion layer of the p-n junction. This process is either thermal diffusion or thermionic emission of electrons into a p-type semiconductor from an n-type semiconductor across a depletion region, which is followed by thermalization of "hot" electrons by optical and acoustic phonon emission. The dynamics of the thermal diffusion and thermionic emission process are very different in the following two cases: (1) The external circuit relaxation time constant CR_s is much shorter than the thermionic emission time $CR_d = kTC/eI$ (constant-voltage operation), and (2) CR_s is much longer than $CR_d = kTC/eI$ (constant-current operation), where R_s and $R_d = \frac{kT}{eI}$ are the source resistance of a driving circuit and the differential resistance of the p-n junction and T is the temperature. In the case of constant-current operation, ($R_s \gg R_d$) a semiconductor laser can be considered as a sub-Poissonian laser with suppressed pump (intensity) fluctuation, which we will discuss in Chapter 15. The principle of the pump noise suppression in a semiconductor p-n junction laser is twofold: current noise suppression in dissipative conductors due to the Pauli exclusion principle and current noise suppression in p-n junctions due to the collective Coulomb blockade effect. This chapter will review the first underlying principle for squeezed state generation by semiconductor lasers (the second principle was discussed in the previous chapter).

Section 13.1 will discuss the suppression of electron transport noise in highly dissipative conductors. It is well known that the current noise in a macroscopic conductor under a finite bias is independent of the bias voltage and is equal to the Johnson-Nyquist thermal noise. However, this formulation was originally derived for equilibrium case with no average current flow. The current noise in a mesoscopic conductor is generally found to be much higher than the Johnson-Nyquist thermal noise and is often equal to the shot noise. We will discuss that the nonequilibrium current noise in a mesoscopic conductor is suppressed by inelastic scattering of electrons and simultaneously the phase coherence of electron wave is lost. In Sec. 13.2 we will discuss the quantum interference effect in the collision of two fermionic particles. Section 13.3 will show the negative correlation between partitioned electrons.

13.1. SUPPRESSION OF ELECTRICAL CURRENT NOISE IN DISSIPATIVE CONDUCTORS

13.1.1. Equilibrium and Nonequilibrium Transport Noise in Mesoscopic Conductors

Let us consider a single-mode electron waveguide shown in Fig. 13-1, in which an electron wavepacket is emitted by one reservoir electrode and propagates toward another reservoir electrode without scattering. The conductance of such a ballistic single electron channel is given by quantum unit of conductance, $G_Q = 2e^2/h$, if the electron spin degeneracy is taken into account. The reason why the ballistic channel without scattering has a finite conductance is that only two electrons (spin up and down) can be accommodated for each electron wavepacket (degree of freedom) that is counted by the minimum time-energy uncertainty product. $\Delta E \Delta t = \frac{\hbar}{2}$ (i.e., the number of degrees of freedom for a given time internal Δt is finite for a given chemical potential difference, $\mu_2 - \mu_1 = \Delta E$, where μ_1 and μ_2 are the chemical potentials of the two electrodes).

If there is no chemical potential difference (zero external bias voltage) between two reservoir electrodes, as shown in Fig. 13-1, the average current is zero but there is a finite noise current. The electron energy distributions in both electrodes obey a Fermi-Dirac distribution f, and electrons in both electrodes near the Fermi energy are stochastically emitted according to the partition theorem with an emission rate of $0 < f < 1$. The variance of the emitted electron number is $\langle \Delta N^2 \rangle = Nf(1-f)$, where N is the total number of wavepackets. This stochastic emission of electrons due to the Fermi-Dirac distribution at a finite temperature T results in thermal noise with a current spectral density $S_i^{\text{th}}(\omega) \cong 4kTG_Q$, if the preceding variance is integrated over all energies and summed over the two reservoirs. There is another noise mechanism that exists even at zero temperature ($T = 0$). When the electron wavepackets above and below the Fermi energy (one empty and one occupied)

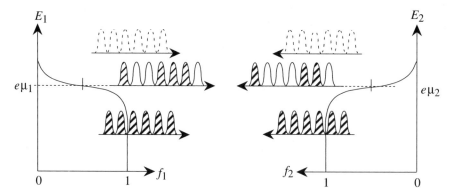

FIGURE 13-1: Equilibrium noise in a ballistic electron channel. The stochastic emission of electrons near the Fermi energy due to Fermi-Dirac distribution results in thermal equilibrium noise, and the beating between high-energy empty electron wavepacket and low-energy occupied electron wavepacket results in quantum noise.

are absorbed by the receiving reservoir, beating between these two wavepackets results in a noise current at a finite frequency determined by the energy difference between the two wavepackets. This beating between an occupied low-energy electron wavepacket and an empty high-energy electron wavepacket results in quantum noise with a current spectral density $S_i^q(\omega) = 2\hbar\omega G_Q$. The total noise current spectral density is identical to generalized Nyquist noise spectral density [1]:

$$S_i(\omega) = 4\hbar\omega \left[\frac{1}{\exp(\hbar\omega/kT) - 1} + \frac{1}{2}\right] G_Q. \tag{13.1}$$

If there is a finite chemical potential difference between two reservoir electrodes, the average current increases with the bias voltage $\langle i \rangle = G_Q(\mu_1 - \mu_2)/e$ but the noise current remains constant. This is because the thermal noise and the quantum noise occur in the two reservoir electrodes independently, so the chemical potential difference does not alter the noise current.

Next let us consider an electron channel with a single elastic scattering center between two reservoir electrodes as shown in Fig. 13-2. An elastic scattering center with a finite transmission T and reflection $(1 - T)$ coefficients can either be a naturally existing ionized impurity located inside or outside the channel or an artificially introduced channel constriction and/or tunnel barrier. When there is no chemical potential difference between the two electrodes ($\mu_1 = \mu_2$), the generalized Nyquist formula (13.1) still holds if the quantum unit of conductance G_Q is replaced by TG_Q [2]. As long as the system is in thermal equilibrium, the generalized Nyquist formula (13.1) holds irrespective of the details of the scattering element (i.e., independent of its material, shape, and size).

However, when there is a finite chemical potential difference between the two electrodes ($\mu_1 > \mu_2$), as shown in Fig. 13-2, nonequilibrium excess noise is generated. This is because the transmitted occupied electron wavepacket (from the left

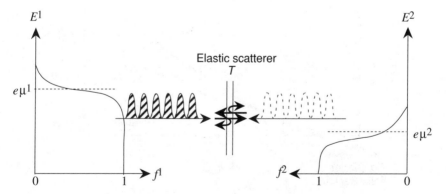

FIGURE 13-2: Nonequilibrium noise in a ballistic electron channel with a single elastic scatterer. A transmitted (reflected) occupied electron wavepacket from the left electrode and reflected (transmitted) empty electron wavepacket from the right electrode beat with each other to produce the nonequilibrium quantum noise.

electrode) and reflected empty electron wavepacket (from the right electrode) beat with each other. Similarly, a reflected occupied electron wavepacket and transmitted empty electron wavepacket produce independent excess beat noise. This nonequilibrium excess noise, often referred to as partition noise, has a quantum mechanical origin similar to that of the high-frequency quantum noise $2\hbar\omega G_Q$ discussed previously. The noise current spectral density for such a single-mode electron channel with a single elastic scatterer is given by [3]

$$S_{\Delta I}(\omega) = 2kTG_Q\mathcal{T}(1-\mathcal{T})\left[\frac{\Omega+\nu}{2}\coth\left(\frac{\Omega+\nu}{2}\right) + \frac{\Omega-\nu}{2}\coth\left(\frac{\Omega\nu}{2}\right)\right.$$
$$\left. -2\frac{\Omega}{2}\coth\left(\frac{\Omega}{2}\right)\right], \qquad (13.2)$$

where $\nu = eV/kT$ and $\Omega = \hbar\omega/kT$.

The solid line in Fig. 13-3 shows the noise current spectral density at 1 KHz normalized by the equilibrium thermal noise $4kT \cdot \mathcal{T}G_Q$ versus normalized bias voltage $\nu = eV/kT$ [3]. The transmission coefficient $\mathcal{T}$ of an elastic scatterer is assumed to be 0.5. When the chemical potential difference ($eV = \mu_1 - \mu_2$) is smaller than the thermal energy kT that determines the region of stochastic emission of electrons due to Fermi-Dirac distribution, the nonequilibrium excess noise of quantum mechanical origin is smaller than thermal noise and the total noise current spectral density is dominated by thermal noise. However, when the chemical potential difference is

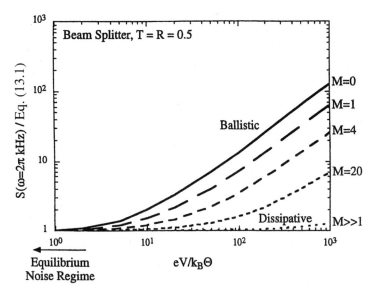

FIGURE 13-3: For an electron 50%-50% beam splitter, the current noise at 1 kHz [normalized by the equilibrium noise, Eq. (13.1) with $G = G_Q/(M+1)$] is plotted as a function of normalized bias voltage. As inelastic scattering increases (increasing the number of voltage probe reservoirs M), the excess noise is suppressed to the generalized Nyquist noise limit.

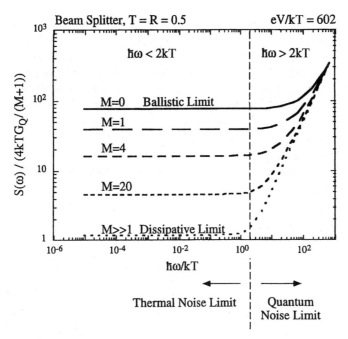

FIGURE 13-4: The normalized current noise for an electron 50%-50% beam splitter is plotted as a function of normalized frequency. For no inelastic scattering ($M = 0$), quantum nonequilibrium noise induces a large deviation from the generalized Nyquist noise. This is suppressed with increasing M to the Nyquist noise limit.

larger than the thermal energy, $eV > kT$, the nonequilibrium excess noise exceeds thermal noise and the total noise current spectral density is just one-half of full shot noise $S_i(\omega) = 2e(1 - T)TG_QV = e\langle i \rangle$. As the transmission coefficient T decreases, the total noise current spectral density approaches the full shot noise [i.e., $S_i(\omega) = 2e(1 - T)TG_QV \longrightarrow 2e\langle i \rangle (T \ll 1)$]. The solid line in Fig. 13-4 shows the noise current spectral density at a bias voltage $eV/kT = 602$ versus normalized frequency $\hbar\omega/kT$ [3]. In the low-frequency region ($\hbar\omega \leq eV$), the noise current spectral density is dominated by the nonequilibrium quantum (partition) noise with a spectral density of $S_i(\omega) = 2e(1 - T)TG_QV = e\langle i \rangle$ (half-shot noise). On the other hand, in the high-frequency region ($\hbar\omega \geq eV$), the noise current spectral density is dominated by the equilibrium quantum (beat) noise with a spectral density of $S_i(\omega) = 2\hbar\omega TG_Q = \hbar\omega G_Q$.

13.1.2. Suppression of Nonequilibrium Partition Noise by Inelastic Scattering

If an electron is scattered by a lattice vibration (phonon), the electron loses part of its kinetic energy by phonon emission; it could also gain energy by phonon absorption. This inelastic scattering process in a single-mode electron channel is macroscopically modeled by a high-impedance voltage probe reservoir [4]. In this voltage probe model, all incident electrons are absorbed by the reservoir and form a thermal equilibrium Fermi-Dirac distribution characterized by a local chemical potential, as shown in Fig. 13-5. The local chemical potential is determined in such a way that the total outgoing current is equal to the total incoming current. The two key features in dissipative electron transport are implemented in this voltage probe model: the Pauli exclusion principle leads to a local chemical potential (i.e., thermal equilibrium electron distribution) and Coulomb interaction between electrons results in local current conservation [4].

When an external voltage V is applied to two terminal electrodes ($\mu_1 - \mu_2 = eV$) and there is no scatterer between the two, the channel conductance is $G_Q = \frac{2e^2}{h}$, the

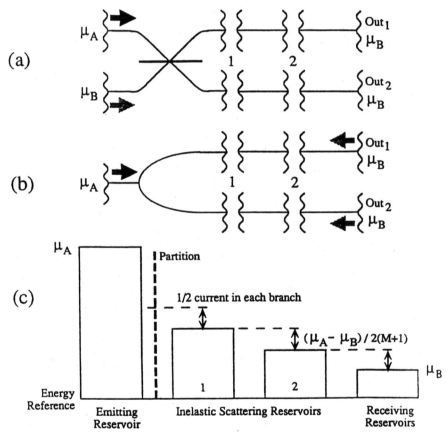

FIGURE 13-5: An electron 50%-50% beam splitter geometry (a) and an impedance matched symmetric 50%-50% Y-branch (b) followed by two inelastic scattering reservoirs in each output branch. Solid arrows indicate the incident occupied electron wavepackets, and the hatched arrows indicate the incident empty electron wavepacket. (c) A schematic of the zero temperature energy distributions for electrons emitted from each of the reservoirs for the aforementioned geometries.

average current is $\langle i \rangle = G_Q V$, and the noise current spectral density $S_i(\omega)$ is given by (13.1). If there is an electron beam splitter with $T = \frac{1}{2}$ and $V \gg \frac{kT\theta}{e}$, the channel conductance is $G = \frac{1}{2} G_Q$ and the average current is $\langle i \rangle = \frac{1}{2} G_Q V$. The noise current spectral density is given by the nonequilibrium partition noise $S_i(\omega) = e\langle i \rangle$ (one-half of full-shot noise). If we add one voltage probe reservoir after the beam splitter, as shown in Fig. 13-5, the average chemical potential of the voltage probe reservoir is $\mu_{\mathrm{VP}} = \frac{1}{4} \mu_1 + \frac{3}{4} \mu_2$. This average chemical potential results in average forward and backward currents with equal magnitudes. The channel conductance is $G = \frac{1}{4} G_Q$ because just one-half of the incident current into the voltage probe reservoir reaches the receiving terminal reservoir and the remaining half is reflected back. The noise current incident into the voltage probe is also halved into the forward and backward directions, so the new noise current spectral density is $S_i(\omega) = \frac{1}{2} e \langle i \rangle$, where $\langle i \rangle = \frac{1}{4} G_Q V$. In general, with M cascade voltage probe reservoirs, the noise current spectral density is reduced to $S_i(\omega) = \frac{e \langle i \rangle}{M+1}$, where $\langle i \rangle = \frac{G_Q}{2(M+1)} V$. With increasing the number of voltage probe reservoirs, the channel conductance and nonequilibrium noise are suppressed by the same factor of $(M+1)^{-1}$. The physical origin of nonequilibrium noise suppression is the self-feedback mechanism produced by the inelastic scattering. When the current incident on the voltage probe reservoir increases, the chemical potential μ_{VP} increases and the backward current increases. The increased backward current counteracts the increased incident current, and this compensation effect results in the suppression of forward current noise. The noise current spectral density with M cascade voltage probe reservoirs is given by

$$S_{\Delta I}(\omega) = \frac{1}{(TM+1)^2} S_{\Delta I_1}(\omega) + 4kT \frac{TG_Q}{TM+1} \cdot \frac{\Omega}{2} \coth\left(\frac{\Omega}{2}\right), \qquad (13.3)$$

where

$$S_{\Delta I_1}(\omega) = 2kT G_Q T(1-T) \left[\frac{\Omega + v_M}{2} \coth\left(\frac{\Omega + v_M}{2}\right) \right.$$
$$\left. + \frac{\Omega - v_M}{2} \coth\left(\frac{\Omega - v_M}{2}\right) - 2 \cdot \frac{\Omega}{2} \coth\left(\frac{\Omega}{2}\right) \right]$$

and $v_M = \frac{v}{TM+1}$.

Figures 13-3 and 13-4 show how the nonequilibrium partition noise is suppressed and the equilibrium noise (generalized Nyquist noise) is recovered by increasing the number M of voltage probe reservoirs [4]. This mathematical model of voltage probe reservoirs has a connection to the real physical situation if the voltage probe reservoirs are replaced by the channel conductance $G = \frac{G_Q}{2(M+1)}$. In the case of a beam splitter configuration shown in Fig. 13-5(a), the suppression of nonequilibrium noise has a linear dependence on the channel conductance. On the other hand, when an electron is split into two output ports by a Y-branch, the suppression of nonequilibrium noise has a (stronger) nonlinear dependence on the channel conductance. This nonlinear suppression in the Y-branch is due to a push-pull compensation process introduced by inelastic back scattering. The chemical potential fluctuations in the voltage probe reservoirs of the two arms are anticorrelated, and the direct return current that flows between the two voltage probe reservoirs suppresses the noise current more efficiently than in the beam splitter case.

13.1.3. Microscopic Model for Suppression of Nonequilibrium Partition Noise and Loss of Electron Wave Coherence

The inelastic scattering process in a single-mode electron channel can be modeled by a Monte Carlo numerical simulation. An external bias voltage V uniquely determines the chemical potential difference between two terminal electrodes. The corresponding electron wavenumber, which contributes to current transport, is divided into small segments with a wavenumber interval of Δk. An electron wavepacket with a wavenumber spread Δk and position spread $\Delta x (\Delta k \Delta x = \frac{1}{2})$ propagates in the channel. The system is divided into many degrees of freedom (DOF), as shown in Fig. 13-6, and each DOF accommodates a single electron (spin degeneracy is neglected). At $T = 0$, every electron wavepacket emitted from the left electrode is occupied. We put an electron beam splitter with a transmission coefficient T to introduce a nonequilibrium partition noise. When an electron moves from one DOF to the next DOF with the same wavenumber, a computer-generated random number q between 0 and 1 is compared with a prescribed phonon scattering probability p. If the random number is greater than the scattering probability and the electron final state is empty, the electron emits a phonon and is back scattered (the process A in Fig. 13-6). Otherwise the electron moves to the next DOF, keeping its wavenumber: This includes the two cases—the random number q is smaller than the scattering probability (the process B in Fig. 13-6) or q is greater than p but the final state is occupied (the process C in Fig. 13-6). The noise current is measured at the input plane of either one of the terminal electrodes.

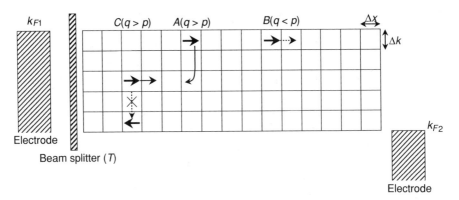

FIGURE 13-6: A Monte Carlo simulation model for an electron transport with inelastic scattering.

Figure 13-7 shows the nonequilibrium noise current normalized by full-shot noise value versus channel conductance with distributed inelastic scattering [5]. In the case of zero inelastic scattering probability, decreasing the beam splitter transmission coefficient decreases the channel conductance according to $G = TG_Q$ and increases the normalized noise current according to $S_i(\omega)/2e\langle i\rangle = (1 - T)$. With increasing inelastic scattering probability, the conductance is further decreased and the nonequilibrium noise is simultaneously suppressed. This noise suppression originates from the final-state occupancy dependent scattering rate. In the case of a very high inelastic scattering probability, both forward and backward propagating electrons obey thermal equilibrium Fermi-Dirac distribution rather than the nonequilibrium distribution originally produced by the beam splitter. In this high inelastic scattering limit, the (macroscopic) voltage probe model and the (microscopic) Monte Carlo simulation model give the same result. However, it is interesting to note that the noise current without elastic scattering center ($T = 1$) increases initially with inelastic scattering: Such a behavior is not predicted by the (macroscopic) voltage probe model.

Figure 13-8 shows the nonequilibrium noise current normalized by full-shot noise value versus channel conductance with distributed elastic scattering [5]. Irrespective of the initial noise value, the final noise current is always reduced to one-third of full-shot noise in the limit of strong elastic scattering. The Pauli exclusion principle is effective in partially suppressing the nonequilibrium noise even if there is no energy dissipation associated with electron back scattering.

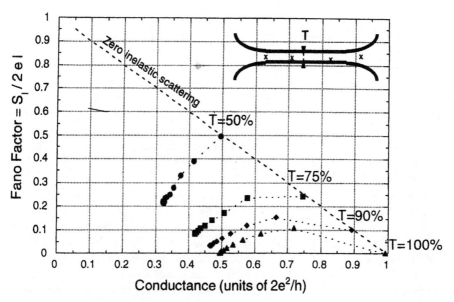

FIGURE 13-7: Normalized noise current spectral density $S_i(\omega)/2eI$ versus channel conductance in a mesoscopic electron channel with a beam splitter with a transmission coefficient T and distributed inelastic scattering.

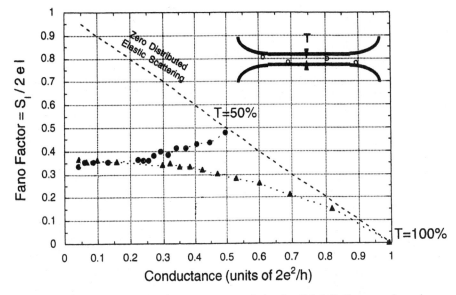

FIGURE 13-8: Normalized noise current spectral density $S_i(\omega)/2eI$ versus channel conductance in a mesoscopic electron channel with and without a 50%-50% beam splitter and distributed elastic scatterings.

Since the nonequilibrium partition noise introduced by a 50%-50% beam splitter originates from the Heisenberg uncertainty relation between the electron number difference and electron phase difference between the two arms, suppression of the nonequilibrium partition noise must accompany an increase in the electron phase difference noise (or loss of phase coherence between the two arms). Figure 13-9(a) shows an electron wave Mach-Zehnder interferometer model in which the upper arm introduces only phase shift but the lower arm couples to phonon reservoirs. The quantum Monte Carlo wavefunction method can be used to study the behavior of this interferometer. Ten electrons are injected one by one into the interferometer and electron numbers are counted in the two output channels as a function of acoustic phonon emission probability. Figure 13-9(b) shows the oscillatory behavior of the electron wave interferometer output $(N_1 - N_2)/(N_1 + N_2)$ for small and large phonon emission probabilities [6]. Figure 13-9(c) shows how visibility is degraded with increasing system-reservoir interaction length [6]. A discrete drop in the visibility by 1/10 is due to the emission of a single phonon in the lower arm, which localizes one electron out of total ten electrons into the lower arm. A continuous decrease in the visibility is due to the absence of phonon emission in the lower arm, which partially localizes an electron into the upper arm. In the example of Fig. 13-9(c), six electrons are found in the lower arm and four electrons must have propagated in the upper arm.

The same mechanism of phonon emission by electrons simultaneously suppresses the electron number partition noise through the Pauli exclusion principle and increases the electron phase partition noise (loss of phase coherence) through local-

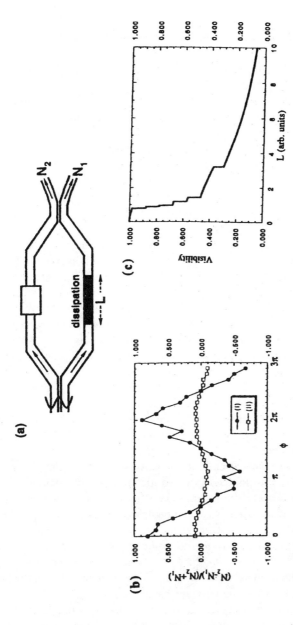

FIGURE 13-9: Suppression of the electron wave intereference by dissipative electron-phonon coupling in an electron interferometer: (a) Schematic structure; (b) the interferometer output versus phase shift of one arm for two values of the length L of the dissipative region: (1) $L = 0.1\hbar k \Gamma^{-1}(k)/m^*$, and (2) $L = 0.5\hbar k \Gamma^{-1}(k)/m^*$; (c) the visibility as a function of L. Each data point is obtained by averaging over 10 injected electrons.

ization of electrons in either one of the two arms. The complementarity between the electron number difference noise ΔN and the electron phase difference noise $\Delta \Phi$ is preserved in this way.

13.2. QUANTUM INTERFERENCE IN ELECTRON COLLISION

The indistinguishability of identical quantum particles can lead to quantum interferences that profoundly affect their scattering [7], [8]. If two particles collide and scatter, the process that results in the detection of the first particle (1) in one direction and the second particle (2) in another direction interferes quantum mechanically with the physically indistinguishable process where the roles of the particles is reversed. For photons, a constructive interference between probability amplitudes can enhance the probability (as compared to classical particles) that both photons are detected in the same direction: This effect is known as *bunching*. For electrons, a destructive interference suppresses this probability: This effect is known as *antibunching* and is the ultimate origin of the Pauli exclusion principle, which states that two electrons can never occupy the same state.

As we have already seen, the scattering of individual particles at a beam splitter is a stochastic process that introduces fluctuations (partition noise) in the output flux, depending on the transmission probability T (Fig. 13-10a). If N identical particles are independently scattered in series, a binomial distribution results for the number of particles transmitted to one of the outputs, N_{out}, and the normalized variance (Fano factor) is equal to $\langle \Delta N_{out}^2 \rangle / \langle N_{out} \rangle = (1 - T)$. In the limit of an extremely small T, the partition noise approaches the Poisson limit (full-shot noise). This single-particle partition process is independent of the quantum statistics of incident particles.

This no longer holds in the case of collisions [Figs. 13-10(b) and (c)]. Assuming a beam splitter with $T = 1/2$, the collision of classical particles (independent partitioning of the two inputs) results in 1/4 probability for two particles to be scattered to the left (right), while the probability for one particle scattered into each output is 1/2. If the incident particle fluxes do not have fluctuation, each output flux exhibits one-half of the shot noise.

In the quantum mechanical case, since particles are described by probability amplitudes, when the wavefunctions of two identical particles initially on the left and right overlap, it becomes impossible to distinguish one particle from the other. Mathematically, the two-particle state can be described by $|1 : \psi_L; 2 : \psi_R\rangle$ (particle 1 in the left and particle 2 in the right) or by $|1 : \psi_R; 2 : \psi_L\rangle$ (vice versa). Only the symmetric and antisymmetric combinations of these states, $|\psi_{\pm}\rangle = (1/\sqrt{2}) [|1 : \psi_L; 2 : \psi_R\rangle \pm |1 : \psi_R; 2 : \psi_L\rangle]$, produce real measurement results. Particles having overall symmetric wavefunctions are called bosons, while those having antisymmetric wavefunctions are called fermions. A photon is a classic example of a boson, whereas an electron is a classic example of a fermion.

The probability amplitudes for the three possible collision outcomes are derived from the unitary evolution of the two-particle wavefunction. The symmetrization or antisymmetrization of the wavefunction yields two contributions to the probability

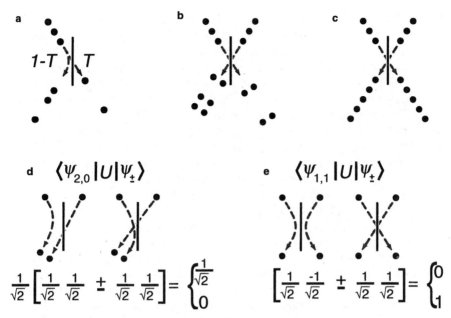

FIGURE 13-10: (a) An ideal beam splitter with transmission T stochastically partitions a constant incident flux into two output ports. (b) Collision (for $T = 1/2$) of quantum particles having a symmetric two particle wavefunction (bosons) results in two particles always in one port or the other. (c) Collision of quantum particles having an antisymmetric wavefunction (fermions) always results in one particle in each port. (d) The symmetrized (upper sign) or antisymmetrized (lower sign) input state $|\psi_\pm\rangle$ is propagated by the beam splitter unitary transformation, U, and projected onto the output state $\langle\psi_{2,0}|$ to calculate the probability amplitude for the case where two particles are in the left port. This results in a constructive interference between the direct and exchange terms for bosons, and a destructive interference for fermions. (e) For the output state where one particle is in each port, $\langle\psi_{1,1}|$, because of the additional minus sign due to reflection from the right input to the right output, a destructive interference results for bosons, and a constructive interference arises for fermions.

amplitude, a direct and exchange term. If we consider the output state where both particles are scattered to the left, then both of these terms have the same magnitude and sign [Fig. 13-10(d)]. For bosons, these terms add to give a probability of 1/2. For fermions, these terms subtract to give completely destructive interference; fermions never have two particles scattering into the same state, which is a manifestation of the Pauli exclusion principle.

If we consider the output state where one particle is in each output, then since the unitary evolution introduces a minus sign for reflection from the right input to the right output in order to satisfy power conservation (or unitarity of the beam splitter scattering matrix), the direct and exchange terms have the same magnitude but opposite signs [Fig. 13-10(e)]. Completely destructive interference is now obtained for bosons, while the probability for fermions equals one. In this ideal beam splitter, while two bosons always scatter into the same port [Fig. 13-10(b)] so that the fluctuations in the output flux exhibit full-shot noise (although not Poissonian), one fermion always scatters into each port [Fig. 13-10(c)] so that the fluctuations in output flux are completely suppressed. Quantum interferences profoundly affect the collision noise.

A mesoscopic electron beam splitter fabricated by electron beam lithography on a GaAs high-mobility, two-dimensional electron gas system demonstrates this quantum interference effect [Fig. 13-11(a)] [9]. The two input and two output ports are defined by point contact constrictions formed by an etched trench and Schottky gates that deplete underlying electrons when negative gate biases are applied. A 40-nm gate finger down the center of the scattering region allows the splitting ratio of the beam splitter to be tuned over a narrow range around 50%-50% divisioning of transmitted electrons. The transport through the device is primarily ballistic since inelastic phonon scattering and elastic ionized impurity scattering length scales at an operating temperature of 1.6 K are much longer than the device size.

The use of ballistic point contacts for the inputs is necessary to achieve streams of single-mode electron wavepackets incident on the beam splitter. Measurements on an isolated point contact (Fig. 13-12 inset) serve to illustrate the expected noise contribution of the Schottky gate point contacts used as inputs in the collision experiment. The conductance through the point contact is proportional to the transmission probability for electrons at the Fermi energy. As the width of a point contact increases to half the de Broglie wavelength of the electrons (typically on the order of 100 nm), partial transmission through its lowest transverse mode becomes possible, and the conductance increases with decreasing partition noise. Once the lowest transverse mode in the point contact is fully transmitting, a plateau is reached in the conductance corresponding to the quantum unit of conductance (with spin degeneracy), $G_Q = 2e^2/h$, and the partition noise is completely suppressed (Fig. 13-12) [9]. Electrons are now steadily injected into the single mode of the point contact and are transmitted without stochasticity. Then, as the width of the point contact increases further, the higher transverse modes open up one by one with partition noise associated only with the opening mode. This is suggested by the fact that the noise normalized by the full-shot noise value corresponding to the total current exhibits peaks with decreasing magnitude as the transmission of the higher opening modes pass through $1/2$.

In the electron collision experiment, similar Schottky gate point contacts form the two inputs of the beam splitter, and both are set at their first conductance plateau, where electrons are steadily injected with unity transmission. The output noise power is plotted in Fig. 13-11(b) against the output current when each input port is biased individually and when both input ports are simultaneously biased [9]. The linear current dependence is a sign of noise due to partitioning. Since the transmissions into the right output are about the same for the two inputs, the slopes for the individual bias cases should be and are approximately the same. On the other hand, it is evident that the slope in the case of collision is smaller. Figure 13-11(c) emphasizes the significance of this by replotting the data as the normalized current noise relative to the expected classical collision noise determined from the sum of the individual partition noise powers. Given the transmission probabilities and using a simple theory that accounts only for the overall transmissions from inputs to outputs that disregards path

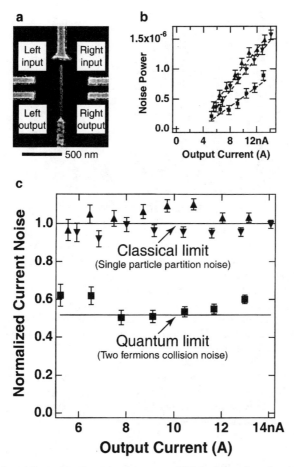

FIGURE 13-11: (a) Scanning electron microscope (SEM) photo of an electron beam splitter device fabricated on a GaAs two-dimensional electron gas system. Schottky gates and an etched trench (dark area near the centre gate) define the inputs (gate-gate point contacts) and the outputs (gate-trench point contacts). (b) The magnitude of the right output's current noise power (arb. units), as a function of its current. A modulation scheme and resonant circuit are employed to improve the discrimination between the stationary background amplifier noise and the current dependent partition noise and thus improve the signal to noise ratio at the 15.6 MHz measurement frequency. When either the left input (open squares) or the right input (open triangles) is biased individually, the single particle partition noises should be and are approximately the same since the transmission probabilities into the right output are similar. Their slopes are used as references to compare the noise during collision (closed squares), which is clearly suppressed. (c) After normalizing the noise by the current (after subtracting zero bias offsets) and scaling by the classical collision noise determined from the weighted average of the slopes of the single particle partition noises, the measured collision noise can be compared to the classical limit. Its suppression indicates a fermion interference. The noise is not completely suppressed due in part to nonidealities in the beam splitter's scattering matrix.

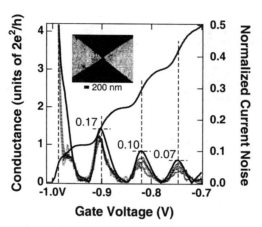

FIGURE 13-12: (Inset) SEM photo of an isolated quantum point contact device. (Main figure) conductance (solid, red) and current noise normalized by the full shot noise value (dotted) (scaled to the expected Fano factor of $1/2$ at $T = 1/2$ in the first plateau), as a function of the voltage applied to the confining Schottky gates. The current through a point contact is carried by discrete transverse modes. As the point contact width is increased with decreasing negative gate bias, a mode's transmission probability increases from zero to unity as its transverse sub-band energy drops below the Fermi energy. Once fully transmitting ($T = 1$), a transverse mode carries a fixed current per unit voltage corresponding to G_Q, and a plateau is reached in the conductance. The partition noise at these plateaux is absent. The current noise was measured for four different source-drain bias voltages of 0.5, 0.75, 1.25, and 1.75 mV. The noise power peaks at the middle of the steps where $T = 1/2$ for the newly opening mode. The theoretical partition noise behavior based on the measured conductance, assuming all fully opened modes have $T = 1$, is also plotted (solid, black) and shows that the trend in experimental noise peaks agrees with theoretical prediction.

length differences, the fermionic collision noise should be 52% of the classical collision noise. The observed suppression of the measured collision noise to an average of 56% of the classical value therefore indicates the presence of fermion quantum interference.

13.3. NEGATIVE CORRELATION IN ELECTRON PARTITION

A thermal boson source tends to emit particles together (bunching) with enhanced particle number fluctuations relative to classical expectations (super-Poisson statistics) [10]. In contrast, a thermal fermion source tends to emit particles separately (anti-bunching) with suppressed particle number fluctuations (sub-Poisson statistics) [11]. As the statistics are different for bosons and fermions, the covariation of these particles under partition also differs, with the cross-covariation between the two output particle number fluctuations being positive for bosons and negative for fermions. The positive cross-covariance for photons was demonstrated more than forty years ago by Hanbury Brown and Twiss [12] using photon intensity interferometry for light from distant stars, a technique which subsequently became an important tool for probing the statistics of photons generated by various types of sources. The fermion complement to the Hanbury Brown and Twiss experiment, where the negative cross-covariance of partitioned electrons from a single-mode, low-temperature, Fermi-degenerate electron source is also demonstrated [13], [14].

In the electron cross-variance measurement, one input port of the device shown in Fig. 13-11(a) is operated as the single-mode electron source, and the other input port is pinched off. The output ports are biased 'open' such that several transverse modes are transmitting, reducing reflections back into the beam splitter.

The results are shown for the experimental normalized cross-covariance as a function of the delay time at four different conductances, $p = G/G_Q = 0.83, 0.77, 0.71$, and 0.61 [Fig. 13-13(a)–(d)], [13]. As the relative delay between the two output channels is increased, the experimental cross-covariance increases toward zero. The characteristic shape of the experimental cross-covariance data is a direct consequence of the bandpass filter (2–10 MHz) used in the measurement circuit. The additional traces shown are from a numerical simulation which accounts for the point contact

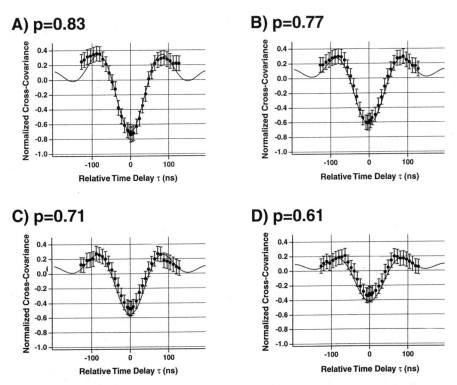

FIGURE 13-13: The normalized cross-covariance is plotted as a function of τ at four values of the input quantum point contact transmission probability: $p = 0.83, 0.77, 0.71, 0.61$. In each case, the minimum cross-covariance occurs at delay ime $\tau = 0$. The solid line represents the simulated cross-covariance for the actual measurement circuit.

transmission p, the beam splitter transmission T, and the entire detection circuit, and closely match the experimental data in all four cases. The negative cross-covariance of the output currents of the mesoscopic electron beam splitter indicates directly the sub-Poisson (anti-bunching) fluctuations of the thermal fermion source.

REFERENCES

[1] H. B. Callen and T. R. Welton, Phys. Rev. **83**, 34 (1951).
[2] R. Landauer, Physica D **38**, 226 (1989).
[3] R. Liu, and Y. Yamamoto, Phys. Rev. B **50**, 17411 (1994); R. Liu and Y. Yamamoto, Physica B **210**, 37 (1995).
[4] M. Buettiker, Phys. Rev. B **33**, 3020 (1986); C. W. J. Beenakker and M. Buettiker, Phys. Rev. B **46**, 1889 (1992).
[5] R. Liu, P. Eastman, and Y. Yamamoto, Solid State Commun. **102**, 785 (1997).
[6] A. Imamōglu and Y. Yamamoto, Phys. Lett. A **191**, 425 (1994).
[7] R. P. Feynman, R. B. Leighton, and M. Sands, *The Feynman Lectures on Physics, vol. 3: Quantum Mechanics* (Addison-Wesley, New York, 1965).
[8] R. Loudon, in *Coherence and Quantum Optics VI*, eds. J. H. Eberly et al. (Plenum Press, New York, 1990), p. 703.
[9] R. Liu, B. Odom, Y. Yamamoto, and S. Tarucha, Nature **391**, 263 (1998).
[10] R. Hanbury Brown and R. Q. Twiss, Nature **178**, 1447 (1956).
[11] E. M. Purcell, Nature **178**, 1449 (1956).
[12] R. Hanbury Brown and R. Q. Twiss, Nature **177**, 27 (1956).
[13] W. D. Oliver, J. Kim, R. C. Liu, and Y. Yamamoto, Science **284**, 299 (1999).
[14] M. Henny, et al., Science **284**, 296 (1999).

14 Nonequilibrium Green's Function Formalism

In Chapters 12 and 13, we discussed the Coulomb blockade effect and the Pauli exclusion principle in mesoscopic systems, where the electron-electron and the electron-phonon interactions play a major role. This is known as the regime of strongly correlated transport and is gaining increasing attention in mesoscopic and condensed matter physics. Due to the complexity of the problem, however, our approach in Chapters 12 and 13 was semiclassical and based on phenomenological models.

In this chapter, we will describe the nonequilibrium Green's function formalism (also referred to as the Keldysh formalism [1]), which provides a microscopic theory for quantum transport with electron-electron and electron-phonon interactions. In Sec. 14.1, we will present the retarded and advanced Green's functions together with the retarded and advanced self-energies. In Sec. 14.2, another set of functions—namely, the electron and hole correlation functions together with the in- and out-scattering functions—will be introduced. In Sec. 14.3, we will formulate the terminal current and relate it to the Landauer–Büttiker formula in the limit of negligible interaction. We end this chapter with a simple analytical example illustrating the application of this formalism to the effect of electron-electron Coulomb repulsion and the Pauli exclusion principle on the current voltage characteristics of resonant tunnel diodes.

The recent text by S. Datta [2] gives an excellent introduction to the subject. Detailed treatments of the material presented in Secs. 14.1 through 14.3 can be found in Refs. [3] and [4].

14.1. GREEN'S FUNCTION AND SELF-ENERGY

The retarded Green's function $G^R(r, r')$ is the (outgoing) Schrödinger wavefunction at point r (or time t) due to the unit excitation at another point r' (or time t') and thus satisfies

$$(E - H + i\eta)G^R(r, r') = \delta(r - r'), \qquad (14.1)$$

where E is the electron energy, H is the system Hamiltonian, and $\eta(\to 0^+)$ is a positive infinitesimal that is introduced to make the solution properly bounded as we move away from the point of excitation. Using an orthonormal set of the eigenfunctions $\psi_\alpha(r)$ of the Hamiltonian H with the eigenenergies ε_α, the retarded Green's function can be expanded as

$$G^R(r, r') = \sum_\alpha \frac{\psi_\alpha^*(r')}{E - \varepsilon_\alpha + i\eta} \psi_\alpha(r). \qquad (14.2)$$

The advanced Green's function is just the conjugate of the retarded Green's function; that is, it is the (incoming) Schrödinger wavefunction at point r due to disappearance of the unit excitation at the point r'. The advanced Green's function is given by

$$G^A(r, r') = \left[G^R(r', r)\right]^* = \sum_\alpha \frac{\psi_\alpha^*(r')}{E - \varepsilon_\alpha - i\eta} \psi_\alpha(r). \qquad (14.3)$$

Therefore, the advanced Green's function in the matrix form is the Hermitian conjugate of the retarded Green's function (i.e., $G^A = \left[G^R\right]^\dagger$).

A common approach to solve the differential equation (14.1) is to discretize the spatial coordinate so that the Green's function becomes a matrix with finite elements $G^R(r, r') \to G^R(i, j)$, where the indices i, j denote points on a discrete lattice. The differential equation becomes a matrix equation, and the Green's function is obtained by inverting the matrix:

$$G^R = [(E + i\eta)I - H]^{-1}, \qquad (14.4)$$

where $[I]$ is the identity matrix. The matrix representation $[H]$ of the Hamiltonian operator is usually evaluated using the tight binding method.

However, the matrix (14.4) is infinite dimensional if we study an open system, which is always the case for a transport problem. A standard procedure to deal with this problem is to separate a small system of primary interest from large systems of secondary interest and to assume that the large systems are in thermal equilibrium conditions. The retarded Green's function for the small system (which is referred to as "system" from now on), is written as

$$G^R = \left[EI - H_s - \tau_p^\dagger g_p^R \tau_p\right]^{-1}. \qquad (14.5)$$

Here H_s is the system Hamiltonian matrix, τ_p is the coupling matrix between the system and the large system p (which is referred to as "reservoir" from now on) and g_p^R is the retarded Green's function of the reservoir p, which we can calculate easily because the reservoir is in thermal equilibrium. The third term on the RHS of (14.5) is called the retarded self-energy. If the system is coupled to several reservoirs, the retarded self-energy and the retarded Green's function are given by

$$\Sigma^R = \sum_p \tau_p^\dagger g_p^R \tau_p, \qquad (14.6)$$

$$G^R = \left[EI - H_s - \Sigma^R\right]^{-1}. \qquad (14.7)$$

The physical meaning of the retarded Green's function G^R in this reservoir model is different from the original one that we gave in the preceding sentence of (14.1). Equation (14.7) expresses the coherent evolution of an electron from the moment it is injected into the system until it loses coherence by its coupling to the reservoirs, which can be external leads, phonon baths, or electron baths. The retarded self-energy Σ^R is considered as an effective potential for an individual electron due to the external leads and the interactions with phonon and electron baths.

The self-energy Σ^R is, in general, a non-Hermitian operator and so the eigenfunctions ψ_α of $H_s + \Sigma^R$ do not form an orthonormal set. The new expression for the retarded and advanced Green's functions can be obtained by introducing the adjoint eigenfunctions:

$$(H_s + \Sigma^R)\psi_\alpha = \varepsilon_\alpha \psi_\alpha, \quad (14.8a)$$

$$(H_s + \Sigma^A)\phi_\alpha = \varepsilon_\alpha^* \phi_\alpha. \quad (14.8b)$$

Here the new *eigenenergy* is $\varepsilon_\alpha = \varepsilon_{\alpha 0} - \Delta_\alpha - i\frac{\gamma_\alpha}{2}$, where $\varepsilon_{\alpha 0}$ is the unperturbed eigenenergy of the system itself, Δ_α is the energy shift, and γ_α is the decay rate due to the coupling to reservoirs. The adjoint eigenfunctions form a bi-orthonormal set:

$$\int \phi_\alpha(r) \psi_\beta^*(r)\, dr = \delta_{\alpha\beta}. \quad (14.9)$$

With these adjoint eigenfunctions, the retarded Green's function is expressed as

$$G^R(r, r') = \sum_\alpha \frac{\phi_\alpha^*(r')}{E - \varepsilon_{\alpha 0} + \Delta_\alpha + i\frac{\gamma_\alpha}{2}} \psi_\alpha(r). \quad (14.10)$$

With this preparation, we are ready to introduce the spectral function and the decay rate. The spectral function A is defined as the imaginary part of the Green's function and is expressed as

$$A \equiv i(G^R - G^A) = \sum_\alpha \psi_\alpha(r)\phi_\alpha^*(r') \frac{\gamma_\alpha}{(E - \varepsilon_{\alpha 0} + \Delta)^2 + \left(\frac{\gamma_\alpha}{2}\right)^2}. \quad (14.11)$$

The local density of states is given by

$$\mathcal{D}(\nabla, \mathcal{E}) = \frac{\infty}{\in \pi} \mathcal{A}(\nabla, \nabla, \mathcal{E}) = -\frac{\infty}{\pi} \mathcal{I}_\updownarrow \left[\mathcal{G}^{\mathcal{R}}(\nabla, \nabla, \mathcal{E}) \right]. \quad (14.12)$$

The decay rate Γ is defined as the imaginary part of the self-energy and is expressed as

$$\Gamma = i\left(\Sigma^R - \Sigma^A\right) = i\left[\frac{1}{G^A} - \frac{1}{G^R}\right]. \quad (14.13)$$

From (14.7) and its adjoint, $G^A = [EI - H_s - \Sigma^A]^{-1}$, the spectral function A and the decay rate Γ are related by

$$A = G^R \Gamma G^A = G^A \Gamma G^R. \tag{14.14}$$

14.2. CORRELATION AND SCATTERING FUNCTIONS

The spectral function A gives all the allowed electronic states, regardless of whether they are occupied or empty. The electron and hole correlation functions describe how many of these states are occupied and how many are empty. In the language of second quantization, they are defined by

$$G^n(k, k'; t, t') = -iG^< = \langle a_{k'}^\dagger(t') a_k(t) \rangle, \tag{14.15}$$

$$G^p(k, k'; t, t') = iG^> = \langle a_k(t) a_{k'}^\dagger(t') \rangle. \tag{14.16}$$

We use the reciprocal-space k-representation instead of r-representation here. We remark that these two-time correlation functions include not only the information about the electron and hole occupation number for a given state ($k = k'$) at time $t = t'$ but also the information about the quantum coherence between different states ($k \neq k'$) at different times ($t \neq t'$). The retarded and advanced Green's functions are expressed similarly by

$$G^R(k, k'; t, t') = -i\theta(t - t') \langle a_k(t) a_{k'}^\dagger(t') + a_{k'}^\dagger(t') a_k(t) \rangle, \tag{14.17}$$

$$G^A(k, k'; t, t') = i\theta(t' - t) \langle a_k(t) a_{k'}^\dagger(t') + a_{k'}^\dagger(t') a_k(t) \rangle. \tag{14.18}$$

Here $\theta(t - t')$ is unity for $t > t'$ and zero for $t < t'$.

In the steady state, the correlation function depends only on the time difference $\tau = t - t'$, and the Fourier transform of (14.15) is given by

$$G^n(k, k'; E) = \int \frac{1}{\hbar} G^n(k, k', \tau) e^{-iE\tau/\hbar} d\tau. \tag{14.19}$$

We remark that the density matrix $\rho(k, k'; t)$ follows from the correlation function by setting $t = t'$ and is given by

$$\rho(k, k'; t) = G^n(k, k'; t, t')_{t=t'} = \frac{1}{2\pi} \int G^n(k, k'; E) \, dE. \tag{14.20}$$

The diagonal element $f(k) = G^n_{k=k', t=t'}$ of (14.15) gives us the number of electrons occupying a particular state k. Similarly, the diagonal element of (14.16), the hole correlation function $G^p_{k=k', t=t'}$, corresponds to the number of holes occupying the state k.

The out-scattering function $\Sigma^{\text{out}}(k, k'; t, t') (= i\Sigma^{>})$ tells us the rate at which electrons are scattered out of state k at time t and appear at state k' at time t', assuming that the state k is initially occupied by an electron. Similarly, the in-scattering function $\Sigma^{\text{in}}(k, k'; t, t') (= i\Sigma^{<})$ is the rate at which holes are scattered out of state k at time t and appear in state k' at time t', assuming that the state k is initially empty. In the steady state, we can obtain the energy-dependent scattering functions $\Sigma^{\text{out}}(k, k'; E)$ and $\Sigma^{\text{in}}(k, k'; E)$ by utilizing the Fourier transform similar to (14.19).

The spectral function A and the decay rate Γ, which are given by (14.11) and (14.13), can be expressed using these new functions:

$$A = G^n + G^p, \tag{14.21}$$

$$\Gamma = \Sigma^{\text{in}} + \Sigma^{\text{out}}. \tag{14.22}$$

From the Schrödinger equation with a source term and the adjoint equation, we can derive the kinetic equations for the electron and hole correlation functions:

$$G^n = G^R \Sigma^{\text{in}} G^A, \tag{14.23}$$

$$G^p = G^R \Sigma^{\text{out}} G^A. \tag{14.24}$$

In general, the preceding kinetic equations (14.23) and (14.24) for G^n and G^p have to be solved simultaneously with (14.7) and its adjoint for G^A. To proceed, however, we need to know the self-energy Σ^R and the scattering functions Σ^{in} and Σ^{out}. This requires the microscopic theory for the system-reservoir coupling (that is, electron-lead, electron-phonon and electron-electron interactions).

14.3. CURRENT FLOW

14.3.1. Noninteracting Transport

Let us consider a small conductor (system) connected to several leads (reservoirs). The terminal current at a particular lead p is given by

$$I_p = 2 \text{ (for spin)} \times \int i_p(E) \, dE, \tag{14.25}$$

where the current density at energy E is expressed in terms of the correlation and scattering functions:

$$i_p(E) = \frac{e}{h} \text{Tr} \left[\Sigma_p^{\text{in}} G^p - \Sigma_p^{\text{out}} G^n \right], \tag{14.26}$$

where trace is taken over the reservoir degrees of freedom. The first term on the RHS of (14.26) is the product of the probability G^p that a particular state k of the system is empty initially, and the rate Σ_p^{in} at which holes are scattered out of the state k of the system and appear in the state k' of the lead p. Thus, this term represents a current flow from the system to the lead p. The second term of LHS of (14.26) is the product of the probability G^n that the state k of the system is occupied initially, and the rate Σ_p^{out} at which electrons are scattered out of the state k of the system and appear in the state k' in the lead p. Therefore, this term corresponds to a current flow from the lead p to the system. The net current is given by the difference of these two current flows.

If we use (14.14), (14.22), and (14.23) in (14.26), the current density is rewritten as

$$i_p(E) = \frac{e}{h}\text{Tr}\left[\Sigma_p^{\text{in}} A - \Gamma_p G^n\right]$$
$$= \frac{e}{h}\text{Tr}\left[\Sigma_p^{\text{in}} G^R \Gamma G^A - \Gamma_p G^R \Sigma^{\text{in}} G^A\right]. \quad (14.27)$$

Here $\Gamma_p = \Sigma_p^{\text{in}} + \Sigma_p^{\text{out}}$ is the decay rate of electrons due to output flow to the lead p. If there are no electron-electron and electron-phonon interactions inside the system, the decay rate Γ and the in-scattering function Σ^{in} are only introduced by the coupling to leads q; that is, $\Gamma = \sum_q \Gamma_q$ and $\Sigma^{\text{in}} = \sum_q (\Sigma_q^{\text{in}})$. We assume that the lead is in thermal equilibrium condition, so the in-scattering functions are given by $\Sigma_p^{\text{in}} = \Gamma_p f_p(E)$ and $\Sigma_q^{\text{in}} = \Gamma_q f_q(E)$, where $f_p(E)$ and $f_q(E)$ are Fermi-Dirac distribution functions. Using these approximations, (14.27) is written as

$$i_p(E) = \frac{e}{h}\sum_q \text{Tr}\left[\Gamma_p G^R \Gamma_q G^A\right][f_p(E) - f_q(E)]$$
$$= \frac{e}{h}\sum_q T_{pq}[f_p(E) - f_q(E)]. \quad (14.28)$$

This is the Landauer-Büttiker formula [5], [6] for coherent transport without interactions inside the system. Here $T_{pq} = \text{Tr}[\Gamma_p G^R \Gamma_q G^A]$ is the (p,q) element of the transmission matrix [2].

14.3.2. Strongly Interacting Transport

Our discussion in this chapter has been based on a simple system-reservoir model, assuming that an individual particle feels an effective potential (the self-energy Σ^R) due to its interaction with the external leads, phonon baths, and electron baths. This simple one-particle picture fails as the system size gets smaller. In mesoscopic scales, we often need to take a many-body approach. This new regime is called the strongly correlated transport, for which the interactions cannot be described by the self-energy Σ^R and scattering functions Σ^{in}, Σ^{out}.

Let us consider the transport problem between two terminals across a single site. This system is described by the Anderson Hamiltonian [7]:

$$H = \sum_{k,\sigma} \varepsilon_{k,\sigma} a_{k,\sigma}^\dagger a_{k\sigma} + \varepsilon_0 \sum_\sigma c_\sigma^\dagger c_\sigma + U n_\uparrow n_\downarrow + \sum_{k,\sigma}(V_{k\sigma} a_{k\sigma}^\dagger c_\sigma + \text{h.c.}). \quad (14.29)$$

The terms of (14.29) represent the unperturbed Hamiltonian of the two reservoir electrodes, the unperturbed Hamiltonian of the single site, the on-site Coulomb repulsive interaction, and the tunnelling Hamiltonian between the single site and the reservoirs, respectively. In the case of a two-terminal system, the current densities at the two leads must satisfy $i_p(E) = -i_q(E)$ due to current continuity condition; thus we can symmetrize (14.27) [8]:

$$i(E) = \frac{1}{2}[i_p(E) - i_q(E)]$$
$$= i\frac{e}{2h}\left\{\text{Tr}[f_p(E)\Gamma_p - f_q(E)\Gamma_q](G^R - G^A) + i\text{Tr}[\Gamma_p - \Gamma_q]G^n]\right\}. \quad (14.30)$$

For the Anderson Hamiltonian (14.29), the couplings to the two leads differ only by a constant factor, $\Gamma_p(E) = \lambda\Gamma_q(E)$, and thus (14.30) is reduced to

$$i(E) = i\frac{e}{h}\left[f_p(E) - f_q(E)\right]\text{Tr}\left[\Gamma(G^R - G^A)\right]$$
$$= \frac{e}{\hbar}\left[f_q(E) - f_q(E)\right]\Gamma\left[-\frac{1}{\pi}I_m(G^R_{\sigma\sigma})\right], \quad (14.31)$$

where $\Gamma = \Gamma_p\Gamma_q/(\Gamma_p + \Gamma_q)$ and $G^R_{\sigma\sigma} = \frac{1}{i\hbar}\theta(t)\langle[c_\sigma(t), c^\dagger_\sigma]_+\rangle$. This innocent-looking formula includes the effect of full nonlinear electron-electron scattering by means of the retarded Green's function $G^R_{\sigma\sigma}$, or more specifically by the spectral function $A_{\sigma\sigma} = i\left(G^R_{\sigma\sigma} - G^A_{\sigma\sigma}\right)$ [8].

The equations of motion for $G^R_{\sigma\sigma}$ are obtained by the following procedure. The Heisenberg equations for c_σ and $a_{\eta k\sigma}$ are given by

$$\frac{d}{dt}c_\sigma = \frac{1}{i\hbar}\left[\varepsilon_0 c_\sigma + Un_{\bar{\sigma}}c_\sigma + \sum_{\eta,k}V^*_{\eta k\sigma}a_{\eta k\sigma}\right], \quad (14.32)$$

$$\frac{d}{dt}a_{\eta k\sigma} = \frac{1}{i\hbar}V_{\eta k\sigma}c_\sigma, \quad (14.33)$$

where η refers to either the right or left electrode. The equation of motion for $G^R_{\sigma\sigma}$ is

$$i\hbar\frac{d}{dt}G^R_{\sigma\sigma} = \delta(t)\langle[c_\sigma(t=0), c^\dagger_\sigma]_+\rangle + \varepsilon_0 G^R_{\sigma\sigma} + \sum_{\eta,k}V^*_{\eta k\sigma}\ll a_{\eta k\sigma}(t); c^\dagger_\sigma \gg$$
$$+ U \ll (c_\sigma n_{\bar{\sigma}})(t); c^\dagger_\sigma \gg . \quad (14.34)$$

Here $\ll a_{\eta k\sigma}(t); c^\dagger_\sigma \gg = \frac{1}{i\hbar}\theta(t)\langle[a_{\eta k\sigma}(t), c^\dagger_\sigma]_+\rangle$ and $\ll (c_\sigma n_{\bar{\sigma}})(t); c^\dagger_\sigma \gg = \frac{1}{i\hbar}\theta(t)\langle[(c_\sigma n_{\bar{\sigma}})(t), c^\dagger_\sigma]_+\rangle$ are the retarded Green's functions for the corresponding dynamic variables. $\bar{\sigma}$ denotes the opposite spin to σ. Taking the Fourier transform of (14.34) and using $[c_\sigma(t=0), c_{\bar\sigma}^\dagger]_+ = 1$ and $\delta(t) = \int \frac{d\omega}{2\pi} e^{-i\omega t}$, we obtain

$$(\hbar\omega - \varepsilon_\sigma) \ll c_\sigma; c_{\bar\sigma}^\dagger \gg_\omega = 1 + \sum_{\eta,k} V_{\eta k\sigma}^* \ll a_{\eta k\sigma}^*; c_\sigma^\dagger \gg + U \ll c_\sigma n_{\bar\sigma}; c_\sigma^\dagger \gg_\omega, \tag{14.35}$$

where $\ll \cdots \gg_\omega$ represents the Fourier transform of the corresponding retarded Green's function. Similarly, we have five other equations:

$$(\hbar\omega - \varepsilon_{\eta k}) \ll a_{\eta k\sigma}; c_\sigma^\dagger \gg_\omega = V_{\eta k\sigma} \ll c_\sigma; c_\sigma^\dagger \gg_\omega, \tag{14.36}$$

$$(\hbar\omega - \varepsilon_\sigma - U) \ll c_\sigma n_{\bar\sigma}; c_\sigma^\dagger \gg_\omega = \langle n_{\bar\sigma} \rangle + \sum_{\eta k} V_{\eta k}^* \ll a_{\eta k\sigma} n_{\bar\sigma}; c_\sigma^\dagger \gg_\omega$$

$$- \sum_{\eta k} V_{\eta k\sigma} \ll c_\sigma a_{\eta k\sigma}^\dagger c_{\bar\sigma}; c_\sigma^\dagger \gg_\omega$$

$$+ \sum_{\eta k} V_{\eta k\sigma}^* \ll c_\sigma c_{\bar\sigma}^\dagger a_{\eta k\bar\sigma}; c_\sigma^\dagger \gg_\omega, \tag{14.37}$$

$$(\hbar\omega - \varepsilon_{\eta k}) \ll a_{\eta k\sigma} n_{\bar\sigma}; c_\sigma^\dagger \gg_\omega = V_{\eta k\sigma} \ll c_\sigma n_{\bar\sigma}; c_\sigma^\dagger \gg_\omega, \tag{14.38}$$

$$(\hbar\omega - \varepsilon_\sigma + \varepsilon_{\eta k} - \varepsilon_{\bar\sigma} - U) \ll c_\sigma a_{\eta k\bar\sigma}^\dagger c_{\bar\sigma}; c_\sigma^\dagger \gg_\omega = -V_{\eta k\sigma}^* \ll c_\sigma n_{\bar\sigma}; c_\sigma^\dagger \gg_\omega$$

$$+ V_{\eta k\sigma}^* f_\eta(\varepsilon_{\eta k}) \ll c_\sigma; c_\sigma^\dagger \gg_\omega, \tag{14.39}$$

$$(\hbar\omega - \varepsilon_\sigma + \varepsilon_{\bar\sigma} - \varepsilon_{\eta k}) \ll c_\sigma c_{\bar\sigma}^\dagger a_{\eta k\bar\sigma}; c_\sigma^\dagger \gg_\omega = -V_{\eta k\sigma} f_\eta(\varepsilon_{\eta k}) \ll c_\sigma; c_\sigma^\dagger \gg_\omega$$

$$+ V_{\eta k\sigma} \ll c_\sigma n_{\bar\sigma}; c_\sigma^\dagger \gg_\omega. \tag{14.40}$$

Here we have introduced the three assumptions:

$$\langle a_{\eta k\bar\sigma}^\dagger c_{\bar\sigma} \rangle = \langle c_{\bar\sigma}^\dagger a_{\eta k\bar\sigma} \rangle = 0, \tag{14.41}$$

which means that there is no quantum coherence between the system and reservoirs;

$$\ll a_{\eta k\sigma} a_{\eta' k'\bar\sigma}^\dagger c_{\bar\sigma}; c_\sigma^\dagger \gg_\omega = \ll a_{\eta k\sigma} c_{\bar\sigma}^\dagger a_{\eta' k'\bar\sigma}; c_\sigma^\dagger \gg_\omega = 0, \tag{14.42}$$

which indicates that the simultaneous tunneling of two electrons (co-tunneling) has negligible probability; and

$$\ll c_\sigma a_{\eta k\sigma}^\dagger a_{\eta' k'\bar\sigma}; c_\sigma^\dagger \gg_\omega = \delta_{\eta\eta'} \delta_{kk'} f_\eta(\varepsilon_{\eta k}) \ll c_\sigma; c_\sigma^\dagger \gg_\omega, \tag{14.43}$$

which means that there is no quantum coherence inside the reservoirs.

Solving (14.35) through (14.40), we have

$$\ll c_\sigma; c_\sigma^\dagger \gg_\omega = \frac{1 - \langle n_{\bar{\sigma}} \rangle}{\hbar\omega - \varepsilon_\sigma - \Sigma_0(\omega) + B} + \frac{\langle n_{\bar{\sigma}} \rangle}{\hbar\omega - \varepsilon_\sigma - U - \Sigma_0(\omega) + C}. \quad (14.44)$$

Here $\Sigma_0(\omega)$ is the retarded self-energy due to tunnel coupling to the reservoirs and is reduced to $-i\delta(\delta \to +0)$ in the weak coupling limit. U is the retarded self-energy due to electron-electron interaction. B and C represent the joint effects of the reservoir coupling and electron-electron scattering.

If it is assumed that $\varepsilon_\sigma = \varepsilon_{\bar{\sigma}}(= \varepsilon_0)$, $\Gamma_p = \Gamma_q = 2\Gamma$, and $\langle n_\sigma \rangle = \langle n_{\bar{\sigma}} \rangle = \frac{1}{2}$, we have

$$-\frac{1}{\pi} \mathrm{Im}\left[G^R_{\sigma\sigma}\right] = \frac{1}{2}[\delta(\varepsilon - \varepsilon_0) + \delta(\varepsilon - \varepsilon_0 - U)]. \quad (14.45)$$

There are two channels for the resonant tunnel current: one at ε_0 corresponding to zero electron in the single site and the other at $\varepsilon_0 + U$ corresponding to one electron in the single site.

Due to the Pauli exclusion principle, there are only two channels. If the one-electron energy ε_0 of the site is equal to $\mu - \frac{U}{2}$, where $\mu = \mu_p = \mu_q$ is the chemical potential of the two electrodes, the tunnel current is blocked within the applied voltage $|V| \leq U/e$ as shown in Fig. 14-1. The system is "insulating" due to the Coulomb blockade effect near $V = 0$. On the other hand, if $\varepsilon_0 = \mu$ or $\varepsilon_0 = \mu - U$, the tunnel current flows at infinitesimal applied voltage as shown in Fig. 14-2: The system is "metallic" in these cases.

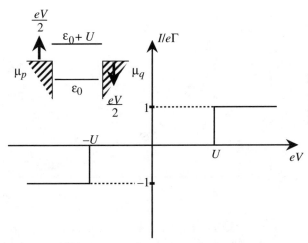

FIGURE 14-1: The current-voltage characteristic of an insulating junction with $\varepsilon_0 = \mu - \frac{U}{2}$.

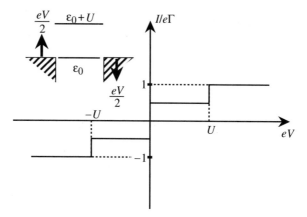

FIGURE 14-2: The current-voltage characteristic of a metallic junction with $\varepsilon_0 = \mu$.

14.3.3. Conductance through a Single Site

The (linear) conductance at zero applied voltage ($V = 0$) is neither zero or infinite in a real junction, because the quantum state of the central site has a linewidth due to the finite decay rate to the reservoirs. Figure 14-3 shows the (linear) conductance at

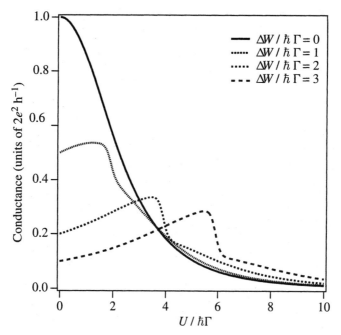

FIGURE 14-3: Linear conductance G/G_Q at $V = 0$ through a single central site versus normalized charging energy $U/\hbar\Gamma$ for different system-reservoir energy level separatoin $\Delta W/\hbar\Gamma$. Here $\hbar\Gamma$ is the energy broadening of the central site.

$V = 0$ through the central site with $\Delta W/\hbar\Gamma = 0, 1, 2$, and 3 as a function of $U/\hbar\Gamma$ [9]. Here $\Delta W = \mu - \varepsilon_0 - \frac{U}{2}$ is the energy difference between the reservoir and the system, $\hbar\Gamma$ is the energy broadening of the central site, and the temperature is $T = 0$. The conductance for $U = 0$ is given by $G_Q \times \frac{(\hbar\Gamma)^2}{(\Delta W)^2+(\hbar\Gamma)^2}$, which accounts for a finite reflection coefficient in the junction due to the energy mismatch ΔW. The conductance has a maximum value at $\frac{U}{2} = |\Delta W|$, where the Fermi level of the reservoir electrodes coincides with one of the two energy levels of the central site (the situation shown in Fig. 14-2). We note that even though electron transfer occurs resonantly through this energy level at $V = 0$, the partial reflection of electrons still exists and the overall conductance is smaller than G_Q.

If a bias voltage V is applied across such a junction, current increases stepwise as shown by the trace ($|\Delta W|/U = 0.5$) in Fig. 14-4 [9]. Except for this singular case, $|\Delta W| = \frac{U}{2}$, the conductance at $V \simeq 0$ is suppressed and current does not flow until the voltage source supplies sufficient energy to compensate for ΔW and U. This is nothing but the Coulomb blockade effect in a resonant tunnel junction.

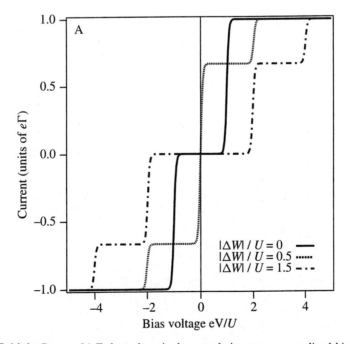

FIGURE 14-4: Current $I/e\Gamma$ through a single central site versus normalized bias voltage eV/U for different system-reservoir energy difference $|\Delta W|/\hbar\Gamma$. Here Γ is the tunnel rate across the junction.

REFERENCES

[1] L. V. Keldysh, Sov. Phys. JETP **20**, 1018 (1965).
[2] S. Datta, *Electronic Transport in Mesoscopic Systems* (Cambridge University Press, Cambridge, UK, 1995).
[3] J. Rammer and H. Smith, Rev. Mod. Phys. **58**, 323 (1986).
[4] G. D. Mahan, Phys. Rep. **145**, 251 (1987).
[5] R. Landauer, IBM J. Res. Dev. **32**, 306 (1988).
[6] M. Büttiker, IBM J. Res. Dev. **32**, 317 (1988).
[7] P. W. Anderson, *Basic Notions of Condensed Matter Physics* (Benjamin, 1984).
[8] Y. Meir and N. S. Wingreen, Phys. Rev. Lett. **68**, 2512 (1992).
[9] F. Yamaguchi and Y. Yamamoto, *Superlattices and Microstructures* **23**, 737 (1998).

15 Quantum Statistical Properties of a Laser

In this chapter we study the quantum statistical properties of a laser oscillator. A laser is a nonequilibrium open system. It generates light with stabilized phase (spectral linewidth much narrower than a cold cavity bandwidth) and stabilized amplitude (photon number distribution much narrower than that of a thermal radiation) at above the oscillation threshold. In general, the macroscopic coherence established in such an open system is determined by the balance between two counteracting forces: the system's ordering force and the reservoir's fluctuating force. In the case of a laser, the system's stabilizing force is the stimulated emission of photons. The amplitude is stabilized around its steady-state value via gain saturation. The phase information is transferred from *photon to photon* by the phase of the coherent stimulated emission. Both laser field and inverted atomic systems couple dissipatively to reservoirs. The laser field decays through output coupling mirrors and internal loss. The atomic dipoles decay due to spontaneous emission of photons and collision with other atoms. The population inversion is subject to dissipation by spontaneous emission and the reverse process (external incoherent pumping). These dissipation processes inevitably introduce fluctuating forces into the system from the reservoirs. Figure 15-1 shows such an open system model for a laser [1].

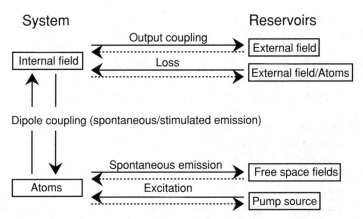

FIGURE 15-1: System-reservoir coupling model of a laser. The solid lines and dashed lines indicate the dissipation and the associated fluctuations, respectively.

We have already formulated the quantum theory for such an open dissipative system in Chapter 7. An important departure of the quantum theory of a laser from that treated in the previous chapter is the onset of nonlinearity (gain saturation). In fact, this gain saturation is a key physical element that separates a linear amplifier and a nonlinear oscillator [1], [2].

We will derive the density operator master equation for a laser in Sec. 15.1 and calculate the photon statistics and the spectral linewidth. The identical results will be obtained by the (quantum) Fokker-Planck equation in Sec. 15.2, and the (quantum) Langevin equation in Sec. 15.3.

It is assumed that the field varies most slowly due to the relation between the decay rates: $\gamma, \gamma_a, \gamma_b \gg \frac{\nu}{Q}$, where $\gamma, \gamma_a, \gamma_b$, and ν/Q are the decay rates for the dipole moment, the upper-level population, the lower-level population, and the laser field. In such a case, the atomic coordinates can be adiabatically eliminated and we have the equation of motion only for the laser field. This is called the *slaving principle*; that is, the dynamics of the whole system follow that of a subsystem with the slowest time constant. In some cases, the aforementioned relation is not satisfied and the adiabatic elimination of atomic coordinates is not justified. Sections 15.4 and 15.5 will treat such a system (i.e., semiconductor laser). Even though the density operator technique we present in the first part of Sec. 15.1 is completely general, the results in the following sections are effective only in the classical regime, where a positive $P(\alpha)$ representation (Glauber-Sudarshon diagonal function of coherent states) exists for the field density operator [2]. This standard quantum theory is not useful for describing the sub-Poissonian lasers, in which pump noise is suppressed to below the ordinary shot noise value. We will treat this problem in Sec. 15.4.

15.1. DENSITY OPERATOR MASTER EQUATION

15.1.1. Derivation of the Master Equation

We start by considering a general four-level laser system in which the two intermediate energy levels are coupled by the laser mode (Fig. 15-2). We assume that the uppermost level may be adiabatically eliminated to give an effective three-level system: This is valid as long as the decay rate γ_{fe} from the uppermost state $|f\rangle$ to the upper laser state $|e\rangle$ is much faster than all the other rates in the atom-field system. In this limit we have an effective incoherent pumping rate R_p from the atomic ground state $|o\rangle$ into $|e\rangle$.

The atom-cavity mode interaction Hamiltonian in the rotating-wave and long-wavelength approximations is given by

$$\hat{H}_{\text{int}} = i\hbar g_l(\hat{\sigma}_{eg}\hat{a} - \hat{a}^\dagger\hat{\sigma}_{ge}), \tag{15.1}$$

where g_l denotes the coupling coefficient given in Chapter 6. Unlike the systems analyzed in Chapter 7, we now have two quantum modes that are interacting coherently. Each of these modes (the cavity and the atom) interacts with independent

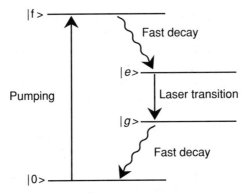

FIGURE 15-2: Energy level diagram of a four-state laser system.

radiation field reservoirs, resulting in cavity decay and spontaneous emission. After eliminating the interaction with these reservoir modes using the techniques developed in Chapter 7, we obtain the density operator master equation for the atom-field system:

$$\frac{d\hat{\rho}}{dt} = g_l(\hat{\sigma}_{eg}\hat{a}\hat{\rho} - \hat{a}^\dagger\hat{\sigma}_{ge}\hat{\rho} - \hat{\rho}\hat{\sigma}_{eg}\hat{a} + \hat{\rho}\hat{a}^\dagger\hat{\sigma}_{ge}) - \frac{\gamma_{\text{dep}}}{2}(\hat{\sigma}_{ee}\hat{\rho}\hat{\sigma}_{gg} + \hat{\sigma}_{gg}\hat{\rho}\hat{\sigma}_{ee})$$

$$- \frac{\kappa}{2}(\hat{a}^\dagger\hat{a}\hat{\rho} + \hat{\rho}\hat{a}^\dagger\hat{a} - 2\hat{a}\hat{\rho}\hat{a}^\dagger) - \frac{R_p}{2}(\hat{\sigma}_{oo}\hat{\rho} + \hat{\rho}\hat{\sigma}_{oo} - 2\hat{\sigma}_{eo}\hat{\rho}\hat{\sigma}_{oe})$$

$$- \frac{\Gamma_g}{2}(\hat{\sigma}_{gg}\hat{\rho} + \hat{\rho}\hat{\sigma}_{gg} - 2\hat{\sigma}_{og}\hat{\rho}\hat{\sigma}_{go}) - \frac{\Gamma_e}{2}(\hat{\sigma}_{ee}\hat{\rho} + \hat{\rho}\hat{\sigma}_{ee})$$

$$+ \Gamma_{eo}\hat{\sigma}_{oe}\hat{\rho}\hat{\sigma}_{eo} + \Gamma_{eg}\hat{\sigma}_{ge}\hat{\rho}\hat{\sigma}_{eg}. \tag{15.2}$$

To obtain (15.2), we have assumed that stimulated emission by thermal photons is negligible except for the pump term, which provides an incoherent pump rate (R_p) from the atomic ground state $|o\rangle$ to upper laser state $|e\rangle$. $\Gamma_e = \Gamma_{eo} + R_p + \Gamma_{eg}$ and $\Gamma_g = \Gamma_{go}$ denote the total decay rate of states $|e\rangle$ and $|g\rangle$, respectively. In (15.2), we have also introduced a dephasing rate γ_{dep} that arises from energy-conserving interactions of atoms with another (unspecified) reservoir. In gas lasers, for example, such a term is introduced by the collision between atoms of different species (foreign gas broadening). In solid-state lasers, low-frequency phonon scattering may be treated as an elastic dephasing process. Finally, κ is the cavity decay rate.

As already mentioned, the master equation of (15.2) describes the dynamics of the coupled atom-cavity system. Even though such a treatment may in principle be used to analyze any atomic laser system, this is not easily done. It is more desirable to obtain a master equation for the cavity mode alone by treating the atoms as a reservoir. In fact, a first-order laser amplifier master equation may be obtained by following the recipes introduced in Chapter 7. However, the fundamental assumption that the reservoir is not affected by the system fails dramatically in the case of a laser oscillator: The onset of nonlinearity or equivalently the gain saturation can be described as a system acting back on the reservoir. We therefore need to be careful in incorporating the gain saturation, which in turn is an essential aspect of the laser dynamics.

The laser master equation for the reduced density operator $\hat{\rho}_f(t)$ is obtained by tracing (15.2) over the atomic parameters. With no further approximation, we obtain

$$\frac{d\hat{\rho}_f}{dt} = g_l(\hat{a}\hat{\rho}_{ge} - \hat{a}^\dagger \hat{\rho}_{eg} - \hat{\rho}_{ge}\hat{a} + \hat{\rho}_{eg}\hat{a}^\dagger) - \frac{\kappa}{2}(\hat{a}^\dagger\hat{a}\hat{\rho}_f + \hat{\rho}_f\hat{a}^\dagger\hat{a} - 2\hat{a}\hat{\rho}_f\hat{a}^\dagger), \quad (15.3)$$

where the operator $\hat{\rho}_{ge} = \langle g|\hat{\rho}|e\rangle$ acts on the laser field Hilbert space. We reemphasize that neither Born nor Markov approximations have been introduced for the atom-cavity interaction. To obtain a closed form equation for the reduced density operator $\hat{\rho}_f$, we need to eliminate the field operators $\hat{\rho}_{ge}$ from (15.3). Using (15.2), we obtain the equation of motion for this field operator:

$$\frac{d\hat{\rho}_{ge}}{dt} = g_l(\hat{\rho}_{gg}\hat{a}^\dagger - \hat{a}^\dagger\hat{\rho}_{ee}) - \frac{\gamma_{\text{tot}}}{2}\hat{\rho}_{ge}, \quad (15.4)$$

with $\gamma_{\text{tot}} = \gamma_{\text{dep}} + \Gamma_e + \Gamma_g$. In the limit $\gamma_{\text{tot}} \gg \kappa$, we can adiabatically eliminate the operators $\hat{\rho}_{ge}$ and $\hat{\rho}_{eg}$ from (15.3), by substituting their steady-state solutions

$$\hat{\rho}_{ge} = \frac{2g_l}{\gamma_{\text{tot}}}(\hat{\rho}_{gg}\hat{a}^\dagger - \hat{a}^\dagger\hat{\rho}_{ee})$$

$$\hat{\rho}_{eg} = \frac{2g_l}{\gamma_{\text{tot}}}(\hat{a}\hat{\rho}_{gg} - \hat{\rho}_{ee}\hat{a}). \quad (15.5)$$

This substitution gives

$$\frac{d\hat{\rho}_f}{dt} = \left\{\frac{2g_l^2}{\gamma_{\text{tot}}}(\hat{a}\hat{\rho}_{gg}\hat{a}^\dagger - \hat{a}^\dagger\hat{a}\hat{\rho}_{gg} + \hat{a}^\dagger\hat{\rho}_{ee}\hat{a} - \hat{a}\hat{a}^\dagger\hat{\rho}_{ee}) + \frac{\kappa}{2}(\hat{a}\hat{\rho}_f\hat{a}^\dagger - \hat{a}^\dagger\hat{a}\hat{\rho}_f)\right\}$$
$$+ \text{h.c.} \quad (15.6)$$

The master equation of (15.6) includes saturation to all orders. Unfortunately, it still depends on the field operators $\hat{\rho}_{ee}$ and $\hat{\rho}_{gg}$. The exact solution at this point requires simultaneous solution of (15.6) along with

$$\frac{d\hat{\rho}_{ee}}{dt} = g_l(\hat{a}\hat{\rho}_{ge} + \hat{\rho}_{eg}\hat{a}^\dagger) - \Gamma_e\hat{\rho}_{ee} + R_p\hat{\rho}_f$$

$$= \frac{2g_l^2}{\gamma_{\text{tot}}}[2\hat{a}\hat{\rho}_{gg}\hat{a}^\dagger - \hat{a}\hat{a}^\dagger\hat{\rho}_{ee} - \hat{\rho}_{ee}\hat{a}\hat{a}^\dagger] - \Gamma_e\hat{\rho}_{ee} + R_p\hat{\rho}_f \quad (15.7)$$

$$\frac{d\hat{\rho}_{gg}}{dt} = -g_l(\hat{a}^\dagger\hat{\rho}_{eg} + \hat{\rho}_{ge}\hat{a}) - \Gamma_g\hat{\rho}_{gg} + \Gamma_{eg}\hat{\rho}_{ee}$$

$$= \frac{2g_l^2}{\gamma_{\text{tot}}}[2\hat{a}^\dagger \hat{\rho}_{ee}\hat{a} - \hat{a}^\dagger \hat{a}\hat{\rho}_{gg} - \hat{\rho}_{gg}\hat{a}^\dagger \hat{a}] - \Gamma_g \hat{\rho}_{gg} + \Gamma_{eg}\hat{\rho}_{ee}, \quad (15.8)$$

where we have assumed $\hat{\rho}_f = \hat{\rho}_{oo} + \hat{\rho}_{gg} + \hat{\rho}_{ee} \simeq \hat{\rho}_{oo}$ (i.e., most of the atoms are in the ground state $|o\rangle$ at all times).

To proceed, we need to eliminate $\hat{\rho}_{ee}$ and $\hat{\rho}_{gg}$ using (15.7) and (15.8), respectively. Recently, a closed form equation for $\hat{\rho}_f$ has been obtained using inverse relaxation super operators [3]. However, due to the operator nature of the variables, it is impossible in general to obtain closed form expressions that are easy to manipulate. We therefore need to concentrate on a particular laser system that allows us to make simplifying assumptions.

15.1.2. Fast Dephasing Case

Most laser systems are dephasing broadened and have a very fast decay rate that allows for the removal of population from the lower laser level $|g\rangle$. If we assume $\gamma_{\text{dep}} \gg \Gamma_g \gg \Gamma_e$, we can discard the terms in (15.6) that are proportional to $\hat{\rho}_{gg}$. In the weak saturation $(1 \gg g_l^2 \langle \hat{a}\hat{a}^\dagger\rangle/(\gamma_{\text{tot}}\Gamma_e))$ regime, we obtain

$$\hat{\rho}_{ee} \simeq \frac{R_p}{\Gamma_e}\hat{\rho}_f - \frac{R_p g_l^2}{\gamma_{\text{tot}}\Gamma_e^2}[\hat{a}\hat{a}^\dagger \hat{\rho}_f + \hat{\rho}_f \hat{a}\hat{a}^\dagger]. \quad (15.9)$$

Here the first term gives linear gain and the second term gives the first-order saturation. Substitution in (15.6) yields

$$\frac{d\hat{\rho}_f}{dt} = \left\{ \frac{2R_p g_l^2}{\gamma_{\text{tot}}\Gamma_e}(\hat{a}^\dagger \hat{\rho}_f \hat{a} - \hat{a}\hat{a}^\dagger \hat{\rho}_f) + \frac{\kappa}{2}(\hat{a}\hat{\rho}_f\hat{a}^\dagger - \hat{a}^\dagger \hat{a}\hat{\rho}_f) \right.$$
$$\left. + \frac{2R_p g_l^4}{\gamma_{\text{tot}}^2\Gamma_e^2}\left[\hat{\rho}_f(\hat{a}\hat{a}^\dagger)^2 + \hat{a}\hat{a}^\dagger \hat{\rho}_f \hat{a}\hat{a}^\dagger - 2\hat{a}^\dagger \hat{\rho}_f \hat{a}\hat{a}^\dagger \hat{a})\right] \right\} + \text{h.c.} \quad (15.10)$$

15.1.3. Lifetime Broadened Laser System

Next we consider the idealized laser system with $\Gamma_e = \Gamma_g = \Gamma$ and $\gamma_{\text{dep}} = \Gamma_{eg} = 0$: This is the scheme analyzed in most other laser physics textbooks, and we consider it here so that the reader can make a direct connection to the previous treatments [1]. Once again, we consider the weakly saturated regime: Since the spontaneous emission on the laser transition is neglected, we find

$$\hat{\rho}_{gg} \simeq \frac{2g_l^2}{\Gamma^2}\hat{a}^\dagger \hat{\rho}_{ee}\hat{a} = \frac{2R_p g_l^2}{\Gamma^3}\hat{a}^\dagger \hat{\rho}_f \hat{a} \quad (15.11)$$

$$\hat{\rho}_{ee} \simeq \frac{R_p}{\Gamma}\hat{\rho}_f - \frac{R_p g_l^2}{\Gamma^3}[\hat{a}\hat{a}^\dagger \hat{\rho}_f + \hat{\rho}_f \hat{a}\hat{a}^\dagger]. \quad (15.12)$$

By substituting (15.11) and (15.12) in (15.6), we obtain

$$\frac{d\hat{\rho}_f}{dt} = \left\{ \frac{A}{2}(\hat{a}^\dagger \hat{\rho}_f \hat{a} - \hat{a}\hat{a}^\dagger \hat{\rho}_f) + \frac{\nu}{2Q}(\hat{a}\hat{\rho}_f \hat{a}^\dagger - \hat{a}^\dagger \hat{a}\hat{\rho}_f) \right.$$
$$\left. + \frac{B}{8}(\hat{\rho}_f(\hat{a}\hat{a}^\dagger)^2 + 3\hat{a}\hat{a}^\dagger \hat{\rho}_f \hat{a}\hat{a}^\dagger - 4\hat{a}^\dagger \hat{\rho}_f \hat{a}\hat{a}^\dagger \hat{a}) + \text{h.c.} \right\}, \quad (15.13)$$

where $\mathcal{A} = 2R_p g_l^2 / \Gamma^2$ and $\mathcal{B} = \mathcal{A} g_l^2 / \Gamma^2$ are the gain and saturation coefficients, respectively [1]. The cavity-Q factor is $Q = \nu/\kappa$.

Comparing the laser master equations (15.10) and (15.13), we observe that the effects of the atomic parameters are (1) to change the gain coefficient $\mathcal{A}$; and (2) to change the form of the saturation term. Naturally, (15.13) is identical to the master equation obtained in Ref. [1].

As mentioned earlier, in general it is not possible to find an analytical expression for $\hat{\rho}_f$ in the intermediate saturation limit. To analyze the laser dynamics in this limit, we concentrate on the matrix elements of the density operator $\hat{\rho}_{nm} = \langle n | \hat{\rho}_f | m \rangle$. At this point, we can introduce the effect of pumping into the lower laser state $|g\rangle$ by modifying the Q-factor as $\frac{\nu}{Q} = \frac{\nu}{Q_e} + \frac{\nu}{Q_o}$. We define

$$N_{nm} = \frac{1}{2}(n+m+2) + \frac{1}{16}(n-m)^2 B/A \quad (15.14)$$

$$N'_{nm} = \frac{1}{2}(n+m+2) + \frac{1}{8}(n-m)^2 B/A \quad (15.15)$$

and use (15.6) through (15.8) to obtain

$$\dot{\rho}_{nm} = -\left(\frac{N'_{nm} A}{1 + N_{nm} B/A}\right)\rho_{nm} + \left(\frac{\sqrt{nm}\, A}{1 + N_{n-1,m-1} B/A}\right)\rho_{n-1,m-1}$$
$$-\frac{1}{2}\frac{\nu}{Q}(n+m)\rho_{nm} + \frac{\nu}{Q}[(n+1)(m+1)]^{1/2}\rho_{n+1,m+1}, \quad (15.16)$$

where $n_{\text{th}} = 0$ is assumed. The diagonal element ρ_{nn} (probability of finding n photons) obeys the equation of motion ($N'_{nn} = N_{nn} = n+1$):

$$\dot{\rho}_{nn} = -\left[\frac{(n+1)A}{1+(n+1)B/A}\right]\rho_{nn} + \left(\frac{nA}{1+nB/A}\right)\rho_{n-1,n-1}$$
$$-\left(\frac{\nu}{Q}\right)n\rho_{nn} + \frac{\nu}{Q}(n+1)\rho_{n+1,n+1}$$
$$\simeq -[A - B(n+1)](n+1)\rho_{nn} + (A - Bn)n\rho_{n-1,n-1}$$
$$-\left(\frac{\nu}{Q}\right)n\rho_{nn} + \frac{\nu}{Q}(n+1)\rho_{n+1,n+1}. \quad (15.17)$$

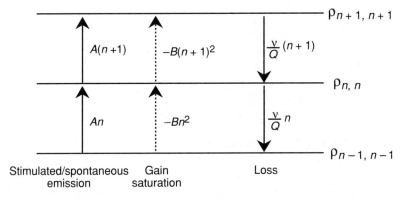

FIGURE 15-3: Flow of probability of n photons in a laser.

The second equality is based on the fourth-order approximation (i.e., $nB/A \ll 1$). This approximation is not valid far above threshold $(n \geq \frac{A}{B})$. Figure 15-3 summarizes the flow of the probability of having n photons. The first term in (15.17) represents the flow from $|n\rangle$ state to $|n+1\rangle$ state due to the stimulated emission (proportional to n) and the spontaneous emission (proportional to 1), where the term $-B(n+1)^2 \rho_{nn}$ represents the gain *saturation* effect caused by the depletion of excited atoms due to strong stimulated emission. The second term in (15.18) represents the flow from $|n-1\rangle$ state to $|n\rangle$ state due to the stimulated emission (proportional to $n-1$) and the spontaneous emission (proportional to 1). The third and fourth terms represent the flows from $|n\rangle$ state to $|n-1\rangle$ state and from $|n+1\rangle$ state to $|n\rangle$ state by either output coupling loss or internal absorption loss.

The field described by (15.17) is a mixed state. This can be seen from the fact that the density matrix element ρ_{nm} couples only with the elements with the same $n-m$ values. This simplification is due to the trace over atomic coordinates. The trace means that we do not measure the atomic state after the interaction with the field. In such a case, the coherence induced between the field and atom is lost and the field density operator is left in a mixed state.

15.1.4. Threshold Characteristics

Let us define the spontaneous emission coefficient β by

$$\beta = \frac{\text{spontaneous emission rate into the single laser mode}}{\text{total spontaneous emission rate}}$$

$$= \frac{A}{\Gamma}$$

$$= 2\frac{g_\ell^2}{\Gamma} \left(= \frac{B}{2A} \right). \tag{15.18}$$

From (15.18) we see that the spontaneous emission coefficient β is identical to the (normalized) saturation coefficient $\frac{B}{2A}$.

The average photon number $\langle \hat{n} \rangle \equiv \sum_n n\rho_{nn}$ satisfies the following equation [using (15.17)]:

$$\frac{d}{dt}\langle \hat{n} \rangle = \frac{A(\langle \hat{n} \rangle + 1)}{1 + 2\beta(\langle \hat{n} \rangle + 1)} - \frac{v}{Q}\langle \hat{n} \rangle. \qquad (15.19)$$

The steady-state solution of this equation is schematically shown in Fig. 15-4.

At the pump rate $R_p = \frac{v/Q}{2\beta}$, the average photon number $\langle \hat{n} \rangle$ is equal to one; here the stimulated emission rate is equal to the spontaneous emission rate into the same lasing mode. If the pump rate exceeds this point (threshold), the stimulated emission rate exceeds the spontaneous emission rate and the differential quantum efficiency $\eta_D \equiv (\Delta \langle \hat{n} \rangle) / (\Delta (R_p/v/Q))$ jumps from β (below threshold) to $\frac{1}{2}$ (above threshold). The average photon number is $\langle \hat{n} \rangle \simeq (R_p)/(2(v/Q))\beta$ below the threshold and $\langle \hat{n} \rangle \simeq (R_p)(2(v/Q))$ well above the threshold.

The reason why η_D does not reach unity but only $\frac{1}{2}$ is simply due to our model of the laser oscillator. In the highly saturated regime, an excited atom experiences a Rabi oscillation many times before being incoherently scattered with a 50%-50% probability to an excited or lower state.

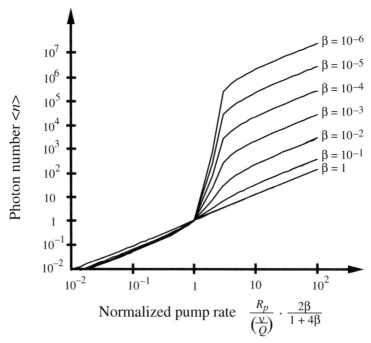

FIGURE 15-4: Average photon number $\langle \hat{n} \rangle$ versus normalized pump rate $\frac{R_p}{\left(\frac{v}{Q}\right)} \cdot \frac{2\beta}{1+4\beta}$.

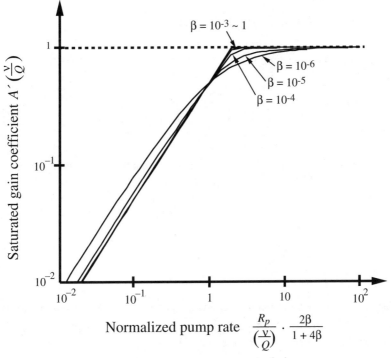

FIGURE 15-5: Normalized spontaneous emission rate $A'/\left(\frac{v}{Q}\right)$ versus normalized pump rate $\frac{R_p}{\left(\frac{v}{Q}\right)} \cdot \frac{2\beta}{1+4\beta}$.

The normalized spontaneous emission rate (saturated gain coefficient) $\frac{A'}{\left(\frac{v}{Q}\right)} = \frac{A/\left(\frac{v}{Q}\right)}{1+2\beta(\langle\hat{n}\rangle+1)} = \frac{\beta R_p/\left(\frac{v}{Q}\right)}{1+2\beta(\langle\hat{n}\rangle+1)}$ is plotted in Fig. 15-5 as a function of $\frac{R_p}{\left(\frac{v}{Q}\right)} \cdot \frac{2\beta}{1+4\beta}$. The normalized spontaneous emission rate is $A'/\left(\frac{v}{Q}\right) \simeq \frac{R_p\beta}{\left(\frac{v}{Q}\right)}$ below the threshold and $A'/\left(\frac{v}{Q}\right) \simeq 1$ well above the threshold. The classical definition of a laser threshold is given by $A = \frac{v}{Q}$ (unsaturated gain = loss), which gives $R^c_{p,\text{th}} = \left(\frac{v}{Q}\right)\frac{1}{\beta}$. The quantum definition of a laser threshold is given by $\langle\hat{n}\rangle = 1$ (photon number = 1), which results in $R^q_{p,\text{th}} = \left(\frac{v}{Q}\right)\frac{1}{2\beta}$. The factor of 2 difference stems from the fact that spontaneous emission is included in the quantum definition but not in the classical definition.

15.1.5. Photon Statistics

The steady-state condition $\dot{\rho}_{nn} = 0$ is satisfied if the net flow of the probabilities, for instance, between the $|n+1\rangle$ state and the $|n\rangle$ state, is zero (see Fig. 15-3):

$$-\frac{(n+1)A}{1+(n+1)B/A}\rho_{nn} + \frac{\nu}{Q}(n+1)\rho_{n+1,n+1} = 0. \tag{15.20}$$

The recursion relation is then given by

$$\rho_{n+1,n+1} = \left\{\frac{A^2}{\frac{\nu}{Q}[A+(n+1)B]}\right\}\rho_{nn}, \tag{15.21}$$

which determines ρ_{nn} for any n in terms of ρ_{00} by mathematical induction:

$$\rho_{nn} = \frac{\left[A^2/\left(\frac{\nu}{Q}B\right)\right]^n}{(n+A/B)!}\rho_{00}. \tag{15.22}$$

ρ_{00} is the probability of finding zero photons, which serves as a normalization constant satisfying $\sum_n \rho_{nn} = 1$.

Figure 15-6 shows the photon statistics ρ_{nn} for $R_p < R^Q_{p,\text{th}}$ (below threshold), $R_p = R^Q_{p,\text{th}}$ (threshold), and $R_p > R^Q_{p,\text{th}}$ (above threshold). The photon statistics below threshold are close to the thermal distribution ρ_{th}:

$$\rho_{nn} \simeq \langle \hat{n} \rangle^n \rho_{00} \longleftrightarrow \rho_{\text{th}} = \frac{1}{1+n_{\text{th}}}\left(\frac{n_{\text{th}}}{1+n_{\text{th}}}\right)^n$$

$$\simeq n^n_{\text{th}}, \tag{15.23}$$

where $\langle \hat{n} \rangle \simeq \frac{R_p}{2\left(\frac{\nu}{Q}\right)}\beta$ and $n_{\text{th}} = \frac{1}{e^{\hbar\omega/k_BT}-1}$, so the effective temperature T increases with the pump rate R_p. At threshold $R_p = R^Q_{p,\text{th}} = \frac{\nu}{Q}\cdot\frac{1}{2}\beta$, the photon statistics become

$$\rho_{nn} = \left(\frac{1}{2}\right)^n \rho_{00} = \frac{1}{2}\left(\frac{1}{2}\right)^n, \tag{15.24}$$

where $\langle \hat{n} \rangle = 1$ and the effective temperature T is

$$T = \frac{\hbar\omega}{k_B \ln 2} \gg 1. \tag{15.25}$$

At an optical frequency, the effective temperature corresponding to the laser threshold is $\sim 10^4$ K.

DENSITY OPERATOR MASTER EQUATION 275

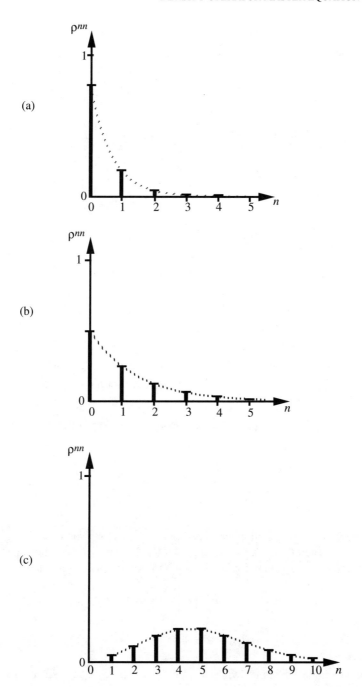

FIGURE 15-6: The photon statistics ρ_{nn} below (a), at (b), and above (c) the threshold.

At well above the laser threshold $R_p \gg R^Q_{p,\text{th}}$, the average photon number $\langle \hat{n} \rangle = \frac{R_p}{2\left(\frac{\nu}{Q}\right)}$ is much larger than $\frac{1}{\beta}$ and therefore we have

$$\rho_{nn} \simeq \frac{\langle \hat{n} \rangle^n}{n!} \rho_{00} = e^{-\langle \hat{n} \rangle} \frac{\langle \hat{n} \rangle^n}{n!}. \tag{15.26}$$

This is the Poisson distribution that we have discussed in Chapter 3.

15.1.6. Spectral Linewidth

The average cavity mode electric field is calculated by

$$\langle E(t) \rangle = i\mathcal{E} \sin kz \, \text{Tr} \left\{ \hat{\rho}(t)(\hat{a} - \hat{a}^\dagger) \right\}$$

$$= i\mathcal{E} \sin kz \sum_{n=0}^{\infty} \langle n|\hat{\rho}(t)(\hat{a} - \hat{a}^\dagger)|n \rangle$$

$$= -i\mathcal{E} \sin kz \sum_{n=0}^{\infty} \sqrt{n+1} \, \rho_{n,n+1}(t) e^{i\nu t} + \text{c.c.}, \tag{15.27}$$

where $\mathcal{E} = \sqrt{\frac{\hbar \omega}{2\epsilon_0 V}}$ is the average electric field amplitude per photon. The equation of motion for $\rho_{n,n+1}$ is derived from (15.16) to fourth-order approximation:

$$\dot{\rho}_{n,n+1} = -\left\{ \left[A - B\left(n + \frac{3}{2}\right) \right] \left(n + \frac{3}{2}\right) + \frac{1}{8}B + \frac{\nu}{Q}\left(n + \frac{1}{2}\right) \right\} \rho_{n,n+1}$$

$$+ \left[A - B\left(n + \frac{1}{2}\right) \right] \sqrt{n(n+1)} \, \rho_{n-1,n}$$

$$+ \frac{\nu}{Q} \sqrt{(n+1)(n+2)} \, \rho_{n+1,n+2}. \tag{15.28}$$

If the laser is operating well above threshold, $\rho_{n,n+1}(t)$ resembles the diagonal solution $\rho_{nn}(t)$. Therefore, we can assume that the elements $\rho_{n-1,n}$ and $\rho_{n+1,n+2}$ are expressed to a very good approximation with the similar recursion relation:

$$\rho_{n-1,n} \simeq \frac{\nu}{Q} \{(A - Bn)[A - B(n+1)]\}^{\frac{1}{2}} \rho_{n,n+1}, \tag{15.29}$$

$$\rho_{n+1,n+2} = \frac{Q}{\nu} \{[(A - B(n+1)][A - B(n+2)]\}^{\frac{1}{2}} \rho_{n,n+1}, \tag{15.30}$$

where the recursion relation (15.21) is used. Using these expressions in (15.28), we obtain

$$\dot{\rho}_{n,n+1} = -\frac{1}{2} D \rho_{n,n+1}, \tag{15.31}$$

where the decay constant

$$D \simeq \frac{1}{2}\frac{A}{\langle \hat{n} \rangle}. \tag{15.32}$$

Even though the diagonal element ρ_{nn} has the steady-state value, the off-diagonal element $\rho_{n,n+1}$ decays to zero. Substituting the integral of (15.31) into (15.27), we find that the average electric field decays according to

$$\langle E(t) \rangle = \langle E(0) \rangle \cos \nu t \exp\left(-\frac{1}{2}Dt\right). \tag{15.33}$$

The field power spectrum is given by the square of the Fourier transform of (15.33),

$$|E(\omega)|^2 = \left| \int_{-\infty}^{\infty} dt \exp(-i\omega t) \langle E(0) \rangle \cos \nu t \exp\left(-\frac{1}{2}Dt\right) \right|^2$$

$$= \frac{|\langle E(0) \rangle|^2}{(\omega - \nu)^2 + \left(\frac{1}{2}D\right)^2}, \tag{15.34}$$

which is a Lorentzian with linewidth (full width at half maximum) D.

At well above the threshold, the lower-level population is created by stimulated emission and the atomic system has an imperfect population inversion:

$$\frac{N_e}{N_e - N_g} = \frac{A}{A - B\langle \hat{n} \rangle} = \frac{A}{\frac{\nu}{Q}}. \tag{15.35}$$

Using this relation in (15.32), we have a new expression for the linewidth D,

$$D = \frac{\frac{\nu}{Q} n_{\rm sp}}{2\langle \hat{n} \rangle} = \frac{\hbar\omega \left(\frac{\nu}{Q}\right)\left(\frac{\nu}{Q_e}\right) n_{\rm sp}}{2P_{\rm out}}, \tag{15.36}$$

where $n_{\rm sp} = \frac{N_e}{N_e - N_g}$ is a population inversion parameter, Q_e is the external cavity Q-value (due to output coupling), and $P_{\rm out} = \hbar\omega \langle \hat{n} \rangle \frac{\nu}{Q_e}$ is the output power. This is the Schawlow–Townes formula. Since $\langle \hat{n} \rangle = \frac{R_p}{2\left(\frac{\nu}{Q}\right)}$ and $R_{p,\rm th}^Q = \left(\frac{\nu}{Q}\right)/2\beta$, (15.36) can be rewritten as

$$D = 2\beta \left(\frac{\nu}{Q}\right) n_{\rm sp} \left(\frac{R_p}{R_{p,\rm th}^Q}\right)^{-1}. \tag{15.37}$$

The linewidth D decreases abruptly at threshold due to the onset of stimulated emission as shown in Fig. 15-7. The linewidth is $D \simeq \frac{1}{2}\left(\frac{\nu}{Q}\right)$ at threshold $R_p = R_{p,\rm th}^Q$ and $D \simeq \beta n_{\rm sp}\left(\frac{\nu}{Q}\right) \ll \frac{\nu}{Q}$ at $R_p = 2R_{p,\rm th}^Q$.

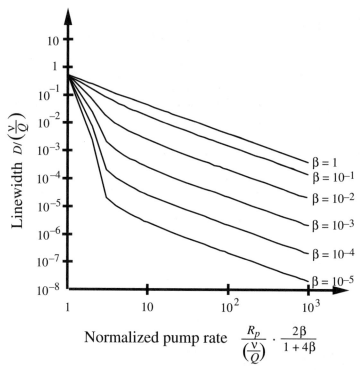

FIGURE 15-7: Normalized linewidth $D/\left(\frac{\nu}{Q}\right)$ versus normalized pump rate $R_p/R_{p,\text{th}}^Q$.

15.2. FOKKER-PLANCK EQUATION

If the field density operator $\hat{\rho}$ in (15.13) is expanded by the diagonal $P(\alpha)$ representation of coherent states, we obtain the corresponding Fokker-Planck equation:

$$\frac{d}{dt}P(\alpha) = -\frac{1}{2}\left\{\frac{\partial}{\partial \alpha}\left[\left(A - \frac{\nu}{Q} - B|\alpha|^2\right)\alpha P(\alpha)\right] + \text{c.c.}\right\} + A\frac{\partial^2}{\partial \alpha \, \partial \alpha^*}P(\alpha).$$

(15.38)

It is convenient to express (15.38) in the polar coordinate (r, θ):

$$\frac{d}{dt}P(r,\theta) = -\frac{1}{2}\frac{1}{r}\frac{\partial}{\partial r}\left[r^2\left(A - \frac{\nu}{Q} - Br^2\right)P(r,\theta)\right]$$

$$+ \frac{1}{4}A\left[\frac{\partial^2}{\partial r^2} + \frac{1}{r}\frac{\partial}{\partial r} + \frac{1}{r^2}\frac{\partial^2}{\partial \theta^2}\right]P(r,\theta). \quad (15.39)$$

The first term on the RHS of (15.39) behaves as a forcelike restoring term, which clusters $P(r)$ around the average value r_0. The steady-state distribution for the field amplitude r is given by

$$\frac{d}{dt} P(r, \theta) = 0. \tag{15.40}$$

If we expand $r = r_0 + \Delta r$ in the first term on the RHS of (15.39), we obtain

$$A - \frac{\nu}{Q} - Br_0^2 = 0, \tag{15.41}$$

$$r^2 \left(A - \frac{\nu}{Q} - Br^2 \right) = -2r Br_0^2 \Delta r = -2r Br_0^2 (r - r_0). \tag{15.42}$$

From (15.39), (15.40), and (15.42), we have

$$\frac{d}{d\theta} P(r, \theta) = 0, \tag{15.43}$$

$$\frac{1}{2} \frac{1}{r} \frac{\partial}{\partial r} [r 2 Br_0^2 (r - r_0) P] + \frac{A}{4} \frac{1}{r} \frac{\partial}{\partial r} \left(r \frac{\partial}{\partial r} P \right) = 0$$

$$\rightarrow \frac{\partial}{\partial r} P = -\frac{4 Br_0^2}{A} (r - r_0) P. \tag{15.44}$$

Equation (15.43) indicates that there is no preferred phase for the laser field. The solution of (15.44) has a Gaussian distribution,

$$P = \frac{1}{\sqrt{2\pi\sigma^2}} \exp\left[-\frac{(r - r_0)^2}{2\sigma^2} \right], \tag{15.45}$$

where the variance σ^2 is given by

$$\sigma^2 = \frac{A}{4 Br_0^2}. \tag{15.46}$$

At well above threshold, the steady-state amplitude r_0 is given by $r_0^2 = \frac{A - \frac{\nu}{Q}}{B} \simeq \frac{A}{B}$. Therefore, the variance is reduced to $\sigma^2 = \frac{1}{4}$. This is the variance of a coherent state, corresponding to a Poissonian photon distribution.

The angular distribution of the laser field $P(\theta)$ obeys the equation

$$\frac{d}{dt} P(\theta) = \frac{A}{4} \frac{1}{r_0^2} \frac{\partial^2}{\partial \theta^2} P(\theta). \tag{15.47}$$

This is the diffusion equation with the diffusion coefficient

$$D = \frac{A}{2r_0^2} = \frac{\frac{\nu}{Q} n_{\text{sp}}}{2 \langle \hat{n} \rangle}, \tag{15.48}$$

which is identical to (15.37). That is, the random-walk phase diffusion determines the spectral linewidth D.

15.3. LANGEVIN EQUATION

15.3.1. Derivation of the Langevin Equation

We will now describe the same laser model in the Heisenberg picture and analyze the time evolution of the field and atom operators. The Langevin equations for the upper- and lower-level population operators, $\hat{\sigma}_e = |e\rangle\langle e|$ and $\hat{\sigma}_g = |g\rangle\langle g|$, are

$$\frac{d}{dt}\hat{\sigma}_e = \Lambda_e - \gamma_e\hat{\sigma}_e + ig[\hat{a}^\dagger\hat{\sigma} - \hat{\sigma}^\dagger\hat{a}] + \hat{F}_e(t), \tag{15.49}$$

$$\frac{d}{dt}\hat{\sigma}_g = \Lambda_g - \gamma_g\hat{\sigma}_g - ig[\hat{a}^\dagger\hat{\sigma} - \hat{\sigma}^\dagger\hat{a}] + \hat{F}_g(t), \tag{15.50}$$

where Λ_e and Λ_g are the (external) pump rates into the upper and lower levels, γ_e and γ_g are the decay rates to other levels (for instance, to the atomic ground state), and $\hat{F}_e(t)$ and $\hat{F}_g(t)$ are the noise operators corresponding to these dissipation processes. The third term on the right-hand side of (15.49) and (15.50) represents the electric dipole coupling of the two levels via the cavity field. The dipole moment operator $\hat{\sigma} = |g\rangle\langle e|$ obeys the following Langevin equation:

$$\frac{d}{dt}\hat{\sigma} = -[\gamma + i(\omega - \nu)]\hat{\sigma} + ig(\hat{\sigma}_e - \hat{\sigma}_g)\hat{a} + \hat{F}_\sigma(t), \tag{15.51}$$

where γ is the dipole decay rate, and ω and ν are the oscillation frequencies of the atomic dipole and the field, respectively. $\hat{F}_\sigma(t)$ stands for the noise operator due to the dipole decay process. Finally, the field operator $\hat{a}$ obeys the Langevin equation

$$\frac{d}{dt}\hat{a} = -\left[\frac{1}{2}\frac{\nu}{Q} + i(\Omega - \nu)\right]\hat{a} - igN\hat{\sigma} + \hat{F}_e(t), \tag{15.52}$$

where Ω is the cold cavity resonance frequency, and N is the number of atoms inside the cavity. The noise operator $\hat{F}_e(t)$ accounts for the fluctuation force due to the field decay with a rate of ν/Q.

We now introduce the population inversion operator $\hat{\sigma}_z \equiv \hat{\sigma}_e - \hat{\sigma}_g$. If there is no pump into the lower level, $\Lambda_g = 0$, and the decay rates of the upper and lower levels are identical, $\gamma_e = \gamma_g = \gamma_z$, we have the following equation:

$$\frac{d}{dt}\hat{\sigma}_z = \Lambda_e - \gamma_z\hat{\sigma}_z + 2ig(\hat{a}^\dagger\hat{\sigma} - \hat{a}\hat{\sigma}^\dagger) + \hat{F}_z. \tag{15.53}$$

The correlation functions of the noise operators are determined by the Einstein relation between the drift and diffusion coefficients as follows:

$$\langle \hat{F}_z(t)\hat{F}_z(t')\rangle = \delta(t-t')[\Lambda_e + \gamma_z(\langle\hat{\sigma}_e\rangle + \langle\hat{\sigma}_g\rangle)], \tag{15.54}$$

$$\langle \hat{F}_\sigma^+(t)\hat{F}_\sigma(t')\rangle = \delta(t-t')[2\gamma\langle\hat{\sigma}_e\rangle + \Lambda_e - \gamma_z\langle\hat{\sigma}_e\rangle], \tag{15.55}$$

$$\langle \hat{F}_\sigma(t)\hat{F}_\sigma^+(t')\rangle = \delta(t-t')[2\gamma\langle\hat{\sigma}_g\rangle - \gamma_z\langle\hat{\sigma}_g\rangle], \tag{15.56}$$

$$\langle \hat{F}_e^+(t)\hat{F}_e(t)\rangle = \delta(t-t')\frac{\nu}{Q}n_{\text{th}}, \tag{15.57}$$

$$\langle \hat{F}_e(t)\hat{F}_e^+(t)\rangle = \delta(t-t')\frac{\nu}{Q}(1+n_{\text{th}}). \tag{15.58}$$

When the atomic dipole decay rate γ is much greater than the population decay rate γ_z and the photon decay rate $\frac{\nu}{Q}$, the dipole operator $\hat{\sigma}$ is given by

$$\hat{\sigma} \simeq \frac{1}{\gamma}[ig\hat{\sigma}_z\hat{a} + \hat{F}_\sigma], \tag{15.59}$$

where we assumed $\omega = \nu$. We can adiabatically eliminate the dipole operator $\hat{\sigma}$ from (15.52) and (15.53) using (15.59):

$$\frac{d}{dt}\hat{a} = -\frac{1}{2}\left(\frac{\nu}{Q}\right)\hat{a} + \frac{g^2}{\gamma}N\hat{\sigma}_z\hat{a} + \hat{F}_e - i\frac{g}{\gamma}N\hat{F}_\sigma, \tag{15.60}$$

$$\frac{d}{dt}\hat{\sigma}_z = -\frac{4g^2}{\gamma}\hat{\sigma}_z\hat{a}^\dagger\hat{a} + \hat{F}_z + i\frac{2g}{\gamma}(\hat{a}^\dagger\hat{F}_\sigma - \hat{F}_\sigma^\dagger\hat{a}). \tag{15.61}$$

If N identical atoms interact with the field $\hat{a}$, (15.61) can be transformed to the equation for a collective population operator $\hat{N}_z \equiv N\hat{\sigma}_z$,

$$\frac{d}{dt}\hat{N}_z = P - \gamma_z\hat{N}_z - \frac{4g^2}{\gamma}\hat{N}_z\hat{a}^\dagger\hat{a} + \hat{G}_p + \hat{G}_{\text{sp}} - 2(\hat{a}^\dagger\hat{G}_\sigma + \hat{G}_\sigma^\dagger\hat{a}), \tag{15.62}$$

where $P = N\Lambda_e$, $\hat{G}_p$, and $\hat{G}_{\text{sp}}$ are the (collective) pump and spontaneous emission noise operators, $\hat{G}_p + \hat{G}_{\text{sp}} = N\hat{F}_z$, and $\hat{G}_\sigma = -i\frac{g}{\gamma}N\hat{F}_\sigma$ is the (collective) dipole noise operator.

15.3.2. Linearization of the Langevin Equations

The quantum Langevin equations (15.60) and (15.62) cannot be solved analytically. However, at well above threshold, both field and population inversion can be split into c-number average values and small fluctuating terms. We consider only small fluctuating terms as operators:

$$\hat{a} = (A + \Delta\hat{A})e^{-i\Delta\hat{\phi}}, \tag{15.63}$$

$$\hat{N}_z = N_0 + \Delta\hat{N}. \tag{15.64}$$

Introducing (15.63) and (15.64) in (15.60) and (15.62), we obtain the linearized Langevin equations:

$$\frac{d}{dt}\Delta\hat{A} = \frac{1}{2A\tau_{st}}\Delta\hat{N} + \frac{1}{2}\left[(\hat{F}_e + \hat{G}_\sigma)e^{i\Delta\hat{\phi}} + e^{-i\Delta\hat{\phi}}(\hat{F}_e^\dagger + \hat{G}_\sigma^\dagger)\right], \quad (15.65)$$

$$\frac{d}{dt}\Delta\hat{\phi} = \frac{i}{2A}\left[(\hat{F}_e + \hat{G}_\sigma)e^{i\Delta\hat{\phi}} - e^{-i\Delta\hat{\phi}}(\hat{F}_e^\dagger + \hat{G}_\sigma^\dagger)\right], \quad (15.66)$$

$$\frac{d}{dt}\Delta\hat{N} = -\left(\frac{1}{\tau_{sp}} + \frac{1}{\tau_{sp}}\right)\Delta\hat{N} - 2\left(\frac{\nu}{Q}\right)A\Delta\hat{A} + \hat{G}_p + \hat{G}_{sp} + \hat{\Gamma}. \quad (15.67)$$

Here the noise operator $\hat{\Gamma} \equiv -2(\hat{a}^\dagger \hat{G}_\sigma + \hat{G}_\sigma^\dagger \hat{a})$ satisfies the following correlation functions:

$$\langle \hat{\Gamma}(t)\hat{\Gamma}(t')\rangle = \delta(t-t')\frac{4g^2 A^2}{\gamma}(\langle \hat{N}_e\rangle + \langle \hat{N}_g\rangle), \quad (15.68)$$

$$\langle \hat{\Gamma}(t)\hat{G}_{\sigma r}(t')\rangle = \langle \hat{G}_{\sigma r}(t)\hat{\Gamma}(t')\rangle = -\delta(t-t')\frac{2g^2 A}{\gamma}(\langle \hat{N}_e\rangle + \langle \hat{N}_g\rangle), \quad (15.69)$$

where $\langle \hat{N}_e\rangle = N\langle \hat{\sigma}_e\rangle$, $\langle \hat{N}_g\rangle = N\langle \hat{\sigma}_g\rangle$ and $\hat{G}_{\sigma r} = \frac{1}{2}(\hat{G}_\sigma + \hat{G}_\sigma^\dagger)$.

We now introduce the Fourier series analysis of period T for the fluctuation operator $\hat{g}(t)$ by

$$\hat{g}(T, \Omega) = \sqrt{\frac{2}{T}} \int_{-\frac{T}{2}}^{\frac{T}{2}} \hat{g}(t)e^{i\Omega t}\, dt. \quad (15.70)$$

The (unilateral) spectral density of $\hat{g}(t)$ is given by Wiener-Khintchin's theorem as

$$S_g(\Omega) = \lim_{T\to\infty} \langle \hat{g}^\dagger(T, \Omega)\hat{g}(T, \Omega)\rangle. \quad (15.71)$$

The amplitude and phase noise spectra are obtained as

$$S_{\Delta A}(\Omega) = \frac{1}{(\Omega^2 + A_2 A_3)^2 + \Omega^2 A_1^2}$$
$$\times \left[A_3^2(S_{G_p} + S_{G_{sp}} + S_\Gamma) \right.$$
$$\left. + (\Omega^2 + A_1^2)(S_{F_{ar}} + S_{G_{\sigma r}}) - 2A_1 A_3 S_{G_{\sigma r}\Gamma}\right], \quad (15.72)$$

$$S_{\Delta\phi}(\Omega) = \frac{1}{\Omega^2 A^2}[S_{F_{ai}} + S_{G_{\sigma i}}], \quad (15.73)$$

where

$$A_1 = \frac{1}{\tau_{sp}} + \frac{1}{\tau_{st}}, \quad (15.74)$$

$$A_2 = 2\left(\frac{v}{Q}\right)A, \qquad (15.75)$$

$$A_3 = \frac{1}{2A\tau_{st}}, \qquad (15.76)$$

$$\frac{1}{\tau_{st}} = \frac{4g^2}{\gamma}A^2. \qquad (15.77)$$

15.3.3. Quantum Noise at Well Above Threshold

At far above threshold $R_p \gg R_{p,\text{th}}^Q$, (15.73) is reduced to the Lorentzian

$$S_{\Delta A}(\Omega) = \frac{\left(\frac{v}{Q}\right)}{\Omega^2 + \left(\frac{v}{Q}\right)^2}. \qquad (15.78)$$

Here the dipole noise terms, S_Γ, $S_{G_{\sigma r}}$, and $S_{G_{\sigma r}\Gamma}$, cancel each other out and the spontaneous emission noise $S_{G_{sp}}$ is negligible. Exactly one-half of the amplitude noise of (15.78) stems from the pump noise S_{G_p} and the remaining half comes from the vacuum fluctuation $S_{F_{ar}}$ associated with the cavity damping.

Using Parseval's theorem, the amplitude noise at well above threshold is calculated as

$$\langle\Delta\hat{A}^2\rangle \equiv \int_0^\infty \frac{d\Omega}{2\pi} S_{\Delta A}(\Omega) = \frac{1}{4}. \qquad (15.79)$$

This result suggests that the cavity internal field approaches a coherent state and that the photon statistics approach a Poissonian distribution.

At far above threshold, (15.73) is reduced to

$$S_{\Delta\phi}(\Omega) = \frac{1}{\Omega^2 A^2}\left[\frac{1}{2}\frac{v}{Q} + \frac{g^2}{\gamma}(\langle\hat{N}_e\rangle + \langle\hat{N}_g\rangle)\right]$$

$$= \frac{\left(\frac{v}{Q}\right)}{\Omega^2 A^2} n_{sp}, \qquad (15.80)$$

where $\frac{1}{2}\frac{v}{Q} = \frac{g^2}{\gamma}(\langle\hat{N}_e\rangle - \langle\hat{N}_g\rangle)$ and $n_{sp} = \frac{\langle\hat{N}_e\rangle}{\langle\hat{N}_e\rangle - \langle\hat{N}_g\rangle}$ are used. The phase noise spectrum has $1/\Omega^2$ dependence on frequency, which indicates that the phase noise is a Wiener-Lévy process. The phase noise spectrum is contributed by the vacuum fluctuation $S_{F_{ai}}$ and the dipole noise $S_{G_{\sigma i}}$.

The field power spectrum is defined as the Fourier transform of the two-time correlation function:

$$I(\omega) = \int_{-\infty}^\infty dt\, e^{-i\omega t}\langle\hat{a}^\dagger(t)\hat{a}(0)\rangle. \qquad (15.81)$$

284 QUANTUM STATISTICAL PROPERTIES OF A LASER

At well above threshold, the amplitude noise $\Delta \hat{A}$ is stabilized around its average value A but the phase noise $\Delta \hat{\phi}$ goes through a random-walk diffusion. Therefore, we have

$$\begin{aligned}\langle \hat{a}^\dagger(t)\hat{a}(0)\rangle &= A^2 \langle \exp\{-i[\hat{\phi}(t) - \hat{\phi}(0)]\}\rangle \\ &= A^2 \langle \exp\left\{-i\int_0^t \frac{d}{dt}\Delta\hat{\phi}\, dt'\right\}\rangle \\ &= A^2 \exp\left(-\frac{\left(\frac{v}{Q}\right)}{2A^2} n_{\rm sp} t\right),\end{aligned} \qquad (15.82)$$

where we used (15.66) in evaluating the time integral in the exponential function. The spectral shape (15.81) is therefore a Lorentzian centered at v and with linewidth (full width at half maximum):

$$D = \frac{\left(\frac{v}{Q}\right)}{2\langle n\rangle} n_{\rm sp}. \qquad (15.83)$$

This is the same result (Schawlow–Townes linewidth) as that obtained by the density operator master equation.

Here we have seen another example of Onsager's regression theorem. That is, the decay of the two-time correlation function $\langle \hat{a}^\dagger(t)\hat{a}(0)\rangle$ obeys the identical form as the decay of the single-time averaged value $\langle \hat{a}(t)\rangle$. A detailed discussion of the quantum regression theorem was already given in Chapter 7.

15.4. SUB-POISSONIAN LASERS

The photon number noise of a laser operating at well above threshold is close to the standard quantum limit of a coherent state (Poisson limit), but the phase noise is considerably higher than that of a coherent state due to the random-walk phase diffusion process. The phase noise spectrum is increased to above the SQL with $(\Omega_c/\Omega)^2$ dependence, where $\Omega_c = \frac{v}{Q}$ is a cavity cut-off frequency and Ω is the fluctuation frequency. Such a phase diffusion noise (i.e., Schawlow–Townes linewidth) is an intrinsic property of a laser oscillator without phase restoring force. On the other hand, there remains a possibility that the photon number noise spectrum can be reduced to below the SQL.

In fact, this turns out to be true. The origin of the photon number noise spectrum of the output field (not the internal cavity field) in the frequency region $\Omega < \Omega_c$ is the (shot-noise-limited) pump fluctuations, while the origin of the photon number noise spectrum of the output field in the frequency region $\Omega > \Omega_c$ is the incident vacuum field fluctuations upon an output coupling mirror. A laser is an incoherently pumped oscillator, in which the phase information of the pump is completely destroyed in the pumping process of a "real" photon absorption process, so the increased pump phase fluctuation does not affect the noise property of a laser. Therefore, if the pump intensity fluctuation is suppressed, the laser output should be a number-phase squeezed state that has a reduced photon number noise that is below the SQL and an enhanced phase noise that is above the SQL.

15.4.1. Standard Quantum Limit of the Output Field

The external output field $\hat{r}$ consists of the transmitted internal field $\hat{a}$ and the reflected vacuum field fluctuation $\hat{f}$:

$$\hat{r} = \sqrt{\frac{\nu}{Q_e}} \hat{a} - \hat{f}. \tag{15.84}$$

Therefore, the in-phase and quadrature phase fluctuations of the output field $\hat{r}$ are

$$\Delta \hat{r}_1 = \sqrt{\frac{\nu}{Q_e}} \Delta \hat{A}_1 - \hat{f}_1, \tag{15.85}$$

$$\Delta \hat{r}_2 = \sqrt{\frac{\nu}{Q_e}} \Delta \hat{A}_2 - \hat{f}_2, \tag{15.86}$$

where $\hat{f}_1 = \frac{1}{2}(\hat{f} + \hat{f}^\dagger)$ and $\hat{f}_2 = \frac{1}{2i}(\hat{f} - \hat{f}^\dagger)$. The noise operator $\hat{F}_e(t)$ in (15.52) can be decomposed as

$$\hat{F}_e(t) = \sqrt{\frac{\nu}{Q_e}} \hat{f} + \sqrt{\frac{\nu}{Q_0}} \hat{f}_0, \tag{15.87}$$

where $\hat{f}_0$ is the noise operator associated with the internal loss. Figures 15-8(a) and (b) compare the internal and external amplitude noise spectra and the internal and external phase noise spectra for $\frac{\nu}{Q_0} \ll \frac{\nu}{Q_e}$ (negligible internal loss). At a pump rate well above the threshold, the amplitude noise spectrum approaches that of a broadband coherent state:

$$S_{\Delta r_1}(\Omega) = \frac{1}{2}, \tag{15.88}$$

which corresponds to the (white) shot-noise-limited photon flux spectrum

$$S_{\Delta N}(\Omega) = 4r_0^2 S_{\Delta r_1}(\Omega) = 2\langle \hat{N} \rangle, \tag{15.89}$$

where $\langle \hat{N} \rangle = \frac{\nu}{Q_e} A^2$ is an average photon flux. The origin of (15.88) in the frequency region $\Omega < \frac{\nu}{Q}$ is the pump noise $\hat{S}_{G_p}$, and in the frequency region $\Omega > \frac{\nu}{Q}$ is the in-phase component $\hat{f}_1$ of the incident vacuum field fluctuation.

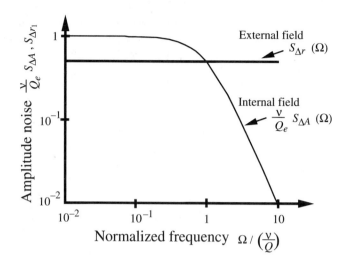

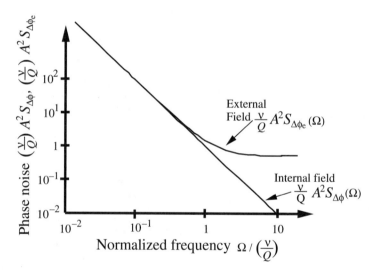

FIGURE 15-8: (a) The amplitude noise spectra of the internal and external fields. (b) The phase noise spectra of the internal and external fields.

The (normalized) phase noise spectrum, with the population inversion parameter $n_{sp} = 1$, is

$$S_{\Delta r_2}(\Omega) = \frac{1}{2}\left(1 + \frac{2\Omega_c^2}{\Omega^2}\right). \tag{15.90}$$

The first term (white noise) in parentheses is due to the quadrature component $\hat{f}_2$ of the reflected vacuum field fluctuation. The second term in parentheses is the phase diffusion term, which is equally contributed by the incident vacuum fluctuation $\hat{f}_2$ and the in-phase component of dipole fluctuation $\hat{\sigma}_1$. The two phase noise contributions, the white noise and the phase diffusion noise with $1/\Omega^2$ dependence, due to the same origin $\hat{f}_2$ are in quadrature phase (90° out of phase) so that they are simply added. On the other hand, the in-phase fluctuation $\sqrt{\frac{v}{Q_e}}\Delta\hat{A}_1$ of the transmitted internal field due to $\hat{f}_1$ and the reflected vacuum field fluctuation $-\hat{f}_1$ have equal magnitudes and opposite phases, so that they cancel out completely.

15.4.2. Photon Number Noise and Phase Noise of Pump-Noise-Suppressed Lasers

If the pump intensity fluctuation is suppressed at least in the low-frequency region—that is,

$$S_{G_p}(\Omega < \Omega_c) = 0, \quad (15.91)$$

by allowing the enhanced pump intensity noise at the high-frequency region—the amplitude noise spectrum $S_{\Delta r_1}(\Omega)$ is reduced to below the SQL in the low-frequency region at a relatively high pump rate, as shown in Fig. 15-9. At far above the threshold, we have

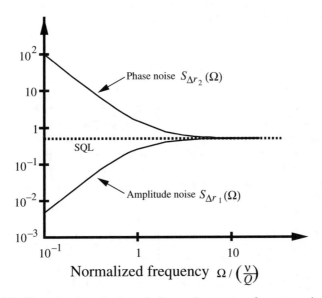

FIGURE 15-9: Normalized amplitude and phase noise spectra of a pump-noise-suppressed laser.

$$S_{\Delta r_1}(\Omega) = \frac{1}{2}\left(1 + \frac{\Omega_c^2}{\Omega^2}\right)^{-1}. \qquad (15.92)$$

Using (15.90) and (15.92), we have the number-phase spectral uncertainty product of a pump-noise-suppressed laser:

$$S_{\Delta r_1}(\Omega) S_{\Delta r_2}(\Omega) = \begin{cases} \dfrac{1}{2} & (\Omega < \Omega_c) \\ \dfrac{1}{4} & (\Omega > \Omega_c) \end{cases}. \qquad (15.93)$$

A number-phase squeezed state is generated at $\Omega < \Omega_c$ (i.e., when the field is measured over a time $T > \Omega_c^{-1}$ in which the uncertainty product is higher than the minimum value by a factor of 2), while a coherent state is generated at $\Omega > \Omega_c$ (i.e., when the field is measured over a time $T < \Omega_c^{-1}$ in which the minimum uncertainty product is exactly satisfied).

A highly saturated laser oscillator operates as a *matched load* near the cavity resonance, $|\omega - \omega_0| < \Omega_c$, where ω is a frequency of the incident wave and ω_0 is the laser cavity resonance frequency. The incident wave is completely absorbed by the cavity and there is no reflection. This is also true for the in-phase component $\hat{f}_1$ of a vacuum field fluctuation. On the other hand, for $|\omega - \omega_0| > \Omega_c$, the cavity has a high input impedance so that the incident wave is simply reflected back. The residual amplitude and phase noise in the high-frequency region shown in Fig. 15-9 are due to the beating of this reflected vacuum field against the coherent excitation of the output field.

15.4.3. Commutator Bracket Conservation

The creation and annihilation operators for the incident wave and the reflected wave from the laser cavity must preserve the proper commutator bracket. The quantum mechanical consistency of this new scheme of generating a number-phase squeezed state using a pump-noise-suppressed laser can be checked by this *commutator bracket test*. For that purpose, we adiabatically eliminate the population inversion operator $\hat{N}$ from the Langevin equation. We also assume $\tau_{st} \ll \tau_{sp}$, which is valid for a pump rate that is far above the threshold. Using the Fourier series analysis of the Langevin equations, we have

$$[\Delta \hat{r}(\Omega), \Delta \hat{r}^\dagger(\Omega)] = \frac{1}{\Omega^4 + \Omega^2 \gamma^2}\left\{-i\Omega^2(\Omega^2 + \gamma^2)[\hat{f}_1(\Omega), \hat{f}_2(\Omega)]\right.$$
$$\left. + i\Omega^2(\Omega^2 + \gamma^2)[\hat{f}_2(\Omega), \hat{f}_1(\Omega)]\right\}$$
$$= [\hat{f}(\Omega), \hat{f}^\dagger(\Omega)]. \qquad (15.94)$$

This is the desired result. The commutator bracket of the output wave is properly preserved only by the commutator bracket of the incident vacuum field $\hat{f}$.

From the commutator bracket conservation viewpoint, both the pump intensity fluctuations and the dipole fluctuations are extrinsic to a saturated laser oscillator. If we can eliminate both, the spectral uncertainty product (15.93) always satisfies the minimum uncertainty product. There are several ways to suppress the pump intensity fluctuation for a laser, such as constant-current injection into a semiconductor p-n junction laser and self-regulated optical pumping of gas lasers [4].

15.5. SQUEEZING IN SEMICONDUCTOR LASERS

In Chapters 12 and 13, we discussed the collective Coulomb blockade effect and the Pauli exclusion principle, which jointly realize suppression of pump noise in semiconductor p-n junction lasers. In the preceding section we showed that light generated by a pump-noise-suppressed laser is a number-phase squeezed state instead of an ordinary coherent state. In this section we will discuss generation of squeezed states from semiconductor lasers and derive the operator Langevin equations for a multimode semiconductor laser. Section 15.6 describes the mode partition noise, which is an important excess noise mechanism in semiconductor lasers.

The Langevin equation for the photon number operator $\hat{n}_i$ of a cavity internal field is given by [5]

$$\frac{d}{dt}\hat{n}_i = \left(\tilde{E}_{cv,i} - \frac{1}{\tau_{ph}}\right)\hat{n}_i + \tilde{E}_{cv,i} + \tilde{G}_i + \hat{g}_i + \hat{f}_i. \qquad (15.95)$$

Here the subscript i denotes the longitudinal, transverse, and polarization modes, $1/\tau_{ph}$ is the cavity photon decay rate, and $\tilde{E}_{cv,i}$ is the spontaneous emission operator into the mode (i), which should be equal to the stimulated emission operator for one photon. The noise operators $\tilde{G}_i$, $\hat{g}_i$, and $\hat{f}_i$ are associated with the stochastic processes of stimulated and spontaneous emission, internal absorption with a rate $1/\tau_{po}$, and output coupling with a rate $1/\tau_{pe}$, respectively. The two-time correlation functions for these noise operators in the relevant frequency/time scale can be approximated by those of Markov processes:

$$\langle \tilde{G}_i(t)\tilde{G}_i(t+\tau)\rangle = \delta(\tau)\langle \tilde{E}_{cv,i}\rangle(\langle \hat{n}_i\rangle + 1), \qquad (15.96)$$

$$\langle \tilde{g}_i(t)\tilde{g}_i(t+\tau)\rangle = \delta(\tau)\frac{\langle \hat{n}_i\rangle}{\tau_{po}}, \qquad (15.97)$$

$$\langle \tilde{f}_i(t)\tilde{f}_i(t+\tau)\rangle = \delta(\tau)\frac{\langle \hat{n}_i\rangle}{\tau_{pe}}. \qquad (15.98)$$

Tilde and hat are used to denote the operators for the electronic and the field systems, respectively.

The Langevin equation for the (total) carrier number operator in the active region is given by [5]

$$\frac{d}{dt}\tilde{N} = P - \frac{\tilde{N}}{\tau_{sp}} - \sum_i \tilde{E}_{cv,i}(\hat{n}_i + 1) + \tilde{\Gamma}_p + \tilde{\Gamma}_{sp} + \tilde{\Gamma}. \tag{15.99}$$

Here P and $1/\tau_{sp}$ are the average rates of pump and spontaneous emission (excluding that into lasing modes i). The associated noise operators are $\tilde{\Gamma}_p$ and $\tilde{\Gamma}_{sp}$, while $\tilde{\Gamma}$ is the noise operator for the stimulated and spontaneous emission processes into the lasing modes i. The two-time correlation function for the pump process is modified by the collective Coulomb blockade effect discussed in Chapter 12. If the fluctuation frequency of interest is much lower than the squeezing bandwidth $f = \frac{eI}{kTC}$ due to the collective Coulomb blockade effect, it is safe to assume that there is no pump fluctuation in the relevant frequency range (i.e., we can neglect the correlation function for $\tilde{\Gamma}_p$). On the other hand, the correlation functions for $\tilde{\Gamma}_{sp}$ and $\tilde{\Gamma}$ are

$$\langle \tilde{\Gamma}_{sp}(t)\tilde{\Gamma}_{sp}(t+\tau) \rangle = \delta(\tau)\frac{\langle \tilde{N} \rangle}{\tau_{sp}}, \tag{15.100}$$

$$\langle \tilde{\Gamma}(t)\tilde{\Gamma}(t+\tau) \rangle = \delta(\tau)\sum_i \langle \tilde{E}_{cv,i} \rangle(\langle \hat{n}_i \rangle + 1), \tag{15.101}$$

$$\langle \tilde{\Gamma}(t)\tilde{G}_i(t+\tau) \rangle = -\delta(\tau)\langle \tilde{E}_{cv,i} \rangle(\langle \hat{n}_i \rangle + 1). \tag{15.102}$$

The nonlinear operator Langevin equations (15.95) and (15.99) for a semiconductor laser cannot be solved analytically in general. At the pump level at well above the threshold, however, we can introduce the linearization technique:

$$\hat{n}_i = \langle \hat{n}_i \rangle + \Delta \hat{n}_i(t) = \langle \hat{a}_i \rangle^2 + 2\langle \hat{a}_i \rangle \Delta \hat{a}_i, \tag{15.103}$$

$$\tilde{N} = \langle \tilde{N} \rangle + \Delta \tilde{N}, \tag{15.104}$$

$$\tilde{E}_{cv,i} = \langle \tilde{E}_{cv,i} \rangle + \frac{d\langle \tilde{E}_{cv,i} \rangle}{d\langle \tilde{N} \rangle}\Delta \tilde{N}. \tag{15.105}$$

Here $\langle \hat{n}_i \rangle$, $\langle \hat{a}_i \rangle$, and $\langle \tilde{N} \rangle$ are the ensemble-averaged values of corresponding operators, and $\Delta \hat{n}_i$, $\Delta \hat{a}_i$, and $\Delta \tilde{N}$ are the small fluctuation operators. The output field fluctuation is expressed in terms of the internal field fluctuation and the noise operators associated with the output coupling [5]:

$$\Delta \hat{r}_i = \sqrt{\frac{1}{\tau_{pe}}}\Delta \hat{a}_i - \sqrt{\tau_{pe}}\,\hat{f}_i. \tag{15.106}$$

This operator boundary condition can be interpreted in such a way that the output field fluctuation is the sum of the transmitted internal field fluctuation and the reflected vacuum field fluctuation.

The standard Fourier analysis of (15.95) and (15.99) with the linearization approximation of (15.103) through (15.104), together with the correlation spectra given by (15.96) through (15.98) and (15.100) through (15.102), allows us to calculate the various noise spectra.

Let us consider an (ideal) single mode laser case first. The amplitude noise spectra $S_{\Delta r_1}(\Omega)$ of the output field from an ordinary laser with shot-noise-limited pump fluctuation and a pump-noise-suppressed semiconductor laser are compared in Figs. 15-10(a) and (b) [6]. The two cases correspond to the situation in which the squeezing bandwidth of the collective Coulomb blockade is much smaller or larger than any other relevant bandwidths of a semiconductor laser. The amplitude noise of the laser operating at just above the threshold ($P/P_{th} \simeq 1$) is much higher than the shot noise value due to *amplified spontaneous emission noise*. The system's ordering force (gain saturation) is weak, and the noise driving forces (spontaneous emission) dominate in this weak pump region. The resonant-like noise peak is due to the so-called relaxation oscillation between the photon number and the population inversion. The amplitude noise of the ordinary laser with shot-noise-limited pump fluctuation operating far above threshold ($P/P_{th} \gg 1$) approaches the standard shot noise value as shown in Fig. 15-10(a). The origin for this shot noise below the cavity cut-off frequency $\left(\Omega < \frac{\omega}{Q}\right)$ is identified as the shot-noise-limited pump noise and that above the cavity cut-off frequency $\left(\Omega > \frac{\omega}{Q}\right)$ is attributed to the reflected vacuum field fluctuation, as we discussed previously. The third noise source of spontaneous emission noise is completely quenched due to strong stimulated emission at such a pump region far above threshold. On the other hand, the amplitude noise of the pump-noise-suppressed laser operating far above threshold is decreased to below the shot noise value for $\Omega < \frac{\omega}{Q}$ but is equal to the shot noise value for $\Omega > \frac{\omega}{Q}$ as shown in Fig. 15-10(b). This is an expected result of the previous identification of the origins of the shot noise of the ordinary laser. The normalized phase noise spectrum of the pump-noise-suppressed laser is also plotted by the dotted line in Fig. 15-10(b). The normalized phase noise increases in proportion to $\left(\frac{\omega}{Q}\right)^2/\Omega^2$ at $\Omega < \frac{\omega}{Q}$ due to the phase diffusion process. The decreased amplitude noise that is below the shot noise value at $\Omega < \frac{\omega}{Q}$ is compensated for by the increased phase (diffusion) noise, and so the Heisenberg uncertainty principle between the amplitude and phase noise is conserved.

The quantum state of a pump-noise-suppressed semiconductor laser operating far above threshold is, therefore, a number-phase squeezed state when the measurement time T_{meas} is much longer than the cavity photon lifetime $\left(\frac{\omega}{Q}\right)^{-1}$, but is a coherent state when T_{meas} is much shorter than $\left(\frac{\omega}{Q}\right)^{-1}$. The effect of the cavity (mirror) is twofold for the complimentarity relation in the generation process of these two kinds of quantum states: it not only provides an optical feedback to induce coherent stimulated emission to stabilize the phase but also introduces a jitter in output coupling of photons to randomize the amplitude. By increasing the cavity photon lifetime $\left(\frac{\omega}{Q}\right)^{-1}$, the phase noise is decreased due to enhanced phase coherent stimulated emission relative to random-phase spontaneous emission, but the amplitude noise is increased due to increased randomness in output coupling of generated photons.

292 QUANTUM STATISTICAL PROPERTIES OF A LASER

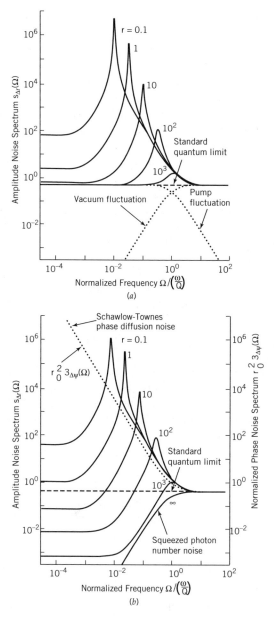

FIGURE 15-10: (a) Amplitude noise spectra for various pump levels, $r \equiv P/P_{th} - 1$, in a laser with shot-noise-limited pump-amplitude fluctuation and the origins for the standard quantum limit (dotted lines). (b) Amplitude noise spectra in a laser with suppressed pump-amplitude fluctuation and normalized phase noise spectrum (dotted line).

The gain bandwidth of a semiconductor laser is usually broader than the longitudinal mode separation, so a few longitudinal modes can oscillate simultaneously even under continuous-wave pump operation. Spontaneously emitted photons primarily go to the particular mode closest to the gain peak, but occasionally spontaneous photons can also go to other longitudinal modes. Such a longitudinal mode competition due to stochastic spontaneous emission coupling was observed in many semiconductor lasers and is known to introduce excess intensity noise. Figure 15-11(a) shows the theoretical lasing characteristics of a semiconductor laser with three longitudinal modes and with homogeneous gain broadening. The curves a and b show the average photon numbers of the central mode and side mode, while the curve c shows the average carrier number. Each mode has a slightly different stimulated emission gain coefficient that is proportional to the spontaneous emission coupling coefficients $\beta_i (i = 1, 2, 3)$, defined by

$$\langle \tilde{E}_{cv,i} \rangle = \beta_i \frac{\langle \tilde{N} \rangle}{\tau_{sp}}. \tag{15.107}$$

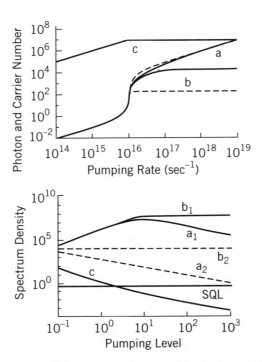

FIGURE 15-11: (a) The average photon numbers of main (a) and side (b) longitudinal modes and the carrier number (c) versus pump rate (electron/s) for a semiconductor laser with a homogeneous gain. (b) The total intensity noise (c) and the intensity noise of individual mode (a, b) versus relative pump rate $P/P_{th} - 1$. Solid lines correspond to $\beta_1 = \beta_3 = (1 - 5 \times 10^{-5})\beta_2$ and dashed lines correspond to $\beta_1 = \beta_3 = (1 - 5 \times 10^{-3})\beta_2$.

The solid lines correspond to a relatively broad gain bandwidth, in which the two side modes have the spontaneous emission coefficients $\beta_1 = \beta_3 = (1 - 5 \times 10^{-5})\beta_2$, where β_2 is the spontaneous emission coefficient of the central mode. The three modes simultaneously oscillate at almost identical threshold, but the central mode starts to dominate with increasing pump rate due to cross-gain saturation. The dashed lines correspond to a relatively narrow gain bandwidth, in which the two side modes have the spontaneous emission coefficients $\beta_1 = \beta_3 = (1 - 5 \times 10^{-3})\beta_2$. A nearly single mode oscillation with large side mode suppression is realized even at a pump rate just above the threshold in this case. As shown in Fig. 15-11(b), the total intensity noise (solid line c) of the three modes is reduced to below the shot noise value at a pump rate $P/P_{th} \geq 3$, but the intensity noise of individual mode is much higher than the shot noise value. This excess intensity noise is caused by the mode competition process. There should exist a strong negative correlation between the central mode intensity and the two side mode intensities, so that the total intensity noise can be squeezed even though the intensity noise of an individual mode is very high.

Injection locking is an efficient way to suppress the side mode intensity and thus the mode competition noise. A detailed analysis of the squeezing characteristics of an injection-locked semiconductor laser using the Langevin equations is given in Refs. [7] through [9].

15.6. OBSERVATION OF MODE PARTITION NOISE

Due to this negative intensity noise correlation, the longitudinal-mode-partition noise can often be suppressed much below the intrinsic intensity noise of individual mode when the total intensity is measured. Actually, intensity squeezing has been observed from a multilongitudinal mode semiconductor laser regardless of its huge longitudinal-mode-partition noise, as shown in Fig. 15-12. The existence of the negative intensity noise correlation between the main mode and the side modes was also directly measured [10], [11]. When the negative intensity noise correlation is not strong enough to suppress the longitudinal-mode-partition noise, the intensity squeezing would be destroyed by the uncanceled mode partition noise. This residual longitudinal-mode-partition noise is considered to be one of the reasons why some semiconductor lasers do not generate number-phase squeezed states even when their quantum efficiencies are very high and they are operating at high pumping levels.

The longitudinal-mode-partition noise has been measured by discriminating the main mode from the other side modes using a monochrometer [10], [11]. However, this method cannot give us the exact magnitude of the longitudinal-mode-partition noise because of a high insertion loss introduced to the measured field. A more precise measurement of the longitudinal-mode-partition noise can be performed using a nonsymmetrical interferometer [12].

The longitudinal mode separation Δf of a typical semiconductor laser is about 300 GHz, which is much smaller than the carrier frequency on the order of 3×10^5 GHz. After traveling through the two arms of a Mach–Zehnder interferometer, the accumulated phase difference between the two beams is $\Delta\phi = 2\pi \Delta l/\lambda$, where

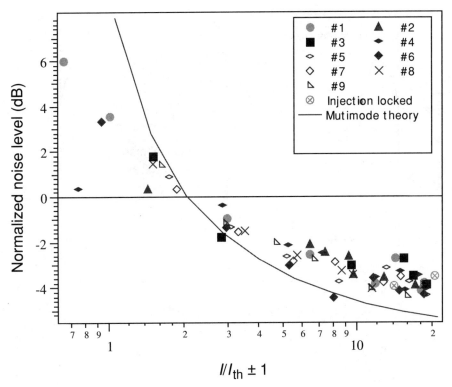

FIGURE 15-12: The total intensity noise normalized by the shot-noise value versus the normalized pump current for nine GaAs transverse junction stripe lasers. The solid line is the theoretical value calculated by the multimode Langevin equations.

Δl is the arm-length difference and λ is the optical wavelength. When one arm length is changed by only a few wavelengths around $\Delta l = 0$, $\Delta \phi$ for different longitudinal modes is almost identical. Thus, the fringe patterns of all the modes of interest (about 100 side modes centered at the main lasing mode) overlap very well. However, when Δl is increased, $\Delta \phi$ for different longitudinal modes begins to differ significantly. For $\Delta l_0 \simeq 0.5 c/\Delta f$, where c is the speed of light, $\Delta \phi$ differs by π for adjacent longitudinal modes. For example, if the interferometer phase difference for the main mode is $\Delta \phi_0$, then the phase difference for the first side modes is $\Delta \phi_0 \pm \pi$ and that for the nth side modes is $\Delta \phi_0 \pm n\pi$. Consequently, at the bright fringe of the interferometer for the main mode ($\Delta \phi_0 = 2k\pi$, k is an integer), all of the even-numbered side modes interfere constructively while all of the odd-numbered side modes interfere destructively. The opposite is true at the dark fringe. Similarly, for $\Delta l = 2\Delta l_0$, $\Delta \phi$ for adjacent modes differs by 2π, and thus the fringes of all the modes overlap again.

Figure 15-13(a) shows the intensity noise and the corresponding shot-noise variation in the output of the interferometer for an arm-length difference of 1 mm, as one arm-length is changed by about one wavelength. The total intensity noise of the free-

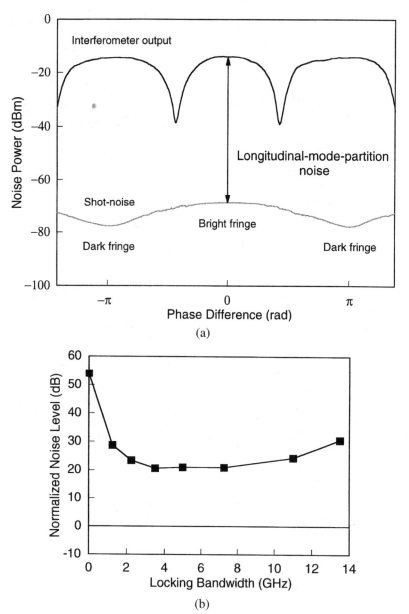

FIGURE 15-13: (a) The intensity noise variation of the interferometer output as one arm-length is changed by about one wavelength. The arm-length difference of the Mach–Zehnder interferometer was 1 mm. The measurement frequency was 20 MHz. The gray curve shows the variation of the corresponding shot-noise value. (b) The longitudinal-mode-partition noise as a function of the locking bandwidth. The longitudinal-mode-partition noise is normalized by the shot-noise level corresponding to the total photocurrent without interferometer.

running laser output is 1.2 dB below the shot-noise level. However, the longitudinal-mode-partition noise (sum of the intensity noise of the main lasing mode and the nonlasing even-numbered side modes) is measured to be 55 dB above the shot-noise level. This indicates that there is a huge longitudinal-mode-partition noise, which is reduced to even below the shot-noise value by negative intensity noise correlation among the modes. From the magnitude of the suppression of the longitudinal-mode-partition noise, we can estimate the strength of the negative intensity noise correlation to be more than 99.9997%! Note that exactly the same excess intensity noise is obtained at the dark fringe, which suggests that the odd-numbered side modes possess the mode partition noise with the same magnitude and opposite phase as the main mode and even-numbered side modes.

Figure 15-13(b) shows the longitudinal-mode-partition noise as a function of the locking bandwidth when the laser is injection locked by the single-mode master laser. The locking bandwidth is changed by changing the injection power into the slave laser. The longitudinal-mode-partition noise is normalized by the shot-noise level. Here, the longitudinal-mode-partition noise at zero locking bandwidth is that of the free-running slave laser output. The longitudinal-mode-partition noise decreases as the locking bandwidth is increased up to about 8 GHz. The minimum longitudinal-mode-partition noise is about 21 dB above the shot-noise level.

Nonlinear gain (hole burning effect) and nonlinear loss (saturable absorber) are known to destroy the negative intensity correlation between different modes and degrade the squeezing characteristics of a semiconductor laser. Detailed theoretical and experimental studies on this point are reported in Ref. [13].

REFERENCES

[1] M. Sargent III, M. O. Scully, and W. E. Lamb, Jr., *Laser Physics* (Addison-Wesley, Reading, Mass., 1974).

[2] D. F. Walls and G. J. Milburn, *Quantum Optics* (Springer-Verlag, Berlin, 1994).

[3] H. M. Wiseman, Phys. Rev. A **47**, 5180 (1993).

[4] Y. Yamamoto, ed. *Coherence, Amplification and Quantum Effect in Semiconductor Lasers* (Wiley, New York, 1991).

[5] Y. Yamamoto, S. Machida, and O. Nilsson, Phys. Rev. A **34**, 4025 (1986).

[6] S. Machida, Y. Yamamoto, and Y. Itaya, Phys. Rev. Lett. **58**, 1000 (1987).

[7] Y. Yamamoto and H. A. Haus, Phys. Rev. A **29**, 1261 (1984).

[8] L. Gillner, G. Björk, and Y. Yamamoto, Phys. Rev. A **41**, 5053 (1990).

[9] S. Inoue, H. Ohzu, S. Machida, and Y. Yamamoto, Phys. Rev. A **48**, 2230 (1993).

[10] S. Inoue, H. Ohzu, S. Machida, and Y. Yamamoto, Phys. Rev. A **46**, 2757 (1992).

[11] F. Marin et al., Phys. Rev. Lett. **75**, 4606 (1995).

[12] S. Inoue and Y. Yamamoto, Opt. Lett. **22**, 328 (1997).

[13] S. Lathi, K. Tanaka, T. Morita, S. Inoue, H. Kan, and Y. Yamamoto, IEEE J. Quantum Electron **35**, 387 (1999).

INDEX

Absorption cross section, 77
Adiabatic elimination, 280
Advanced Green's function, 254
Amplified spontaneous emission noise, 291
Anderson Hamiltonian, 258
Angular momentum eigenstate, 103, 108
Angular momentum quantization, 102
Annihilation operator, 37, 39
Antibunching, photon, 84, 160, 250
Antinormal ordering, 71

Back action evasion, 24
Baker-Hausdorff relation, 56, 74
Bernoulli distribution, 89
Bloch state, 109, 134
Bogoliubov transformation, 53, 204
Bohr radius, 198
Boltzmann distribution, 146
Born approximation, 146
Born-Markov approximation, 146
Bose-Einstein distribution, 149
Bosonization, 211
Bose condensation of excitons, 212
Bulk exciton polariton, 204
Bunching, photon, 83
Bra vector, 2

Canonical quantization, 33
Cauchy-Schwarz inequality, 6
Cavity polariton, 206
Cavity QED, 126
Characteristic function, 73
Charging energy, 221, 229
Coherent atomic state, 109, 134
Coherent state, 11, 53, 91
Collapse of wavefunction, 15
Collapse operator, 164
Collective angular momentum operator, 106
Collective Rabi oscillation, 122

Commutation relation, 5, 32, 287
Conductance quantization, 250
Cosine operator, 63
Coulomb blockade, 220
Coulomb enhancement, 196
Coulomb gauge, 35
Creation operator, 37, 39
Cumming's collapse and revival, 128

De Broglie wavelength, 8, 97
Density operator, 19, 145, 171, 178, 266
Detailed balance, 149
Diagonal P(α) representation, 71, 152
Dicke state, 108, 134
Displacement operator, 56
Dressed boson state, 124
Dressed fermion state, 124
Drift and diffusion coefficients, 144, 278
Dynamical variable, 3

Einstein's relation, 144, 280
Elastic scattering, 244
Electron collision, 246
Electron tunneling, 229
Entangled state, 70
EPR paradox, 71
Equilibrium current noise, 236
Exciton, 197, 202
Exciton-exciton interaction, 199
Exciton polariton, 203

Fermi-Dirac distribution, 236
Fermi's golden rule, 76
First kind measurement, 18
First-order coherence function, 77
Fluctuation-dissipation theorem, 142
Fock state, 45, 96
Fokker-Planck equation, 152, 277
Four-wave mixing, 62

299

INDEX

Generalized Nyquist noise, 237
Generalized projection operator, 22

Hamilton-Jacobi equation, 31
Hanbury Brown-Twiss experiment, 76, 81, 250
Harmonic oscillator, 11
Hartree-Fock approximation, 187
Heavy hole exciton polariton, 210
Heisenberg equation, 32, 140, 175, 280
Heisenberg uncertainty principle, 4
Hermitian operator, 2
Heterodyne detection, 70
Homodyne detection, 70

Indirect measurement, 18
Inelastic scattering, 239

Jaynes-Cummings Hamiltonian, 118

Keldysh formalism, 253
Kerr effect, 176
Ket vector, 1

Lack of causality, 3
Lagrange equation, 30
Lagrangian, 29, 114
Lagrangian density, 33
Langevin equation, 280
Laser threshold, 271
Light hole exciton polariton, 210
Linear superposition, 2
Liouville-von Neumann equation, 19, 145
Longitudinal-transverse splitting, 205
Long wavelength approximation, 117

Mach-Zehnder interferometer, 92, 176
Macroscopic Coulomb blockade, 224
Magnetoexciton, 215
Markov approximation, 142, 146
Master equation, 266
Minimum action principle, 29
Minimum uncertainty state, 8, 53, 60, 68, 109, 178, 287
Mixed state, 19
Mode partition noise, 294
Mollow's triplet, 127, 158
Monte Carlo wavefunction method, 163

Non-Hermitian operator, 8, 164
Noise operator method, 140
Nonequilibrium current noise, 237
Nonequilibrium Green's function, 253
Normal mode splitting, 121
Normal ordering, 72

Observable, 3
Optical Bloch equation, 195
Orbital angular momentum, 105

Parametric amplifier, 51, 62, 67
Partition noise, 58, 239, 294
Pauli exclusion principle, 247, 262
Pauli spin operator, 105
Pegg-Barnett formalism, 90
Phase eigenstate, 90
Phase operator, 63, 89
Phase space filling, 199
Photon counting, 85
Photon number eigenstate, 45, 85, 96, 228
Photon statistics, 274
Photonic de Broglie wave, 97
Poisson bracket, 31
Poisson distribution, 54, 86
Power-Zienau-Wooley transformation, 114
Probability distribution function, 71
Probability interpretation, 1
Projection operator, 2, 22, 164
Pump-noise-suppressed laser, 287
Pure state, 2

QND observable, 24
Quantum entanglement, 67
Quantum interference, 8, 48, 246
Quantum jump, 163
Quantum nondemolition (QND) measurement, 24, 175
Quantum regression theorem, 85, 154
Quantum state diffusion method, 171
Quantum unit of resistance, 221
Quantum well exciton, 206
Quasi-probability denstiy, 49, 72

Radiation trapped state, 128
Ramsey atom interferometer, 182
Reservoir, 137
Retarded Green's function, 254

Rotating wave approximation (RWA), 118, 140, 146

Scalar potential, 34
Schawlow-Townes linewidth, 276, 285
Schrodinger equation, 39, 163, 171
Schrodinger wavefunction, 1
Second kind measurement, 18
Second-order coherence function, 81
Second quantization, 38
Self-energy, 253
Self-phase modulation, 65
Semiconductor Bloch equation, 186
Simultaneous measurement, 25
Sine operator, 63
Single photon turnstile device, 228
Soliton, 181
Spectral function, 255
Spin angular momentum, 105
Spontaneous emission coefficient, 271
Squeezed state, 11, 60, 63, 288
Squeezed vacuum state, 94
Squeezing parameter, 8
Standard quantum limit, 82, 285
State reduction, 15, 68, 164, 178
State vector, 1

Stochastic differential equation, 153
Stochastic wavefunction method, 162
Sub-Poissonian laser, 283
Superradiance, 134

Thermal light, 82, 87
Thermionic emission time, 225
Transverse delta function, 36
Twin-photons, 67
Two-band approximation, 191
Two-time correlation function, 166

Uncertainty principle, 4

Vacuum Rabi oscillation, 119
Vector potential, 34
Voltage probe reservoir, 239
Von Neumann's projection postulate, 14, 68, 164

Wave-particle duality, 8
Wigner distribution function, 74
Wigner-Weisskopf approximation, 131

Young's interference, 77